Organic Chemistry of Drug Degradation

RSC Drug Discovery Series

Editor-in-Chief:
Professor David Thurston, *London School of Pharmacy, UK*

Series Editors:
Dr David Fox, *Pfizer Global Research and Development, Sandwich, UK*
Professor Salvatore Guccione, *University of Catania, Italy*
Professor Ana Martinez, *Instituto de Quimica Medica-CSIC, Spain*
Professor David Rotella, *Montclair State University, USA*

Advisor to the Board:
Professor Robin Ganellin, *University College London, UK*

Titles in the Series:

1: Metabolism, Pharmacokinetics and Toxicity of Functional Groups
2: Emerging Drugs and Targets for Alzheimer's Disease; Volume 1
3: Emerging Drugs and Targets for Alzheimer's Disease; Volume 2
4: Accounts in Drug Discovery
5: New Frontiers in Chemical Biology
6: Animal Models for Neurodegenerative Disease
7: Neurodegeneration
8: G Protein-Coupled Receptors
9: Pharmaceutical Process Development
10: Extracellular and Intracellular Signaling
11: New Synthetic Technologies in Medicinal Chemistry
12: New Horizons in Predictive Toxicology
13: Drug Design Strategies: Quantitative Approaches
14: Neglected Diseases and Drug Discovery
15: Biomedical Imaging
16: Pharmaceutical Salts and Cocrystals
17: Polyamine Drug Discovery
18: Proteinases as Drug Targets
19: Kinase Drug Discovery
20: Drug Design Strategies: Computational Techniques and Applications
21: Designing Multi-Target Drugs
22: Nanostructured Biomaterials for Overcoming Biological Barriers
23: Physico-Chemical and Computational Approaches to Drug Discovery
24: Biomarkers for Traumatic Brain Injury
25: Drug Discovery from Natural Products
26: Anti-Inflammatory Drug Discovery
27: New Therapeutic Strategies for Type 2 Diabetes: Small Molecules
28: Drug Discovery for Psychiatric Disorders
29: Organic Chemistry of Drug Degradation

How to obtain future titles on publication:
A standing order plan is available for this series. A standing order will bring delivery of each new volume immediately on publication.

For further information please contact:
Book Sales Department, Royal Society of Chemistry, Thomas Graham House, Science Park, Milton Road, Cambridge, CB4 0WF, UK
Telephone: +44 (0)1223 420066, Fax: +44 (0)1223 420247,
Email: booksales@rsc.org
Visit our website at http://www.rsc.org/Shop/Books/

Organic Chemistry of Drug Degradation

Min Li
Ringoes, New Jersey
Email: minli88@yahoo.com

RSC Publishing

RSC Drug Discovery Series No. 29

ISBN: 978-1-84973-421-9
ISSN: 2041-3203

A catalogue record for this book is available from the British Library

Published by The Royal Society of Chemistry,
Thomas Graham House, Science Park, Milton Road,
Cambridge CB4 0WF, UK

Registered Charity Number 207890

For further information see our web site at www.rsc.org

Printed in the United Kingdom by CPI Group (UK) Ltd, Croydon, CR0 4YY, UK

This book is dedicated to the memory of my parents,
Shaohua Li and Ruiying Yang,
for their love and inspiration.

Preface

For some time, I have had the desire to write a book on the area of drug degradation chemistry, partly because of the need for a book with in-depth coverage of the mechanisms and pathways of "real" drug degradation, which is defined (in this book) as drug degradation that tends to occur under long term storage and stability conditions. During the 2010 Pittcon in Orlando, while visiting the exhibition booth of RSC Publishing, I met Ms. Roohana Khan, then Regional Business Manager of RSC Publishing for the US, and expressed my idea for the book. She was very interested in the idea and promptly forwarded my initial proposal to the editors of *RSC Drug Discovery Series*, particularly Dr. David Rotella and Mrs. Gwen Jones, which eventually led to the book publishing contract.

The vast majority of drugs are organic and increasingly biological molecular entities. Control or minimization of drug degradation requires a clear understanding of the underlying organic chemistry of drug degradation, which is not only critically important for developing a drug candidate but also for maintaining the quality, safety, and efficacy of an approved drug product over its product life cycle. Specifically, the knowledge of drug degradation is not only vital for developing adequate dosage forms that display favorable stability behavior over the registered product shelf life, but also critical in assessing which impurities would be most likely to be significant or meaningful degradants so that they should be properly controlled and monitored. This book discusses various degradation pathways with an emphasis on the underlying mechanisms of the degradation that tends to occur under the real life scenarios, that is, the long term storage conditions as represented by the stability conditions recommended by the International Conference on Harmonisation of Technical Requirements for Registration of Pharmaceuticals for Human Use (ICH) and the World Health Organization (WHO). The utility and limitation of using stress studies or forced degradation in "predicting" real life drug degradation chemistry is clearly discussed in the book and the reader is alerted

RSC Drug Discovery Series No. 29
Organic Chemistry of Drug Degradation
By Min Li
© Min Li 2012
Published by the Royal Society of Chemistry, www.rsc.org

to the stressing conditions that tend to produce artificial degradation products. Organic reactions that are significant in drug degradation are discussed and illustrated with examples of drug degradation from commercialized drug products as well as drug candidates in various stages of pharmaceutical and manufacturing development. This book consists of nine chapters, with Chapters 2 and 3 devoted to hydrolytic and oxidative degradations, the two most commonly observed types of drug degradation, the latter being perhaps the most complex of all. In Chapter 3, the Udenfriend reaction is discussed in detail with regard to its significant, but little yet known role in the auto-oxidative degradation of drugs. Chapters 4 and 5 cover the remaining vast majority of drug degradation reactions except for photochemical degradation, which is discussed in Chapter 6. Chapter 7 covers the chemical degradation of biological drugs. The book finishes with two chapters, respectively, on strategies for rapid elucidation of drug degradants and control of drug degradation according to current regulatory requirements and guidelines. With the increasing regulatory requirements on the quality and safety of pharmaceutical products, I hope this book will be a handy resource for pharmaceutical and analytical scientists as well as medicinal chemists. A good understanding of drug degradation chemistry should also facilitate lead optimization and help to avoid the degradation pathways that may lead to potentially toxic degradants.

Completing this book has been a laborious but fulfilling experience. As one reviewer of the book proposal put it, "it is potentially a Herculean task". Fortunately, I have been able to complete the book largely on schedule, partly due to the encouraging and constructive comments by the two reviewers. The subject of drug degradation chemistry involves multidisciplines. It requires knowledge and experience in organic chemistry, medicinal chemistry, separation sciences, mass spectrometry, and NMR spectroscopy. Throughout my professional life, I have been fortunate to have gained knowledge and experience in the above disciplines. I am forever indebted to my mentors during the early phases of my career for their advice, passion for science, and example of hard work and integrity. My undergraduate major was polymer chemistry at Fudan University, followed by two years of a master program in the same subject. During that period, I had the opportunity to study the photochemistry and photophysics of polymers under Professor Shanjun Li. This experience triggered my interest in photochemistry, which has enabled me to write Chapter 6, Photochemical Degradation. During my PhD study in the laboratory of Professor Emil H. White at Johns Hopkins University, I learned the principles of organic chemistry, and protein and peptide chemistry with extensive hands-on experience. In this period, I also started to learn the basics of mass spectrometry, particularly fast atom bombardment (FAB) ionization, the technique of choice for mass spectrometric analysis of biological molecules at the time. Use of FAB-MS turned out to be crucial in identifying the exact location of a chemical probe attached to an active-site peptide of a protease, which was one of my main research projects at that time. During my postdoctoral research in Professor Michael E. Johnson's laboratory at University of Illinois at Chicago, Center for Pharmaceutical Biotechnology, I had the opportunity to learn the basic principles of medicinal chemistry, particularly in the field of structure-based drug design. I am also deeply indebted to many of my colleagues in various biotechnical and pharmaceutical companies, particularly Merck and formerly Schering-Plough where I have spent the majority of my career, for their encouragement and support. I would like to thank especially Dr. Zi-Qiang Gu, Dr. Abu M. Rustum, and the members of my research groups at different times. Over a span of more than ten years, my research groups have performed hundreds of investigations related to various drug degradation mechanisms and pathways; a minority of these investigation results were published, many of which are cited in this book. The successful resolution of these challenging investigations would have not been possible without the contribution from the members of my research groups, most notably Dr. Bin Chen, Dr. Xin Wang, Dr. Xin (Jack) Yu, Dr. Mingxiang Lin, and Dr. Russell Maus.

Special thanks go to Dr. Russell Maus who reviewed the manuscript of Chapter 2, Dr. Gary Martin for a constructive discussion on the topic of two-dimensional NMR spectroscopy, and to the editors of RSC Publishing who have done a superb job in the production of the book. Finally, my grateful thanks go to my family, particularly my wife Beihong, for her love, support, and unwavering confidence in me over the past 20 years.

Min Li
Ringoes, New Jersey
27 May 2012
minli88@yahoo.com

Contents

Chapter 1   Introduction   1

    1.1   Drug Impurities, Degradants and the Importance of
           Understanding Drug Degradation Chemistry   1
    1.2   Characteristics of Drug Degradation Chemistry and
           the Scope of this Book   3
    1.3   Brief Discussion of Topics that are Outside the Main
           Scope of this Book   5
         1.3.1   Thermodynamics and Kinetics of Chemical
               Reactions   5
         1.3.2   Reaction Orders, Half-lives and Prediction
               of Drug Product Shelf-lives   7
         1.3.3   Key Elements in Solid State Degradation   9
         1.3.4   Role of Moisture in Solid State
               Degradation and pH in the
               Microenvironment of the Solid State   10
    1.4   Organization of the Book   11
    References   14

Chapter 2   Hydrolytic Degradation   16

    2.1   Overview of Hydrolytic Degradation   16
    2.2   Drugs Containing Functional Groups/Moieties
           Susceptible to Hydrolysis   20
         2.2.1   Drugs Containing an Ester Group   20
         2.2.2   Drugs Containing a Lactone Group   23
         2.2.3   Drugs Containing an Amide Group   24
         2.2.4 β-Lactam Antibiotics   26
         2.2.5   Carbamates   30

RSC Drug Discovery Series No. 29
Organic Chemistry of Drug Degradation
By Min Li
© Min Li 2012
Published by the Royal Society of Chemistry, www.rsc.org

2.2.6   Phosphates and Phosphoramides   32
2.2.7   Sulfonamide Drugs   34
2.2.8   Imides and Sulfonylureas   35
2.2.9   Imines (Schiff Bases) and Deamination   36
2.2.10   Acetal and Hemiacetal Groups   40
2.2.11   Ethers and Epoxides   41
2.3   Esterification, Transesterification and Formation
       of an Amide Linkage   43
References   44

Chapter 3   Oxidative Degradation   48

3.1   Introduction   48
3.2   Free Radical-mediated Autooxidation   49
       3.2.1   Origin of Free Radicals: Fenton Reaction
                and Udenfriend Reaction   49
       3.2.2   Origin of Free Radicals: Homolytic Cleavage
                of Peroxides by Thermolysis and Heterolytic
                Cleavage of Peroxides by Metal Ion Oxidation   53
       3.2.3   Autooxidative Radical Chain Reactions and
                Their Kinetic Behavior   54
       3.2.4   Additional Reactions of Free Radicals   56
3.3   Non-radical Reactions of Peroxides   57
       3.3.1   Heterolytic Cleavage of Peroxides and
                Oxidation of Amines, Sulfides, and
                Related Species   57
       3.3.2   Heterolytic Cleavage of Peroxides and
                Formation of Epoxides   59
3.4   Carbanion/enolate-mediated Autooxidation
       (Base-catalyzed Autooxidation)   61
3.5   Oxidation Pathways of Drugs with Various Structures 62
       3.5.1   Allylic- and Benzylic-type Positions
                Susceptible to Hydrogen Abstraction by
                Free Radicals   62
       3.5.2   Double Bonds Susceptible to Addition by
                Hydroperoxides   68
       3.5.3   Tertiary Amines   71
       3.5.4   Primary and Secondary Amines   76
       3.5.5   Enamines and Imines (Schiff Bases)   79
       3.5.6   Thioethers (Organic Sulfides), Sulfoxides,
                Thiols and Related Species   80
       3.5.7   Examples of Carbanion/enolate-mediated
                Autooxidation   83
       3.5.8   Oxidation of Drugs Containing Alcohol,
                Aldehyde, and Ketone Functionalities   87

3.5.9 Oxidation of Aromatic Rings: Formation of
Phenols, Polyphenols, and Quinones 92
3.5.10 Oxidation of Heterocyclic Aromatic Rings 96
3.5.11 Miscellaneous Oxidative Degradations 99
References 101

**Chapter 4 Various Types and Mechanisms of
Degradation Reactions 110**

4.1 Elimination 110
4.1.1 Dehydration 110
4.1.2 Dehydrohalogenation 114
4.1.3 Hofmann Elimination 116
4.1.4 Miscellaneous Eliminations 117
4.2 Decarboxylation 118
4.3 Nucleophilic Conjugate Addition and
Retro-nucleophilic Conjugate Addition 121
4.4 Aldol Condensation and Retro-aldol 124
4.4.1 Aldol Condensation 124
4.4.2 Retro-aldol Reaction 126
4.5 Isomerization and Rearrangement 127
4.5.1 Tautomerization 127
4.5.2 Racemization 128
4.5.3 Epimerization 129
4.5.4 *Cis-trans* Isomerization 129
4.5.5 *N,O*-Acyl Migration 132
4.5.6 Rearrangement *via* Ring Expansion 133
4.5.7 Intramolecular Cannizzaro Rearrangement 136
4.6 Cyclization 137
4.6.1 Formation of Diketopiperazine (DKP) 137
4.6.2 Other Cyclization Reactions 138
4.7 Dimerization/Oligomerization 139
4.8 Miscellaneous Degradation Mechanisms 144
4.8.1 Diels–Alder Reaction 144
4.8.2 Degradation *via* Reduction or
Disproportionation 145
References 146

Chapter 5 Drug–Excipient Interactions and Adduct Formation 150

5.1 Degradation Caused by Direct Interaction
between Drugs and Excipients 150
5.1.1 Degradation *via* the Maillard Reaction 150
5.1.2 Drug–Excipient Interaction *via* Ester
and Amide Linkage Formation 153

5.1.3   Drug–Excipient Interaction *via*
        Transesterification   154
5.1.4   Degradation Caused by Magnesium
        Stearate   154
5.1.5   Degradation Caused by Interaction
        between API and Counter Ions and
        between Two APIs   156
5.1.6   Other Cases of Drug–Excipient
        Interactions   157
5.2   Degradation Caused by Impurity of Excipients   158
5.2.1   Degradation Caused by Hydrogen Peroxide,
        Formaldehyde, and Formic Acid   158
5.2.2   Degradation Caused by Residual Impurities
        in Polymeric Excipients   159
5.3   Degradation Caused by Degradants of Excipients   160
5.4   Degradation Caused by Impurities from Packaging
      Materials   161
References   162

Chapter 6   Photochemical Degradation   **165**

6.1   Overview   165
6.2   Non-oxidative Photochemical Degradation   166
6.2.1   Photodecarboxylation: Photodegradation of
        Drugs Containing a 2-Arylpropionic Acid
        Moiety   167
6.2.2   Photoisomerization   170
6.2.3   Aromatization of 1,4-Dihydropyridine Class
        of Drugs   174
6.2.4   Dehalogenation of Aryl Halides   176
6.2.5   Cyclization in Polyaromatic Ring Systems   180
6.2.6   Photochemical Elimination   182
6.2.7   Photodimerization and Photopolymerization   184
6.2.8   Photochemistry of Ketones: Norris Type I
        and II Photoreactions   185
6.3   Oxidative Photochemical Degradation   187
6.3.1   Type I Photosensitized Oxidation:
        Degradation *via* Radical Formation and
        Electron Transfer   188
6.3.2   Type II Photosensitized Oxidation:
        Degradation Caused by Singlet Oxygen   189
6.3.3   Degradation Pathways *via* Reaction with
        Singlet Oxygen   190
References   194

Chapter 7   Chemical Degradation of Biological Drugs   198

    7.1   Overview   198
    7.2   Chemical Degradation of Protein Drugs   199
        7.2.1   Hydrolysis and Rearrangement of Peptide
               Backbone Caused by the Asp Residue   199
        7.2.2   Various Degradation Pathways Caused by
               Deamidation and Formation of Succinimide
               Intermediate   202
        7.2.3   Hinge Region Hydrolysis in Antibodies   204
        7.2.4   Oxidation of Side Chains of Cys, Met, His,
               Trp, and Tyr   204
        7.2.5   Oxidation of Side Chains of Arg, Pro,
               and Lys   209
        7.2.6   β-Elimination   211
        7.2.7   Crosslinking, Dimerization, and
               Oligomerization   213
        7.2.8   The Maillard Reaction   214
        7.2.9   Degradation *via* Truncation of a *N*-Terminal
               Dipeptide Sequence through DKP Formation   215
        7.2.10 Miscellaneous Degradation Pathways   215
    7.3   Degradation of Carbohydrate-based Biological Drugs   216
    7.4   Degradation of DNA and RNA Drugs   218
        7.4.1   Hydrolytic Degradation of Phosphodiester
               Bonds   218
        7.4.2   Oxidative Degradation of Nucleic
               Acid Bases   220
    References   222

**Chapter 8   Strategies for Elucidation of Degradant Structures
            and Degradation Pathways   227**

    8.1   Overview   227
    8.2   Practical Considerations of Employing LC-MS^n for
        Structural Elucidation of Degradants at Trace Levels   229
        8.2.1   Conversion of MS-unfriendly HPLC
               Methods to LC-MS Methods   230
        8.2.2   Nomenclature, Ionization Modes and
               Determination of Parent Ions   230
        8.2.3   Fragmentation and LC-MS^n Molecular
               Fingerprinting   233
    8.3   Brief Discussion of the Use of Multi-dimensional
        NMR in Structure Elucidation of Trace Level
        Impurities   239

8.4   Performing Meaningful Stress Studies   240
       8.4.1   Generating Relevant Degradation Profiles   241
8.5   Effective Use of Mechanism-based Stress Studies in
       Conjunction with LC-MS^n Molecular Fingerprinting
       in Elucidation of Degradant Structures and
       Degradation Pathways: Case Studies   245
       8.5.1   Outline of General Strategy   245
       8.5.2   Proposing Type of Degradation Based on
                LC-MS^n Analysis   245
       8.5.3   Design of Stress Studies According to
                Presumed Degradation Type   247
       8.5.4   Tracking and Verification of Unknown
                Degradants Generated in Stress Studies Using
                LC-MS^n Molecular Fingerprinting   248
       8.5.5   Case Study 1: Elucidation of a Novel
                Degradation Pathway for Drug Products
                Containing Betamethasone Dipropionate and
                Similar Corticosteroidal 17,21-Diesters   248
       8.5.6   Case Study 2: Rapid Identification of Three
                Betamethasone Sodium Phosphate Isomeric
                Degradants – Use of Enzymatic
                Transformation When a Direct MS^n
                Fingerprint Match is not Available   251
       8.5.7   Case Study 3: Identification of an Impurity in
                Betamethasone 17-Valerate Drug
                Substance – Structure Prediction When an
                Exact MS^n Fingerprint Match is not Available   256
       References   258

Chapter 9   Control of Drug Degradation   262

9.1   Overview   262
9.2   Degradation Controlling Strategies *Versus* Multiple
       Degradation Pathways and Mechanisms   262
9.3   Design and Selection of a Drug Candidate
       Considering Drug Degradation Pathways and
       Mechanisms   263
9.4   Implication of the Udenfriend Reaction and
       Avoidance of a Formulation Design that may Fall
       into the "Udenfriend Trap"   265
9.5   Control of Oxygen Content in Drug Products   267
9.6   Use of Antioxidants and Preservatives   268
9.7   Use of Chelating Agents to Control Transition Metal
       Ion-mediated Autooxidation   268

9.8	Control of Moisture in Solid Dosage Forms	269
9.9	Control of pH	270
9.10	Control of Photochemical Degradation Using Pigments, Colorants, and Additives	270
9.11	Variability of Excipient Impurity Profiles	271
9.12	Use of Formulations that Shield APIs from Degradation	271
9.13	Impact of Manufacturing Process on Drug Degradation	272
9.14	Selection of Proper Packaging Materials	272
9.15	Concluding Remarks	273
	References	274

Subject Index **278**

CHAPTER 1

Introduction

1.1 Drug Impurities, Degradants and the Importance of Understanding Drug Degradation Chemistry

A drug impurity is anything that is not the drug substance (or active pharmaceutical ingredient, API) or an excipient according to the definition by the US Food and Drug Administration (FDA).[1] Impurities can be categorized into process impurities, drug degradation products (degradants or degradates), and excipient and packaging-related impurities. Process impurities are produced during the manufacture of the drug substance and drug product, while degradants are formed by chemical degradation during the storage of the drug substances or drug products. The storage conditions are typically represented by the International Conference on Harmonisation (ICH)- and World Health Organization (WHO)-recommended stability conditions which simulate different climatic zones of the world.[2,3] Certain process impurities can also be degradants, if they continue to form in storage under stability conditions. Packaging-related impurities, also called leachables, are typically various plasticizers, antioxidants, UV curators, and residual monomers that leach out of the plastic or rubber components and labels of the package/container of a drug product over time.

Those process impurities that are not degradants may be controlled or eliminated by modifying or changing the process chemistry. On the other hand, control or minimization of drug degradants requires a clear understanding of the drug degradation chemistry, which is not only critically important for developing a drug candidate but also for maintaining the quality, safety, and efficacy of an approved drug product. Specifically, knowledge of drug degradation is not only vital for developing adequate dosage forms that display favorable stability behavior over the registered product shelf-life, but is also critical in assessing which impurities would be most likely to be significant or meaningful degradants so that they can be included in the specificity

RSC Drug Discovery Series No. 29
Organic Chemistry of Drug Degradation
By Min Li
© Min Li 2012
Published by the Royal Society of Chemistry, www.rsc.org

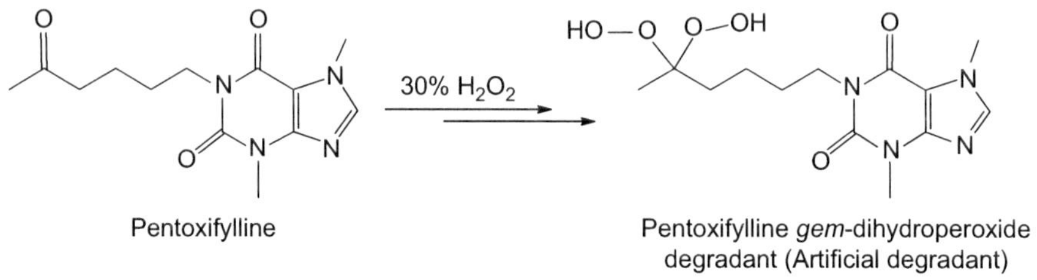

Scheme 1.1

mixture when developing and validating stability-indicating analytical methodologies. A common problem in the development of stability-indicating HPLC methods using stress studies (or forced degradation) is a lack of proper evaluation if the stress-generated degradants would be real degradants or not. From a practical point of view, the real degradants are those that can form under long term storage conditions such as the International Conference on Harmonisation (ICH) stability conditions.[2] On the other hand, various artificial degradants can be generated during stress studies, in particular when excessive degradation is rendered or the stress conditions are not consistent with the degradation pathways of the drug molecule under the usual stability conditions. For example, forced degradation of a ketone-containing drug, pentoxifylline, using 30% hydrogen peroxide at room temperature for eight days produced a geminal dihydroperoxide degradation product (Scheme 1.1).[4] This compound is highly unlikely to be a real degradant of the drug product.

This book is devoted to increasing our understanding and knowledge of the organic chemistry of drug degradation. The knowledge derived from this endeavor should also be beneficial for the elucidation of drug metabolite structures and bioactivation mechanisms. Most drugs undergo at least certain level of metabolism,[5] that is, chemical transformation catalyzed by various enzymes. Except in the case of pro-drugs, drug metabolites can be considered as drug degradants formed *in vivo*. Chemical degradation and drug metabolism can produce the same degradants, even though they may go through different reaction intermediates or mechanisms. *In vitro* chemical reactions have been used to mimic enzyme-catalyzed drug metabolism processes, in order to help elucidate the enzymatic mechanisms for the catalysis.[6] On the other hand, understanding the mechanisms of drug metabolism may also facilitate the elucidation of drug degradation pathways *in vitro*.

Regardless of their origins, certain drug degradants can be toxic, which is one of the main contributors to undesirable side effects or adverse drug reactions (ADR) of drugs.[7] In the early stage of drug development, the degradants (including metabolites) and degradation pathways (or bioactivation pathways in the case of reactive metabolites) of a drug candidate need to be elucidated, followed by toxicological evaluation of these degradants. Dependent upon the outcome of the evaluation, the structure of the drug candidate may have to be modified to avoid the formation of a particular toxicophore based on the understanding of the degradation chemistry (or bioactivation pathways) elucidated. Failure to uncover toxic degradants, usually the low level ones, in the early development stage can lead to hugely costly failure in later stage clinical studies or even withdrawal of an approved drug product from the market.

1.2 Characteristics of Drug Degradation Chemistry and the Scope of this Book

The vast majority of therapeutic drugs are either organic compounds or biological entities. The latter drugs include protein and nucleic acid (RNA and DNA)-based drugs which are biopolymers comprising small molecule building blocks. This book focuses on the organic chemistry aspect of drug degradation, in particular, the mechanisms and pathways of the chemical degradation of both small and large molecule drugs under real life degradation scenarios, as represented by the usual long term stability conditions. Stress studies or forced degradation can help elucidate the structures of real degradants and the degradation pathways of drugs. Nevertheless, caution needs to be taken in differentiating the real and artificial degradants. This subject will be discussed in detail in Chapter 8, Strategies for Elucidation of Degradant Structures and Degradation Pathways.

Drug degradation chemistry differs from typical organic chemistry in several ways. First, the yield of a drug degradation reaction is usually very low, from approximately 0.05% to a few percentage points at the most. Dependent upon the potencies and maximum daily dosages of the drugs, ICH guidelines require that the impurities and/or degradants of a drug be structurally elucidated, once they exceed certain thresholds, which are typically between 0.05% and 0.5%, relative to the drug substances.[8,9] For potential genotoxic impurities, they need to be characterized and controlled at a daily maximum amount of 1.5 μg for drugs intended for long term usage.[10] Such low yields would be meaningless from the perspective of the regular organic chemistry. Second, due to the low yields and limited availability of samples, particularly stability samples of formulated drugs, the quantity of a drug degradant is usually extremely low, posing a serious challenge for its isolation and/or characterization. Despite the advent of sensitive and powerful analytical methodologies such as high resolution tandem liquid chromatography-mass spectrometry (LC-MS/MS), liquid chromatography-nuclear magnetic resonance (LC-NMR), and cryogenic micro NMR probes, the identification of drug degradants remains one of the most challenging activities in pharmaceutical development.[11] Third, the typical conditions and "reagents" of drug degradation reactions are limited in scope. For example, the ICH long term stability conditions for different climatic zones specify the requirements for heat and moisture (relative humidity, RH), for example, 25 °C/60% RH and 30 °C/65% RH, while the ICH accelerated stability condition requires heating at 40 °C under 75% RH. In addition to moisture, the other most important "reagent" in drug degradation reactions is molecular oxygen. Since molecular oxygen is ubiquitous and difficult to remove from drug products, oxidative degradation of drugs is one of the most common degradation pathways. Often, the impact of molecular oxygen can be indirect. For example, a number of polymeric drug excipients such as polyethylene glycol (PEG), polysorbate, and povidone, are readily susceptible to autooxidation, resulting in the formation of various peroxides including hydrogen peroxide.[12–14] These peroxides can cause significant drug degradation once formulated with drug substances containing oxidizable moieties. In contrast, reductive degradation is rarely seen in drug degradation reactions owing to the lack of a reducing agent in common drug excipients that is strong enough to cause meaningful reductive degradation. Other possible "reagents" in drug degradation reactions are usually limited to drug excipients and their impurities. For example, excipients consisting of oligosaccharides and polysaccharides with reducing ends, such as lactose and starch, are frequently used in drug formulation. The aldehyde functionality of these excipients can react with the primary and secondary amine groups of drugs to undergo degradation via the Maillard reaction. This topic will be covered in Chapter 5, Drug–Excipient Interaction and Adduct Formation.

As indicated above, this book focuses on the organic chemistry of drug degradation, in particular, the mechanisms and pathways of the chemical degradation of both small and large molecule drugs under real life degradation scenarios. Owing to the variety of dosage forms of formulated drugs, degradation of drugs can occur in various states including solid (tablets, capsules, and powders), semi-solid (creams, ointments, patches, and suppositories), solution (oral, ophthalmic, and optic solutions, nasal sprays, lotions, injectables), suspension (suspension injectables), and gas phase (aerosols). Obviously, a drug molecule can exhibit different degradation pathways and kinetics in different dosage forms. Nevertheless, as the emphasis of this book is on drug degradation chemistry with regard to mechanisms and pathways in general, we will not discuss in too much detail in which state a particular degradation pathway occurs. For readers who are interested specifically in drug degradation in the solid state, the book *Solid-state Chemistry of Drugs* by Byrn, Pfeiffer, and Stowell is a good resource, in which an in-depth treatment of a drug's degradation behavior *versus* its polymorphism is presented.[15]

Additionally, the topic of drug degradation kinetics is outside the main scope of this book, although kinetic parameters such as activation energy, E_a, reactant half-life, and reaction rate constant, are used extensively in Chapter 2, Hydrolytic Degradation, for the purpose of comparing the hydrolytic lability of various functional groups on a semi-quantitative basis. Those who are interested in drug degradation kinetics are referred to the book by Yoshioka and Stella, *Stability of Drugs and Dosage Forms*,[16] in which various kinetics models of drug degradation are described. Note that the topic of process impurities of drugs is also out of the scope of this book. There are a number of publications on process chemistry development and control of process impurities.[17–19] Last, this book tries to focus mainly on the major degradation pathways and mechanisms of drugs, rather than to be all-inclusive.

1.3 Brief Discussion of Topics that are Outside the Main Scope of this Book

Although there will not be a detailed discussion of topics that are outside the main scope of this book, such as those mentioned above, a brief overview of some of these topics is beneficial for a better overall understanding of drug degradation chemistry and this is given here.

1.3.1 Thermodynamics and Kinetics of Chemical Reactions

A change in Gibbs free energy, ΔG, of a chemical reaction governs the propensity of the reaction to proceed. ΔG is defined as follows:

$$\Delta G = \Delta H - T\Delta S \tag{1.1}$$

where ΔH is the change in the reaction enthalpy, T is the reaction temperature (in Kelvin), and ΔS is the change in the reaction entropy.

For a thermodynamically favored reaction, that is, a reaction that occurs spontaneously, if allowed by the reaction kinetics, the ΔG of the reaction is negative. In other words, the free energy of the products is lower than that of the reactants in such a case. A schematic diagram of a thermodynamically favored reaction is presented in Figure 1.1. In contrast, a thermodynamically unfavorable reaction has a positive ΔG.

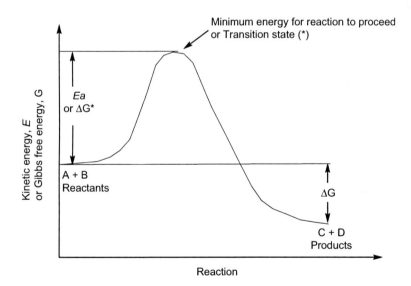

Figure 1.1 Schematic diagram of a thermodynamically favored reaction, where the Gibbs free energy of the reaction, ΔG, is negative. E_a is the activation energy per the collision theory, while ΔG^* is Gibbs free energy of activation according to transition theory.

ΔG determines if the reaction of $A + B \rightarrow C + D$ is favored or not, but it does not determine how fast the reaction, whether thermodynamically favored or not, would take place. The rate of the reaction or its kinetics is governed by the energy that is necessary to activate the reactants to a certain state so that they can convert to their products. There are two theories describing this process: collision theory and transition state theory. Collision theory is embodied in the well-known Arrhenius equation (equation (1.2)), which was first proposed by van't Hoff in 1884 and later justified and interpreted by Arrhenius in 1889:[20]

$$k = Ae^{-Ea/RT} \tag{1.2}$$

where k is the reaction rate constant, A is the pre-exponential (or frequency) factor which can generally be approximated as a temperature-independent constant, E_a is the activation energy which is defined as the minimum energy the reactants must acquire through collision in order for the reaction to occur, R is the gas constant, and T is the reaction temperature (in Kelvin).

According to the Arrhenius equation, the rate constant of a reaction is temperature dependent and by taking the natural logarithm of equation (1.2), the Arrhenius equation takes the following format (equation (1.3)):

$$\ln k = \frac{-E_a}{R}\frac{1}{T} + \ln A \tag{1.3}$$

This expression shows that the higher the temperature, the faster the reaction rate. Additionally, if one measures the reaction rate constants (k) at different temperatures (T), one should get a linear relationship by plotting $\ln k$ *versus* $1/T$. Hence, the activation energy, E_a, can be obtained from the slope ($-E_a/R$) of the linear plot and $\ln A$ from the y-intercept.

Despite its widespread use, the Arrhenius equation and its underlying collision theory have been challenged over time. The major competing theory appears to be transition state theory which was developed independently by Eyring, and Evans and Polanyi in 1935.[21] The equation derived according to transition state theory is the Eyring equation, also called the Eyring–Polanyi equation (1.4):

$$k = \frac{k_B T}{h}e^{-\Delta G^*/RT} \tag{1.4}$$

where ΔG^* is Gibbs free energy of activation, k_B is the Boltzmann constant, and h is Planck's constant.

This equation bears some resemblance to the Arrhenius equation in that the $k_B T/h$ item corresponds to the pre-exponential factor, A, and ΔG^* corresponds to the activation energy, E_a. Nevertheless, in the Eyring equation, ΔG^*, in addition to $k_B T/h$, is temperature dependent, as $\Delta G^* = \Delta H^* - T\Delta S^*$. Hence, the Erying equation can be written as equation (1.5) after taking natural logarithm and rearrangement:

$$\ln k/T = (-\Delta H^*/R)(1/T) + \ln k_B/h + \Delta S^*/R \qquad (1.5)$$

where ΔH^* is enthalpy of activation and ΔS^* is entropy of activation.

Hence, ΔH^* can be obtained from the slope $(-\Delta H^*/R)$ of a linear plot of $\ln k/T$ *versus* $1/T$, while ΔS^* can be obtained from the y-intercept $(\ln k_B/h + \Delta S^*/R)$. Therefore, one can obtain both E_a (from equation (1.3)) and ΔH^* and ΔS^* (from equation (1.5)) from a single dataset of reaction rate constant, k, *versus* reaction temperature, T. Although application of the Eyring equation enables one to obtain both ΔH^* and ΔS^* values and the ΔS^* value should help elucidate the reaction mechanism,[22] it appears that the use of Arrhenius equation exceeds the use of Eyring equation, at least in the hydrolytic stability studies of drugs. With respect to the numeric difference between the values of E_a and ΔH^*, we can again rearrange the Eyring equation (1.5) into the following format (equation (1.6)):

$$\ln k = (-\Delta H^*/R)(1/T) + \Delta S^*/R + \ln k_B/h + \ln T \qquad (1.6)$$

Among the last three items of the equation, only $\ln T$ is a variable of reaction temperature, while the other two are constants. However, for reactions that are studied within a relatively narrow window of temperature, say no greater than 100 K above room temperature (298 K), a temperature change of 100 K with regard to $\ln T$ does not appear to have too much impact on the overall value of the summation for the last three items. Hence, the Arrhenius equation may be considered a simplified version of the Eyring equation when reactions are studied within a relatively narrow range of temperature; the vast majority of the degradation reactions of drugs fall into this category. Therefore, numerically the value of E_a would not be too much different from that of ΔH^*. Indeed, in a hydrolysis study of a group of sulfamides, the difference between the two values is no more than 1 kcal mol^{-1}.[23]

1.3.2 Reaction Orders, Half-lives and Prediction of Drug Product Shelf-lives

If a reaction only involves a single reactant, A, and the rate of this reaction is proportional to its concentration, the reaction order of this unimolecular reaction is said to be 1 with regard to A and the reaction is a first order reaction. This relationship can be expressed by equation (1.7):

$$K = k[A] \qquad (1.7)$$

where K is the reaction rate, k is the reaction rate constant, and [A] is the concentration of A.

For a first order reaction as illustrated above, K can be expressed as:

$$K = -\mathrm{d}[A]/\mathrm{d}t \tag{1.8}$$

where t is the reaction time. As a result, the first order reaction equation can be rewritten as:

$$-\mathrm{d}[A]/\mathrm{d}t = k[A] \quad \text{or} \quad \mathrm{d}[A]/[A] = -k\mathrm{d}t \tag{1.9}$$

Integration of the equation results in the following:

$$[A] = -[A]_0\, e^{kt} \quad \text{or} \quad [A]/[A]_0 = -e^{kt} \tag{1.10}$$

where $[A]_0$ is the initial concentration of A. The reaction time when half of A is consumed, that is, $[A]/[A]_0 = 1/2$, is the half-life of A, $t_{1/2}$. The equation now becomes:

$$e^{kt_{1/2}} = -1/2 \tag{1.11}$$

By taking the natural logarithm and rearranging the resulting equation, $t_{1/2}$ can be calculated by equation (1.12):

$$t_{1/2} = \ln 2/k = 0.693/k \tag{1.12}$$

Thus, for a first order reaction, the half-life of the reactant can be readily calculated from the reaction rate constant, k. While a true unimolecular reaction is not very common, a great number of reactions are bimolecular reactions such as the one illustrated in Scheme 1.1. The rate of the latter reaction can be expressed by equation (1.13), if the reaction order for either A or B is 1:

$$K = k[A][B] \tag{1.13}$$

where K is the reaction rate, k is the reaction rate constant, and $[A]$ and $[B]$ are the concentrations of reactants A and B, respectively.

For a dimerization reaction of reactant A, $K = k[A]^2$, and the reaction order for A is 2. Frequently during studies of the kinetics of bimolecular reactions, the concentration of one reactant, for example $[B]$, can either be kept constant experimentally or at a large excess with respect to the other reactant. The latter category includes the hydrolysis of drug molecules in aqueous solutions, where water is reagent B in large excess. Consequently, $[B]$ becomes or can be approximated as a constant and the bimolecular rate expression can be written as $K = k'[A]$, where $k' = k[B]$. In such a case, the bimolecular reaction becomes a pseudo first order reaction and the half-life of A can be calculated using the formula for first order reaction shown in equation (1.12) above.

In order to "calculate" the shelf-lives of drug products, more meaningful reaction times would usually be when 5% or 10% of the drug substances are degraded.[24] Frequently, it is desirable to perform an accelerated stability study at a higher temperature, T_1, from which the degradation rate constant of the accelerated stability study, k_1, is obtained, to "predict" the shelf-life of the drug product at a regular stability temperature (*e.g.* 298 K), T_2. In principle, this can be readily done for drug products that follow first or pseudo first order degradation kinetics, since the degradation rate constant at the regular stability temperature, k_2, can be calculated according to the following formula, equation (1.14), based on the Arrhenius equation (equation (1.2)):

$$\frac{k_2}{k_1} = \frac{e^{-E_a/RT_1}}{e^{-E_a/RT_2}} \quad \text{or} \quad k_2 = k_1 e^{\frac{E_a}{R}\left(\frac{1}{T_2} - \frac{1}{T_1}\right)} \tag{1.14}$$

Hence, the shelf-life, t, can be calculated or "predicted" based on a calculation using equation (1.14). Nevertheless, in many cases, such predication produces tremendous errors, rendering this approach of no or little practical value. This is due to a number of factors. For example, the degradation mechanism may not be the same at different temperatures. Thus, the dependence of k on T would deviate from the Arrhenius equation. In addition, the exponential relationship between k and T means that a small error in the k_1 value at T_1, could propagate into a huge variation in the k_2 value at T_2.[25]

Because of the limitation of the above approach, various non-linear statistical models for "predicting" drug product shelf-lives have been developed with varying degrees of success.[26–29] Some of these models take into consideration the degradation types of reaction orders at or greater than two using a polynomial degradation model.[28] Last, it needs to be pointed out that in the past decade or so, shelf-lives of drug products have been increasingly constrained by the occurrence of degradants rather than by the loss of active ingredient potency due to the advancement of analytical methodologies and tightening of regulatory requirements.[29]

1.3.3 Key Elements in Solid State Degradation

Solid exists in different forms called polymorphs, which can be generally divided into crystal and non-crystal (or amorphous) forms. Among the crystal forms are anhydrates, solvates, and co-crystals. The most important solvates are hydrates; anhydrates and hydrates can interconvert. For example, high temperature and low moisture should promote the loss of crystal water from hydrates, while the opposite combination should promote the hydration of anhydrates. Acidic and basic drug molecules can be presented in final drug dosage forms in their native forms or in various salt forms. Both the native and salt forms may be capable of producing different polymorphs, that is, an amorphous state, different crystal forms, and/or different hydration states including anhydrates. Depending upon which form is chosen as the physical form of a drug substance in its solid, semi-solid, or other dosage form, where the physical form of the drug molecule matters, the conversion of the selected physical form to other forms is physical degradation, which may result in changes in the solubility and chemical stability of the drug molecule. Such changes are very likely to have an impact on the bioavailability and toxicity profiles of the drug product. Therefore, selection of an appropriate physical form is essential for maintaining drug product quality, safety, and efficacy.[30–32]

Different polymorphs usually have different stabilities or different rates of degradation. In general, crystalline materials are more stable than amorphous ones owing to more restrictive molecular mobility of the former. Likewise, molecules tend to be more stable in solid dosage forms than in liquid ones in most cases, albeit it may not always be the case, since molecules in solid state are more restricted in their mobility. For example, the activation energy of the cyclization of aspartame in the solid state forming its diketopiperazine (DKP) degradant is $268\,kJ\,mol^{-1}$ $(64.1\,kcal\,mol^{-1})$,[33] which is comparable to other solid state reactions.[34] On the other hand, the activation energy for the same degradation in solution is only $70\,kJ\,mol^{-1}$.[35] In certain cases, a particular degradation pathway may only occur in the solid state or associated with particular crystal forms; such a degradation is categorized as a true solid state degradation. A classic example is the solid state photooxidation of 21-cortisol tert-butylacetate to the corresponding 21-cortisone ester.[36,37] Among the five crystal forms obtained, only forms I and IV are reactive. The crystal structure of form I was resolved and its susceptibility to photooxidation was attributed to the ease of penetration by oxygen into a channel along the axis of the helix of the crystal.

1.3.4 Role of Moisture in Solid State Degradation and pH in the Microenvironment of the Solid State

Moisture present in solids is generally categorized as bound (crystal water) and/ or unbound or surface absorbed water. Nevertheless, the bound water molecules can also be mobile and migrate within the crystal lattice or along the solid surface over a long time scale.[38,39] Hence, with regard to the role of moisture in solid state degradation, Ahlneck and Zografi concluded that it is preferable to think of it as a plasticizer rather than causing surface dissolution of the solid.[40] Furthermore, moisture tends to be absorbed more favorably in minor amorphous defects or disordered regions present in crystalline materials. This causes a further increase in molecular mobility in these already "activated" or "hot" spots, which then triggers drug degradation in these hot spots. A large number of studies have confirmed the correlation between increased molecular mobility caused by the plasticizing effect of water and drug degradation.[37,41,42] In most solid state degradation reactions such as hydrolysis and oxidation, water can act as both a plasticizer and a reactant.[40]

The concept of pH is usually associated with aqueous solutions. With drug products formulated in solid dosage forms, the concept of pH may be used in two ways: first, lyophilized solid dosage forms, such as most protein-based drugs, are typically made from pH buffered solutions. Hence, the pH of the buffered solutions are used as the "pH" of the corresponding lyophilized solids or lyophiles.[43] A solid state stability study of lyophilized insulin displayed a profile of degradation rate *versus* "pH" that is very similar to insulin degradation in solutions in the same pH range.[44] Second, for other solid dosage forms, the concept of microenvironment pH has been used, which is defined as the pH value of the slurry resulting from the disintegration of the solid forms by water.[45] The impact of pH in such solid dosage formulations also depends on the manufacturing process: a buffer in a formulation resulting from a wet granulation tends to be more effective in stabilizing the drug substance than that from a direct compression.[46] It appears that wet granulation enables a more uniform distribution of the buffer and perhaps also renders closer contact between the buffer and the drug substance than does the direct compression.

1.4   Organization of the Book

This book is divided into nine chapters. The previous sections of this chapter stipulate the objective and scope of the book. In addition, this chapter also covers briefly a number of topics that are useful for the discussion in later chapters but are outside the main scope of this book. These topics include thermodynamics and kinetics of chemical reactions and several key concepts in solid state chemistry.

Chapters 2 and 3 discuss hydrolytic and oxidative degradations, respectively, which are the two most commonly observed types of drug degradation, owing to the universal presence of moisture and oxygen. In Chapter 2, Hydrolytic Degradation, several hydrolytic mechanisms are discussed first along with electrical and steric effects that impact hydrolysis, followed by more than 30 examples of drugs containing functional groups susceptible to hydrolytic degradation. The susceptibilities of these drugs are compared, whenever possible, by their hydrolysis activation energies, on a semi-quantitative basis in most cases. In cases where activation energies were not reported in the literature, enthalpy of activation and/or reaction rate constants are used for the comparison.

In Chapter 3, Oxidative Degradation, several major types of autooxidation mechanisms are discussed first, followed by discussion of specific oxidation pathways of drugs with various functional groups and structures in relation to each type of major autooxidation mechanism. Among the latter, the ubiquitously known Fenton reaction and the little known, but more relevant Udenfriend reaction, are discussed. The discussion focuses on their roles in free radical-mediated autooxidation by activating molecular oxygen into several reactive oxygen species (ROS), that is, $O_2^{-\bullet}/HO_2^{\bullet}$, H_2O_2, and $HO^{\bullet}$. The radicals ROS then trigger radical chain reactions, in which process organic peroxyl radicals and hydroperoxides are predominant intermediates. Hydroperoxides can undergo homolytic cleavage, owing to their relatively low O–O bond dissociation energies, as well as metal ion-catalyzed heterolytic cleavage. Homolytic cleavage generates alkoxyl and hydroxyl radicals, while heterolytic cleavage reproduces peroxyl radicals. Non-radical reactions of peroxides are also discussed, in particular those responsible for the formation of *N*-oxide, sulfoxide, and epoxide degradants. Lastly, the general mechanism for a non-radical autooxidation pathway, carbanion/enolate-mediated autooxidation (base-catalyzed autooxidation), is also discussed. This mechanism is well known in synthetic organic chemistry but much less known for its role in degradation chemistry. Examples are used to demonstrate that this pathway can be significant for those drug molecules containing somewhat "acidic" carbonated CH_n moieties, particularly when the drugs are formulated in liquid dosage forms. Overall, more than 60 examples of oxidative drug degradation in real life scenarios are presented. These examples cover the functional groups, moieties, and structures that are commonly seen in drug molecules.

Chapter 4, Various Types and Mechanisms of Degradation Reactions, covers most of the commonly occurred drug degradation reactions, except for hydrolysis, oxidation, degradation caused by interaction with excipients, and photochemical degradation. The latter two categories of drug degradation are discussed in Chapters 5 and 6, respectively. The degradation reactions covered in Chapter 4 include elimination, decarboxylation, nucleophilic conjugate addition and its reverse process, aldol condensation and the retro-aldol process, rearrangement and isomerization, cyclization, dimerization and oligomerization, and a few examples of degradation via miscellaneous mechanisms and pathways. While some of the above classifications are relatively specific in scope, for example, decarboxylation, nucleophilic conjugate addition, and aldol condensation, others are more complicated as they can involve many sub-types or different types of degradation pathways and mechanisms. For example, cyclization and dimerization can involve many different types of degradation mechanisms. More than 50 examples of drug degradation are discussed in this chapter.

Chapter 5, Drug–Excipient Interaction and Adduct Formation, is organized according to the types of interaction between a drug substance and excipients: direct interaction between the drug substance and excipients, interaction of the drug substance with impurities of excipients, interaction of the drug substance with degradants of excipients, and interaction of the drug substance with impurities from packaging materials. In the first category, the various pathways of the well-known Maillard reaction are reviewed first, followed by discussion of the interaction between carboxyl-containing drug substances and excipients possessing hydroxyl and amino groups through the formation of ester and amide linkages, respectively. Drug–excipient interaction due to transesterification is also covered. In the second part of the chapter, drug degradation caused by reaction with impurities and degradants originating from excipients and packaging materials is described.

Chapter 6, Photochemical Degradation, starts with a concise overview of photophysical and photochemical events upon irradiation of a molecule with a chromophore that absorbs in the wavelength of irradiation. Afterwards, various non-oxidative photochemical degradation pathways are discussed, including photochemical decarboxylation, isomerization, aromatization, dehalogenation, cyclization, elimination, dimerization, and Norrish type I and II photoreactions. In the second part of the chapter, type I and II photo-sensitized oxidative reactions are covered. Type I photosensitized oxidative degradation is mediated by free radicals and it appears that $HO^{\bullet}$ free radical is frequently involved, while type II photosensitized oxidative degradation involves reaction with singlet oxygen. Approximately 30 examples are presented that illustrate the mechanistic details of various photochemical degradation pathways.

Chapter 7, Chemical Degradation of Biological Drugs, discusses chemical degradation pathways of drugs derived from proteins, carbohydrates, and nucleic acids. Owing to the fact that the vast majority of biological drugs are protein and peptide drugs, the major emphasis of the chapter is devoted to the degradation chemistry of proteins and peptides with a specific discussion on each of the amino acid residues containing labile side chains. Some discussion of degradation involving the peptide backbone is also conducted. The last two sections of the chapter provide an overview of the degradation pathways commonly seen in carbohydrates and nucleic acids, including hydrolysis of glycosidic linkages, hydrolysis of phosphorodiester linkages, and oxidation of nucleobases.

In Chapter 8, Strategies for Elucidation of Degradant Structures and Degradation Pathways, strategies for elucidating degradant structures and degradation pathways are discussed. The emphasis is placed on the use of the LC-MS^{n} molecular fingerprinting technique (n is typically 1 to 4) in conjunction with mechanism-based stress studies and NMR spectroscopy. Typically, samples are first analyzed using LC-MS^{n}, through which parent ions, major fragments, and their structural relationship *versus* those of the API can be established. Based on the LC-MS^{n} molecular fingerprinting results, a possible degradation mechanism may be inferred. A particular type of stress study (forced degradation) of the API is then designed and carried out accordingly. The formation of the degradant in the stress study can be confirmed by LC-MS^{n} through matching LC-MS^{n} molecular fingerprints of the stress-generated degradant and the one observed in the original sample. Sufficient amounts of the degradant can usually be generated for various 1D and 2D NMR analyses for further structure elucidation or verification if necessary. The degradation pathways of the drug can be proposed using the elucidated structures of the degradants and their intermediates. This effective approach is demonstrated through three in-depth case studies.

The last chapter, Chapter 9, Control of Drug Degradation, provides a high level overview of the strategies for controlling drug degradation based on understanding of the degradation pathways and mechanisms. The overview consists of 13 discussion topics, which cover the strategies for controlling drug degradation from the perspectives of (i) early phase drug design and development, (ii) consideration of multiple degradation pathways, (iii) formulation

development with regard to the use of antioxidants, preservatives, chelating agents, control of pH, variability of excipient impurities, and the use of excipients that shield or protect APIs from degradation, (iv) the impact of the manufacturing process, and (v) selection of proper packaging materials with regard to control of moisture, oxygen, and light induced degradation.

References

1. http://www.fda.gov/downloads/Drugs/GuidanceComplianceRegulatory-Information/Guidances/ucm079235.pdf, last accessed 07 July 2012.
2. International Conference on Harmonisation, *ICH Harmonised Tripartite Guideline: Stability Testing of New Drug Substances and Products, Q1A(R2)*, dated 6 February 2003.
3. World Health Organization, *Stability Testing of Active Pharmaceutical Ingredients and Finished Pharmaceutical Products*, WHO Technical Report Series, No. 953, 2009, Annex 2.
4. M. K. Mone and K. B. Chandrasekhar, *J. Pharm. Biomed. Anal.*, 2010, **53**, 335.
5. D. P. Williams, N. R. Kitteringham, D. J. Naisbitt, M. Pirmohamed, D. A. Smith and B. K. Park, *Curr. Drug Metab.*, 2002, **3**, 351.
6. Y. Nagatsu, T. Higuchi and M. Hirobe, *Chem. Pharm. Bull.*, 1990, **38**, 400.
7. A. S. Kalgutkar, I. Gardner, R. S. Obach, C. L. Shaffer, E. Callegari, K. R. Henne, A. E. Mutlib, D. K. Dalvie, J. S. Lee, Y. Nakai, J. P. O'Donnell, J. Boer and S. P. Harriman, *Curr. Drug Metab.*, 2005, **6**, 161.
8. International Conference on Harmonisation, *ICH Harmonised Tripartite Guideline: Impurities in New Drug Substances, Q3A(R2)*, dated 25 October 2006.
9. International Conference on Harmonisation, *ICH Harmonised Tripartite Guideline: Impurities in New Drug Products, Q3B(R2)*, dated 2 June 2006.
10. *Guideline on the Limits of Genotoxic Impurities*, European Medicines Agency, London, 28 June 2006.
11. M. J. Frank, *Anal. Chem.*, 1985, **57**, 68A.
12. J. W. McGinity, T. R. Patel and A. H. Naqvi, *Drug Dev. Commun.*, 1976, **2**, 505.
13. T. Huang, M. E. Garceau and P. Gao, *J. Pharm. Biomed. Anal.*, 2003, **31**, 1203.
14. W. R. Wasylaschuk, P. A. Harmon, G. Wagner, A. B. Harman, A. C. Templeton, H. Xu and R. A. Reed, *J. Pharm. Sci.*, 2007, **96**, 106.
15. S. R. Byrn, R. R. Pfeiffer and J. G. Stowell, *Solid-state Chemistry of Drugs*, SSCI Inc., West Lafayette, Indiana, 2nd edn, 1999.
16. S. Yoshioka and V. J. Stella, Stability of Drugs and Dosage Forms, Kluwer Academic Publishers, New York, 2002.
17. M. D. Argentine, P. K. Owens and B. A. Olsen, *Adv. Drug Delivery Rev.*, 2007, **59**, 12.
18. S. Ahuja and K. M. Alsante (eds), *Handbook of Isolation and Characterization of Impurities in Pharmaceuticals*, Academic Press, San Diego, CA, 2003.

19. Z. Cimarosti, F. Bravo, P. Stonestreet, F. Tinazzi, O. Vecchi and G. Camurri, *Org. Process Res. Dev.*, 2010, **14**, 993.
20. http://en.wikipedia.org/wiki/Arrhenius_equation. Last accessed 22 April 2012.
21. http://en.wikipedia.org/wiki/Transition_state_theory. Last accessed 10 April 2012.
22. M. D. Pluth, R. G. Bergman and K. N. Raymond, *J. Org. Chem.*, 2009, **74**, 58.
23. M. A. Portnov, M. N. Vaisman, T. A. Dubinina and V. A. Zasosov, *Pharm. Chem. J.*, 1974, **8**, 381.
24. M. E. Gil-Alegre, J. A. Bernabeu, M. A. Camacho and A. I. Torres-Suarez, *Il Farmaco.*, 2001, **56**, 877.
25. R. T. Darrington and J. Jiao, *J. Pharm. Sci.*, 2004, **93**, 838.
26. S. Y. P. King, M. S. Kung and H. I. Fung, *J. Pharm. Sci.*, 1984, **73**, 657.
27. S. Ebel, M. Ledermann and B. Reyer, *Eur. J. Pharm. Biopharm.*, 1991, **37**, 80.
28. R. T. Magari, K. P. Murphy and T. Fernandez, *J. Clin. Lab. Anal.*, 2002, **16**, 221.
29. K. C. Waterman and R. C. Adami, *Int. J. Pharm.*, 2005, **293**, 101.
30. P. H. Stahl and C. G. Wermuth (eds), *Handbook of Pharmaceutical Salts: Properties, Selection, and Use*, Verlag Helvetica Chimica Acta, Zurich, Switzerland, 2008.
31. G. G. Z. Zhang and D. Zhou, in *Developing Solid Oral Dosage Forms: Pharmaceutical Theory and Practice*, eds Y. Qiu, Y. Chen, L. Liu and G. G. Z. Zhang, Academic Press, Burlington, MA, 2009, Chapter 2.
32. R. J. Bastin, M. J. Bowker and B. J. Slater, *Org. Process Res. Dev.*, 2000, **4**, 427.
33. S. S. Leung and D. J. W. Grant, *J. Pharm. Sci.*, 1997, **86**, 64.
34. H. Kissinger, *Anal. Chem.*, 1957, **29**, 1702.
35. J. Stamp, *Kinetics and Analysis of APM Decomposition Mechanisms in Aqueous Solutions Using Multiresponse Methods*, Ph.D. thesis, University of Minnesota, Minneapolis, MN, 1990.
36. C.-T. Lin, P. Perrier, G. G. Clay, P. A. Sutton and S. R. Byrn, *J. Org. Chem.*, 1982, **47**, 2978.
37. S. R. Byrn, W. Xu and A. W. Newman, *Adv. Drug Delivery Rev.*, 2001, **48**, 115.
38. L. W. Jelinski, J. J. Dumais, R. E. Stark, T. S. Ellis and F. E. Karasz, *Macromolecules*, 1983, **16**, 1019.
39. G. Zografi, *Drug Dev. Ind. Pharm.*, 1988, **14**, 1905.
40. C. Ahlneck and G. Zografi, *Int. J. Pharm.*, 1990, **62**, 87.
41. E. Y. Shalaev and G. Zografi, *J. Pharm. Sci.*, 1996, **85**, 1137.
42. S. Yoshioka and Y. Aso, *J. Pharm. Sci.*, 2007, **96**, 960.
43. R. G. Strickley and B. D. Anderson, *Pharm. Res.*, 1996, **13**, 1142.
44. R. G. Strickley and B. D. Anderson, *J. Pharm. Sci.*, 1997, **86**, 645.
45. A. T. M. Serajuddin, A. B. Thakur, R. N. Ghoshal, M. G. Fakes, S. A. Ranadive, K. R. Morris and S. A. Varia, *J. Pharm. Sci.*, 1999, **88**, 696.
46. S. I. F Badawy, R. C. Williams and D. L. Gilbert, *J. Pharm. Sci.*, 1999, **88**, 428.

CHAPTER 2

Hydrolytic Degradation

2.1 Overview of Hydrolytic Degradation

Hydrolysis is a chemical reaction during which a bond in the hydrolysis substrate is cleaved by a water molecule. Hydrolytic degradation is probably the most commonly observed drug degradation pathway for the following two reasons: First, a great number of functional groups and/or structural moieties of drug molecules are susceptible to hydrolysis. These functional groups and/or structural moieties include, but are not limited to, ester, lactone, amide, lactam, carbamate, phosphate, phosphamide, sulfamide, imine, guanidine, and ether. Second, water molecules are ubiquitously present in the form of moisture (as in solid dosage forms), liquid (as in aqueous liquid formulations), or crystal water (as in crystalline drug substances and excipients).

The hydrolysis of carbonyl compounds (esters, lactones, amides, lactams, and carbamates) constitutes the majority in the overall hydrolytic degradation of drugs. The mechanisms of specific acid-catalyzed (*i.e.* hydronium) and specific base-catalyzed (*i.e.* hydroxide ion) hydrolysis are generally illustrated in Scheme 2.1.[1]

In terms of drug stability under real life scenarios, for example, the stability during long term storage conditions and stability during drug passage through the acidic environment of the stomach, the specific acid-catalyzed mechanism may be directly relevant at pH 1 to 3, which is the typical pH range in the stomach.[2] In other cases, hydrolysis via catalysis by general acid, base, and nucleophilic attack may be more relevant. In the general acid- and base-catalyzed hydrolysis, the catalyst functions as a proton transferring agent in the activated complex,[3] as illustrated in Scheme 2.2.

In neutral hydrolysis in which no apparent acid or base is present, water can act as a general base catalyzing the hydrolysis.[3] In cases where nucleophilic species are present in drug formulations, hydrolysis via nucleophilic attack may

RSC Drug Discovery Series No. 29
Organic Chemistry of Drug Degradation
By Min Li
© Min Li 2012
Published by the Royal Society of Chemistry, www.rsc.org

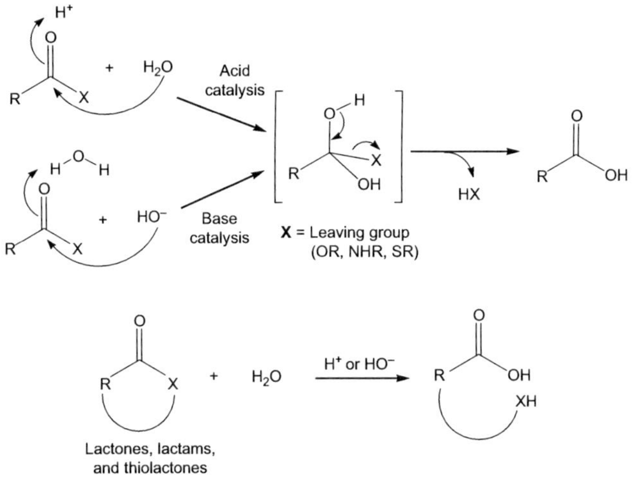

Acid catalysis
Base catalysis
X = Leaving group (OR, NHR, SR)
HX
H⁺ or HO⁻
Lactones, lactams, and thiolactones

Scheme 2.1

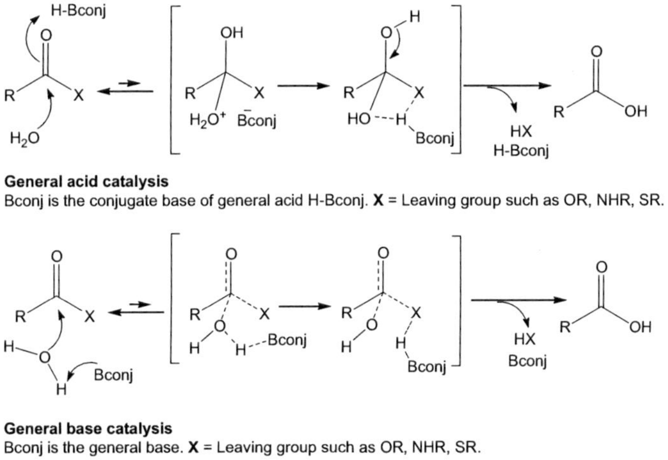

General acid catalysis
Bconj is the conjugate base of general acid H-Bconj. X = Leaving group such as OR, NHR, SR.
General base catalysis
Bconj is the general base. X = Leaving group such as OR, NHR, SR.

Scheme 2.2

occur if the acylated nucleophile intermediates are labile towards hydrolysis. Hydrolysis via nucleophilic attack is shown in Scheme 2.3.

Certain metal ions, in particular divalent ions, such as Zn^{2+}, Cu^{2+}, Ni^{2+}, and Co^{2+}, can also catalyze hydrolysis of esters,[4,5] amides,[6] and acetals.[7] It appears that the complexation of the metal ions to the carbonyl group causes further polarization on the carbonyl carbon, making the latter more susceptible to hydrolytic attack in a similar way to the specific acid-catalyzed hydrolysis.

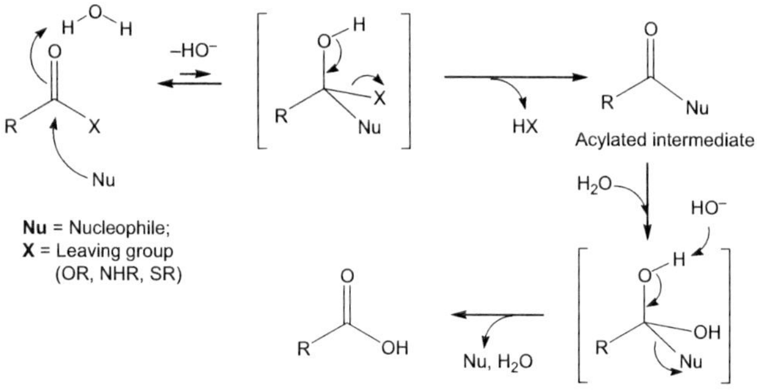

Scheme 2.3

It is important to note that the actual hydrolytic degradation of a drug molecule in a particular formulation may proceed via one or a combination of the mechanisms shown above. Since hydrolytic degradation involves attack of the hydrolysis substrate by a water molecule, the degradation reaction is generally of second order. Nevertheless, for hydrolytic degradation in aqueous formulations, the degradation kinetics can usually be approximated as pseudo-first order, since the amount (or concentration) of water far exceeds the hydrolysis substrate, that is, the hydrolysable drug molecule. Based on the pseudo-first order kinetics treatment and use of the Arrhenius equation, half-lives of the hydrolysis substrates and the activation energies of the hydrolytic reactions can be easily determined (see Section 1.3.1), which in turn could help to estimate the shelf-lives of the formulated drug products by extrapolation.[8]

A number of factors have an impact on the rate and/or mechanism of hydrolysis, such as the temperature, pH, steric hindrance, electronic properties of the hydrolysable moiety, and nature of the leaving group. While an increase of steric hindrance in the hydrolysis substrate slows down the hydrolysis, the presence of electron-withdrawing groups in the acyl part of the substrate (which further polarizes the carbonyl group) and good leaving groups promotes the hydrolysis. While higher temperature accelerates the hydrolysis as it does for all chemical reactions, the impact of pH is not straightforward within the pH range 1 to 13. Usually, the hydrolytic stability of drug molecules is evaluated within such a pH range. These factors will be discussed in the following sections with examples of specific drug molecules. The hydrolytic susceptibility of each type of drug molecules will be compared semi-quantitatively, wherever possible, based on their activation energies for hydrolysis. All the drug molecules with activation energy data, which are discussed in the chapter, are summarized in Table 2.1.

In this chapter, degradation caused by esterification, transesterification, and formation of amide linkage will be briefly discussed. These degradation pathways are related to the hydrolytic degradation: esterification and formation of amide linkage are the reverse reactions of hydrolytic degradation. On the other hand, an alcohol and amine play the role of water in the hydrolytic degradation, respectively, in transesterification (alcoholysis) and aminolysis.

Table 2.1 Summary of Activation Energies for Hydrolytic Degradation of Drug Molecules Discussed in this Chapter.

Drug molecule/model compound	E_a (kcal mol^{-1})^a	Rate constant or half life ($t_{1/2}$)	Comments	Ref.
Ester				
Ethyl acetate	9.2			9
Hexyl acetate	11.4			9
Methylphenidate	16.0; 12.4		Acidic; basic	10
Aspirin	16.7; 12.5		Acidic; basic	15
Procaine	16.8		Acidic	14
Benzocaine	18.6		Acidic	14
t-Butyl acetate	27			9
Diloxanide furoate		$t_{1/2}$ 12 min	1% aq. NaHCO$_3$, pH ~8.5 RT	11
Lactone				
Lovastatin	12–13		pH 2	21
Simvastatin	12–13		pH 2	21
Daptomycin	13.61		pH 10	22
Amide				
Acetamide	18.2			9
Chloramphenicol	21		Acidic 10% aq. propylene glycol	25
Chloramphenicol	24	7.5×10^{-9} s^{-1}	pH 6; $t_{1/2}$ at 37 °C	10
Indomethacin	24.26		pH 7	27
Lidocaine	33.8; 26.3		Acidic; basic	28
Prazosin		0.0096;0.99 h^{-1}	0.1 M HCl; 0.1 M NaOH. 80 °C	26
Terazosin		0.097;1.75 h^{-1}	0.1 M HCl; 0.1 M NaOH. 80 °C	26
Doxazosin		0.042;15.71 h^{-1}	0.1 M HCl; 0.1 M NaOH. 80 °C	26
Lactam				
Amoxicillin	17.35		pH 5	32
Ampicillin	18.3; 9.2		pH 5; 9.78	31
Latamoxef	21		12% aq. manitol	36
Cefepime	22.2			40
Cefaclor	25.95; 17.08		pH 7.20; 9.95	35
Carbamate				
Estramustine	21.3		Enthalpy of activation, pH ~1–9	49
Carzelesin	26.7; 31.2		pH 1.5; 7.2	54
Phosphate				
6α-Methylprednisolone sodium phosphate	27		pH 7.5	56
Prednisolone sodium phosphate	30.16		pH 8	55
Phosphoramide				
Fosaprepitant dimeglumine	22		ACN/0.1% H$_3$PO$_3$ (v/v, 1/1)	58

Table 2.1   (*Continued*)

Drug molecule/model compound	E_a (kcal mol^{-1})^a	Rate constant or half life ($t_{1/2}$)	Comments	Ref.
Sulfonamide			Sulfonamide bond is extremely stable. E_a should be > 30 kcal mol^{-1}	
N,N-Dimethyl-methanesulfonamide		$< 2 \times 10^{-9}$ M^{-1}s^{-1}	This rate constant is $\sim 10^4$ times lower than a carboxamide	62
Cyclic β-sultam			Rate constant of small ring β-sultam is $\sim 10^7$ times greater than a typical sulfonamide	63
Imide				
Phenobarbital	18.87		pH 10.12	69, 70
Phenobarbital glycosy-lated metabolites (2)	19.0/19.1; 16.1/16.3		pH 7; 9.95	69, 70
Urea bond				
Urea		$t_{1/2}$ 3.6 years	Aq. Soln, 38 °C	71
Oxime				
Obidoxime	26.2			76
Glycosidic bond				
Doxorubicin	22.0		0.5 N HCl	80
Tobramycin	32; 15		1N HCl; 1N NaOH	77
Epoxide				
Propylene oxide	~ 19.0–19.5			86, 87
Propylene oxide		6.9×10^{-7} s^{-1}		88
Butadiene oxide		1.4×10^{-5} s^{-1}	A vinylepoxide	88
Cycloheptadiene oxide		2.7×10^{-5} s^{-1}		88
Cyclohexadiene oxide		2.6×10^{-4} s^{-1}		88
Cyclopentadiene oxide		5.2×10^{-3} s^{-1}		88

^{a}If the values reported in the literature are in the unit of kJ mol^{-1}, they are converted into kcal mol^{-1} by dividing the reported values by 4.184.

2.2 Drugs Containing Functional Groups/Moieties Susceptible to Hydrolysis

2.2.1 Drugs Containing an Ester Group

In general, the ester linkage is quite prone to hydrolytic degradation due to the weak ester bond. For example, in neutrally buffered solution, the activation energy of hydrolysis of ethyl acetate, which can be considered a typical, simple ester, is only 38.4 kJ mol^{-1} (9.2 kcal mol^{-1}), while the activation energy of hydrolysis of acetamide is 76.0 kJ mol^{-1} (18.2 kcal mol^{-1}).[9] With longer and sterically more hindered groups, the esters become more resistant toward hydrolysis: the activation energies for hydrolysis of hexyl acetate and t-butyl acetate are 47.5 and 113 kJ mol^{-1} (11.4 and 27 kcal mol^{-1}), respectively. Because

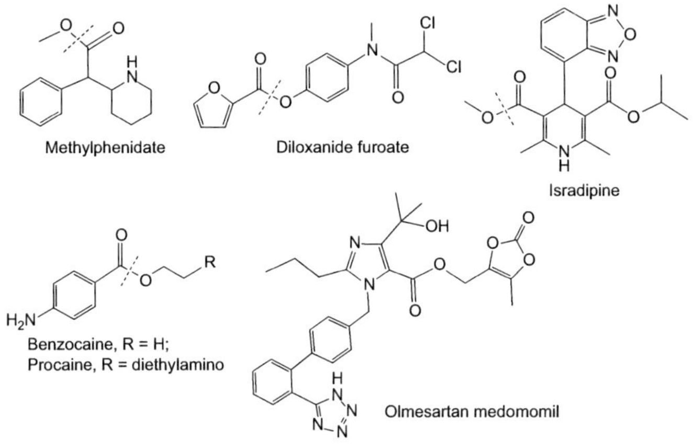

Figure 2.1 Examples of drugs containing an ester linkage. The labile ester linkage is indicated by a dashed line.

of this general labile nature, ester functionality of appropriate chain lengths has been frequently employed in the design of pro-drugs.

The stability of esters in solutions is strongly dependent upon the pH of the solution. The rate of ester hydrolysis becomes significant under both strongly acidic and basic conditions. In general, a base-catalyzed hydrolysis is more efficient than an acid-catalyzed one. For example, the activation energies for the hydrolysis of methylphenidate, a psychostimulant drug containing a methyl ester moiety (Figure 2.1), are 16.0 and 12.4 kcal mol^{-1}, respectively, under acid and base catalysis.[10] Gadkariem *et al.* studied the hydrolytic stability of diloxanide furoate, a furoyl pro-drug used as a luminal amoebicide, which contains both an ester and an amide linkage (Figure 2.1).[11] In a simulated gastric fluid at 37 °C, the drug was found to be stable for at least 2 hours. On the other hand, in 1% aqueous NaHCO$_3$ solution at room temperature, it underwent hydrolysis through the cleavage of the furoyl ester linkage, with a half-life of only 12 min. The 1% aqueous NaHCO$_3$ solution should have a pH of approximately 8.5; nevertheless, at the end of the hydrolysis study, the pH of the solution became 10.6. Another example demonstrating esters' different stability behaviors under acidic and basic catalysis can be found in the hydrolytic stress study of isradipine,[12] a potent calcium channel antagonist in the dihydropyridine family. Isradipine contains both a methyl and an isopropyl ester linkage. When stressed in 0.1 N HCl at 60 °C for 6 hours, 4% of isradipine was degraded, while stress in 0.1 N NaOH under the same temperature and duration of time caused it to degrade by 15%. Under both the acidic and basic conditions, the only degradation product is the hydrolytic degradant resulting from the loss of the methyl ester group. The sterically more hindered isopropyl ester linkage remained intact.

Olmesartan medomomil (Figure 2.1) is an ester of imidazolyl acid, which was found to hydrolyze approximately 40% more under acidic (0.01 N HCl) than under basic (0.01 N NaOH) catalysis.[13] This somewhat unusual behavior may be attributed to the fact that under the acidic condition, the imidazole ring is protonated which could make the ester carbonyl carbon more positive, causing the ester linkage more susceptible to hydrolysis.

Benzoates and related compounds are commonly encountered moieties in drugs and excipients. Benzocaine and procaine are two 4-aminobenzoate-based local anesthetic drugs, with the former containing a simple ethyl ester and the latter a substituted ethyl ester (Figure 2.1). The activation energies for acid-catalyzed hydrolysis of the two drugs are 18.6 and 16.8 kcal mol^{-1}, respectively.[14]

Aspirin, one of the oldest synthetic drugs, has been frequently cited as an example for the hydrolytic degradation of drugs. Chemically, it is acetylsalicylic acid or 2-(acetyloxy)benzoic acid. In its hydrolysis, the leaving group is an *ortho*-carboxylated phenol, which is a good leaving group. Its activation energies were found to be 16.7 and 12.5 kcal mol^{-1} under acid and base catalysis, respectively.[15] Owing to the presence of the *ortho*-carboxyl group, the mechanism of aspirin hydrolysis becomes somewhat complicated, resulting in different hypotheses during the course of elucidating its mechanism. It appears that it all started from the observation made by Edwards that hydrolysis of aspirin is independent of pH in the region of pH 4 to 8.[16,17] To explain this behavior, the mechanism of intramolecular nucleophilic catalysis was proposed, in which the neighboring ionized carboxyl group acts as the nucleophile (Scheme 2.4).[18]

In this nucleophilic mechanism, the formation of the anhydride intermediate resulting from the initial nucleophilic attack would be the rate determining step, followed by a quick hydrolysis step. This mechanism appeared to be supported by a hydrolysis study conducted in H$_2$^{18}O-enriched water, in which 6% ^{18}O was found to be incorporated in the hydrolysis product.[19] Nevertheless, Fersht and Kirby suspected the validity of the nucleophilic mechanism based on the fact that the proposed anhydride is a quite reactive species and as such, the 6% incorporation of ^{18}O seems to be too low.[20] Furthermore, they found that no significant incorporation of ^{18}O occurred in a hydrolysis study of aspirin conducted in 20 atom % ^{18}O-enriched water at 39 °C. Only about 2% ^{18}O incorporation was observed at 100 °C. Nevertheless, the authors attributed the 2% incorporation to a slow acid-catalyzed exchange with the carbonyl group of salicylic acid, which occurs after the hydrolysis of aspirin is completed. Based on these results and others that contradict the nucleophilic mechanism, Fersht and Kirby compared aspirin hydrolysis with that of phenyl acetate

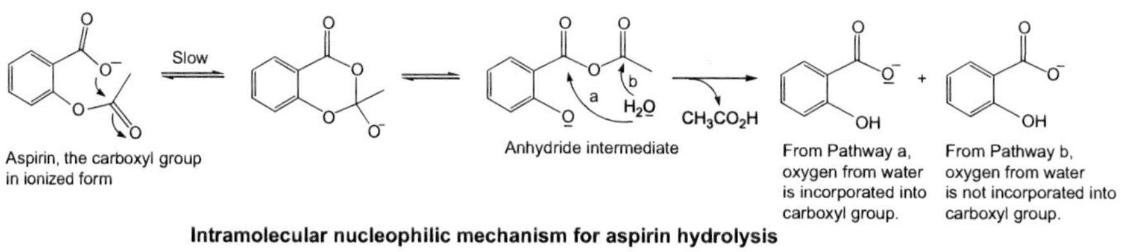

Aspirin, the carboxyl group in ionized form   Anhydride intermediate   From Pathway a, oxygen from water is incorporated into carboxyl group.   From Pathway b, oxygen from water is not incorporated into carboxyl group.

Intramolecular nucleophilic mechanism for aspirin hydrolysis

Scheme 2.4

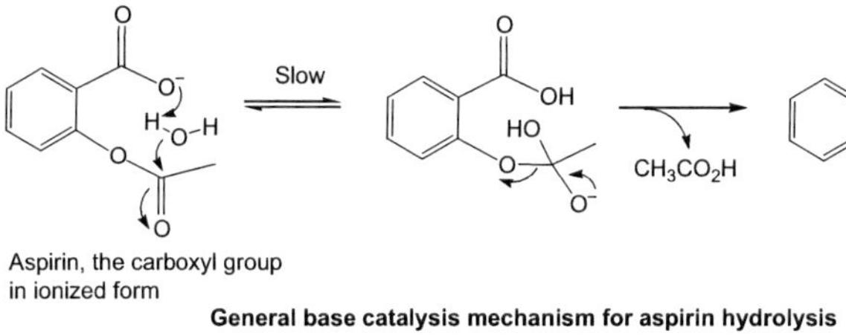

Aspirin, the carboxyl group
in ionized form

General base catalysis mechanism for aspirin hydrolysis

Scheme 2.5

that is known to involve general base-catalyzed hydrolysis. Based on the comparison, they concluded that aspirin undergoes intramolecular general base-catalyzed hydrolysis with the *ortho*-carboxyl group acting as the general base (Scheme 2.5). The two authors further concluded that the hydrolysis is a classic general base catalysis as opposed to a kinetically equivalent mechanism of general acid-specific base catalysis.

2.2.2 Drugs Containing a Lactone Group

Lactones are cyclic esters. Lovastatin and simvastatin, two members of the enormously successful family of HMG-CoA reductase inhibitors for hyper-cholesterol treatment, contain a six-membered lactone moiety which is subject to hydrolysis, giving rise to the active forms of the drugs, δ-hydroxyl acids. Kaufman studied the hydrolysis kinetics and thermodynamics of the two drugs along with two model compounds.[21] Since the structures of the two drugs only differ by a methylene group which is away from the lactone moiety, the two drugs were found to have very similar kinetic and thermodynamic parameters. The activation energies for hydrolysis of the two drugs should be in the neighborhood of 12–$13\,\mathrm{kcal\,mol^{-1}}$ at pH 2.0, based on the values determined from the two closely related model compounds. These values of the activation energies do not seem to differ much from those for hydrolysis of regular esters. Nevertheless, a notable difference between lactones and regular esters is that the hydrolysis products of the former, such as the δ-hydroxyl acids in the cases of lovastatin and simvastatin, can undergo lactonization (*i.e.* reforming the cyclic esters) almost as equally well as the hydrolysis. The lactonization becomes particularly significant for lactones with small to moderate ring sizes. In the same study, Kaufman demonstrated that both the hydrolysis (of the lactones) and lactonization (of the δ-hydroxyl acids) produced the same reaction mixture under acidic catalysis, indicating that acid-catalyzed hydrolysis of the lactones is reversible. This reversibility would be true for the hydrolysis of drugs con-taining regular ester functionality, should the leaving group (an alcohol) remain in close proximity to the resulting acid group. However, in drug degradation of regular esters, the acid and alcohol parts resulting from the hydrolytic degra-dation are typically very low in concentration and tend to diffuse away from each other, making the reverse reaction (esterification) insignificant.

A large number of macrocyclic drug molecules contain an ester linkage. Since the ring size is usually quite large, the hydrolysis of these drugs appears to be similar to that of the linear esters with similar numbers of carbons. For example, daptomycin is a lipopeptide antibiotic that contains a macrocyclic ring connected by nine amide bonds and one ester bond. The activation energy for hydrolysis of the ester bond at pH 10 was reported to be 13.61 kcal mol^{-1}.[22]

2.2.3 Drugs Containing an Amide Group

In general, amides are much more stable towards hydrolysis than esters of similar structures, as evidenced by the significant difference in their activation energies (see Section 2.2.1). This stability is usually explained by the following two facts. First, nitrogen is less electronegative than oxygen and hence the amide carbonyl is less electrophilic than the ester carbonyl. Second, the resonance between the lone pair electrons of the nitrogen and the carbonyl group gives the amide bond partial double bond character.[23] The amide linkage is the major chemical bond in protein and peptide drugs, as well as in many small molecule drugs. In this section, we will only discuss the hydrolysis of non-peptide drugs containing primarily one amide linkage. Hydrolytic degradation of peptide drugs will be discussed along with protein drugs in Chapter 7, Chemical Degradation of Biological Drugs.

Chloramphenicol is the first broad-spectrum antibiotic discovered in 1947. Owing to its toxicity, its use nowadays has been mostly limited to topical treatment of eye infections. It contains a dichloroacetylamide linkage which can undergo hydrolytic degradation in its pharmaceutical preparation.[24] At pH 6, the activation energy for the amide hydrolysis was found to be 24 kcal mol^{-1}.[10] At this pH and 25 °C, the pseudo-first order rate constant for the hydrolysis was determined to be 7.5×10^{-9} s^{-1}, corresponding to a half-life of approximately 3 years. Under acid catalysis, the activation energy for the hydrolysis carried out in an aqueous solution containing 10% propylene glycol was found to be 21 kcal mol^{-1}.[25]

Piperazine is a frequently used scaffold in drug design and, as such, an amide linkage to the piperazine nitrogen is often employed to connect the moiety to other parts of the drug molecule. Ojha *et al.* performed stress studies on three α_1-adrenergic receptor antagonists containing a piperazinylamide moiety (Figure 2.2).[26]

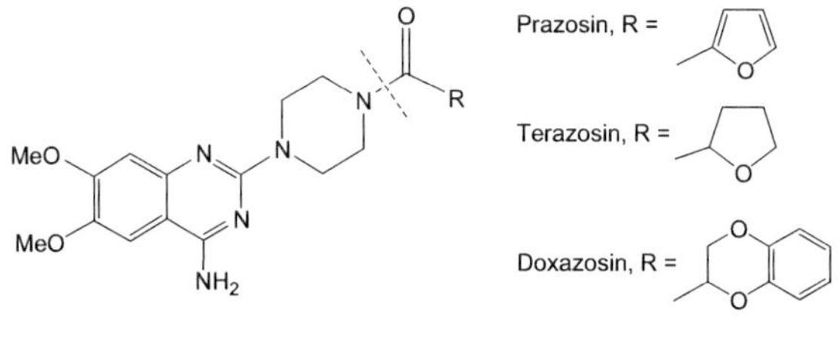

Figure 2.2 Three α_1-adrenergic receptor antagonists containing a piperazinylamide moiety.

This amide linkage was found to hydrolyze much more efficiently under basic than acidic conditions. The rate constants of the pseudo-first order hydrolysis of prazosin, terazosin, and doxazosin in 0.1 M NaOH at 80 °C were determined to be 0.99, 1.75, and 15.71 h^{-1}, respectively. These values are 103, 18, and 370 times higher than the rate constants of the three drugs in 0.1 M HCl at 80 °C, respectively. Under both acidic and basic conditions, prazosin is most resistant toward hydrolysis, which probably can be attributed to the stabilization of the amide bond by the conjugation of the amide carbonyl group to the aromatic furoyl ring.

Indomethacin, a non-steroidal anti-inflammatory drug (NSAID), contains an indole in its core structure (Figure 2.3). The nitrogen of the indole ring is connected to a 4-chlorobenzoyl group and the resulting amide bond is susceptible to hydrolysis. Krasowska studied the hydrolytic stability of indomethacin in aqueous solutions containing varying concentrations of polysorbates (2.5 to 10% wt/v).[27] It was found that in the pH range 2.2 to 8.0, the stability of indomethacin increased as the concentration of the surfactant was raised. For example, at pH 7.0 and room temperature, the 10% degradation time ($t_{0.1}$) of indomethacin was found to increase from 45 days in the absence of polysorbates to 316 days in the presence of 2.5% polysorbates. Accordingly, the hydrolysis activation energies were also found to increase from 101.50 kJ mol^{-1} (24.26 kcal mol^{-1}) at 0% polysorbates to 116.82 kJ mol^{-1} (27.92 kcal mol^{-1}) at 10% polysorbates. These results clearly show that the use of polysorbates stabilizes indomethacin from hydrolytic degradation.

Lidocaine, a local anesthetic, is an acylated aniline derivative. It is quite stable in both acidic and basic conditions,[28] probably attributable to the steric hindrance provided by the two methyl groups at the 2- and 6-positions of the aniline ring. The activation energies for hydrolysis of protonated and free base lidocaine were 33.8 and 26.3 kcal mol^{-1}, respectively. The more facile hydrolysis of the latter was presumed to be due to intramolecular general base catalysis shown in Scheme 2.6.

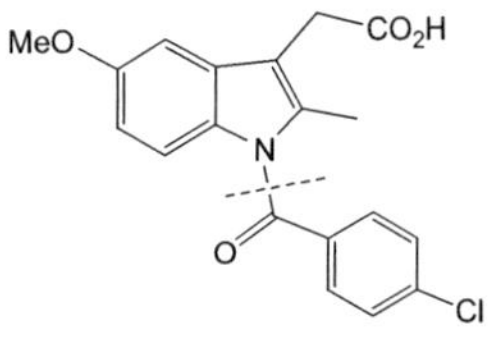

Figure 2.3   Indomethacin.

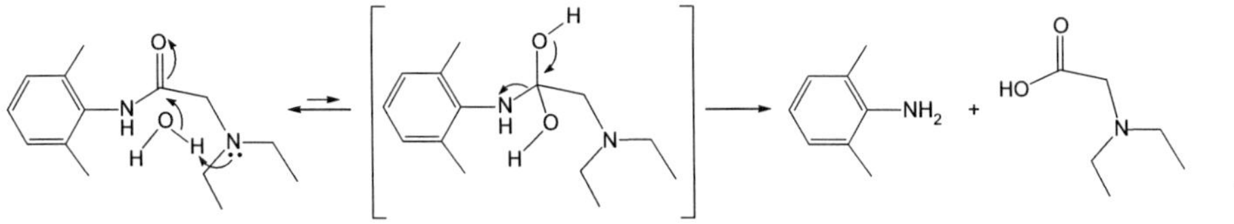

Scheme 2.6

2.2.4 β-Lactam Antibiotics

Lactams are cyclic amides and the most important class of drugs based on lactams are probably β-lactam antibiotics which include the families of penicillins and cephalosporins. Penicillins contain a β-lactam ring fused to a five-membered thiazolidine ring, where one of the atoms in the fully saturated ring is sulfur. On the other hand, the β-lactam in cephalosporins is fused to a six-membered ring. β-Lactam antibiotics are suicide inhibitors of bacterial peptidases. They inhibit the enzymes by acylating the hydroxyl group of the active site serine residue. The acylation occurs via a nucleophilic attack by the serine hydroxyl group, resulting in the opening of the four-membered lactam ring. A similar process also occurs during the hydrolytic degradation of β-lactam antibiotics both *in vitro* and *in vivo*, in which water becomes the nucleophile. It is estimated that the acylating capability of a β-lactam is about 100 times more powerful than a regular amide; fusion of the β-lactam to the thiazolidine ring causes another 100 times increase in its reactivity.[29]

Davis *et al.* studied alcohol-catalyzed hydrolysis of benzylpenicillin, also known as penicillin G, under basic conditions.[30] It was found that the rate-limiting step is not the formation of the tetrahedron intermediate, but rather the subsequent ring opening step. In the latter step, water acts as a general acid providing a proton for the leaving amine group. The penicilloyl ester intermediate then undergoes hydrolysis to give the penicillic acid. Prior to the hydrolysis, the penicilloyl ester intermediate is in equilibrium with the enamine form resulting from an intramolecular elimination (or retro-nucleophilic addition) process. All the degradation pathways are summarized in Scheme 2.7.

The above mechanism may be meaningful in cases where a penicillin drug is formulated with excipients that contain a hydroxyl group. Obviously, water and/or hydroxide ion can also attack the lactam carbonyl group directly, producing the penicillic acid/penicillate in one step.

Both ampicillin and amoxicillin are amino derivatives of benzylpenicillin. The difference between the two aminopenicillins is that amoxicillin has an additional hydroxyl group at the 4-position of the benzyl group. The introduction of the amino group at the benzylic position broadens the antibacterial

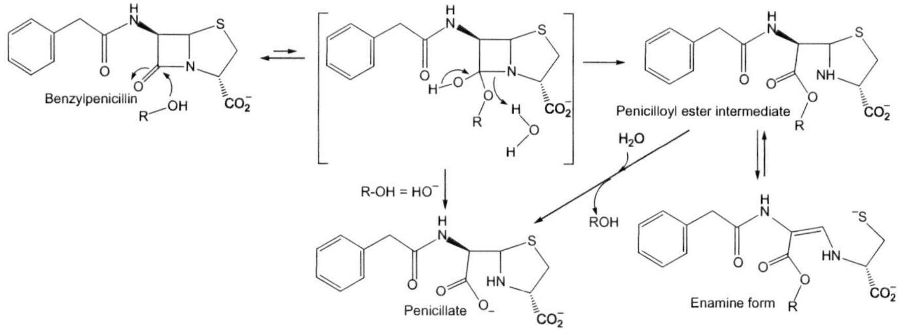

Scheme 2.7

activities of both drugs. On the other hand, the presence of the amino group also has an impact on the stability and degradation pathways of the two drug molecules. For example, aminopenicillins have increased stability under acidic conditions; at pH 5, the activation energies of the degradation of ampicillin and amoxicillin are 18.3 kcal mol^{-1}[31] and 17.35 kcal mol^{-1} (72.59 kJ mol^{-1}),[32] respectively. The enhanced stabilities are attributed to the presence of the predominant zwitterion forms of both drugs at pH 5. At pH below 5, amoxicillin degrades mainly via hydrolysis of its β-lactam ring. At pH 5 to 7, the formation of diketopiperazine (DKP) becomes its major degradation pathway, resulting from the intramolecular attack of the β-lactam ring by the benzylic amino group. Above pH 8, hydroxide ion-catalyzed hydrolysis predominates. Overall, the stability of amoxicillin reaches its maximum around pH 5, where it is mostly present in the zwitterion form. Consequently, its pH–degradation rate curve displays a U-shape within the pH range 1 to 10.[32] The degradation pathways of amoxicillin are summarized in Scheme 2.8. It is anticipated that ampicillin should follow very similar pathways owing to the high similarity in their structures; for example, the pH–rate curve of ampicillin also shows a similar U-shape.[31,33]

Because of their broad antibacterial activities, enhanced stabilities, and improved pharmacokinetic profiles, ampicillin and in particular amoxicillin, are among the most prescribed antibiotic drugs today.

Cephalosporins are another very important class of β-lactam-based antibiotics which emerged after the penicillin family was introduced. They differ from the penicillins in that a six-membered ring replaces the five-membered thiazolidine ring of the penicillins. In general, the degradation pathways of cephalosporins are similar to those of penicillins:[34] at both low and high pH

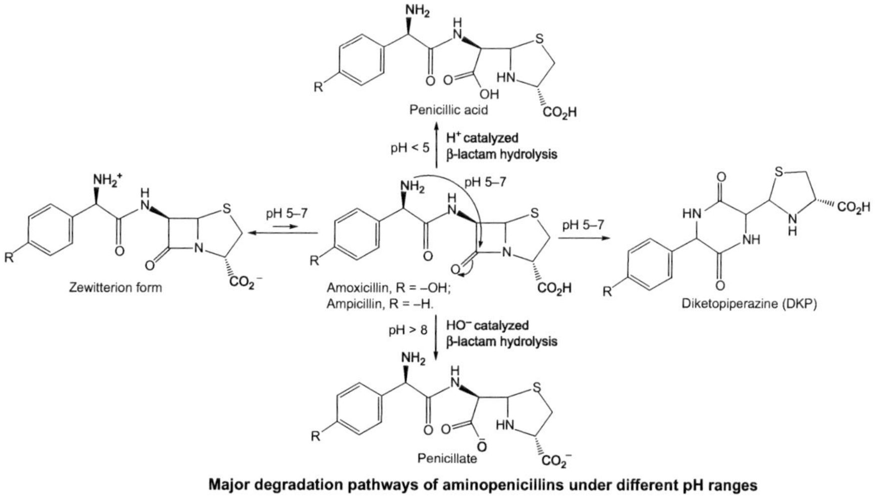

Major degradation pathways of aminopenicillins under different pH ranges

Scheme 2.8

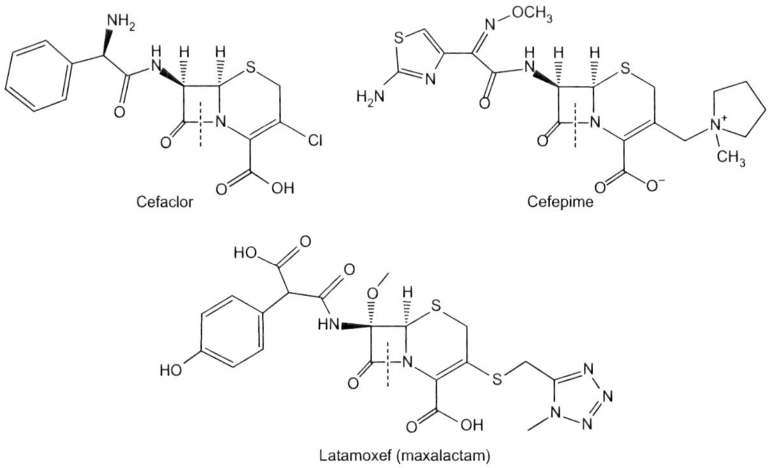

Figure 2.4   Cefaclor, cefepime, and latamoxef.

ranges, the hydrolysis of the β-lactam ring is usually the major pathway. In the mid-range pH, formation of the DKP degradant becomes a significant pathway. An example can be found in the case of cefaclor, a second generation cephalosporin antibiotic that is still widely used today. Similar to aminopenicillins, it contains the benzylamino moiety (Figure 2.4).

As such, it also displays a U-shaped pH–rate profile according to a kinetic study performed by Dimitrovska *et al.*,[35] in which the maximum stability of cefaclor occurred between pH 3 and 4. In this pH range, the majority of cefaclor molecules should exist in zwitterion form. The activation energies of cefaclor degradation were found to be 25.95 and 17.08 kcal mol^{-1}, respectively, at pH 7.20 and 9.95. At basic pH, cefaclor appears to be much more stable than ampicillin, which is an aminopenicillin, judging from the value of the latter's activation energy of only 9.2 kcal mol^{-1} at pH 9.78.[31]

Latamoxef (maxalactam, Figure 2.4) is an oxacephem antibiotic. Its bicyclic nucleus is identical to that of a cephalosporin except that the sulfur atom of the latter is replaced by an oxygen atom. In an experimental solid formulation of amorphous latamoxef with 12% (wt/wt) manitol, the drug molecule degraded via two comparable degradation pathways: hydrolysis of the lactam ring and decarboxylation.[36] The activation energies of the two pathways were found to be very similar, both at approximately 88 kJ mol^{-1} (21 kcal mol^{-1}).

The hydrolysis of penicillins and cephalosporins can also be facilitated by transition metal ions such as Cu^{2+} and Zn^{2+}; it was reported that the hydrolysis of benzylpenicillin and cephaloridine could be enhanced by a 1:1 molar ratio of Cu^{2+} at ~10^8 and ~10^4 folds, respectively.[37,38] The dramatic rate enhancement was attributed to the stabilization of the tetrahedral transition state of the hydrolysis by the metal ions. For example, it was determined that Cu^{2+} can lower the energy of the transition state by 13.9 kcal mol^{-1} at 30 °C.[39] Nevertheless, in the case of cefaclor, the degradation rates were found to increase only by a very moderate 25% and 49% at pH 4.08 and 6.00, respectively.[35] This observation does not seem to fully support the hypothesis of transition state stabilization by metal ions.

Cefepime is a fourth generation cephalosporin that contains an alkoxylamine moiety rather than the benzylic amino group that is typically present in second generation cephalosporins (Figure 2.4). In the side chain attached to its six-membered ring, cefepime has a quaternary ammonium functional group. The pH–rate profile of the drug molecule also shows a U-shaped curve,[40] which is typical for many cephalosporins,[34,41] with maximum stability occurring at pH 4 to 6. The activation energy for the spontaneous degradation (presumably hydrolysis) of cefepime was determined as 22.2 kcal mol^{-1}, which is comparable to the activation energy for the spontaneous hydrolysis of a typical cephalosporin.

It has been frequently observed that the initial hydrolytic degradation products expected from the hydrolysis of cephalosporins, that is, cephalosporoates, are difficult to isolate, most likely owing to further degradation.[42] For example, cepirome, cefsulodin, and a number of other structurally similar cephalosporins undergo further degradation via hydrolysis and decarboxylation after the β-lactam ring is cleaved.[43–45] The initial β-lactam hydrolysis-triggered degradation pathways for these cephalosporins are summarized in Scheme 2.9 using ceftazidime as an example.[40]

Imipenem was the first drug to be approved in the β-lactam family of carbapenems, which are variations of penicillins in which the sulfur atom of the penicillin core structure is replaced by a methylene group. The carbapenems have extremely broad anti-bacterial activity[46] and are usually co-administered clinically with cilastatin, an inhibitor of the renal brush border enzyme dehydropeptidase-I (DHP-I). Ratcliffe *et al.* studied the degradation behavior of imipenem under acidic conditions.[47] At dilute concentrations, the lactam ring of imipenem was quickly cleaved to produce a 2-pyrroline degradant, which then isomerized to a diastereomeric 1-pyrroline mixture upon neutralization. At high concentrations, dimerization of the drug molecule occurred, which ultimately led to the formation of an intermolecular diketopiperazine (DKP) degradant. Meanwhile, the formamidinium side chain of imipenem was found to undergo facile rotational isomerization, apparently due to the hindered rotation about the partial double bond between the carbon and nitrogen. The degradation pathways are summarized in Scheme 2.10.

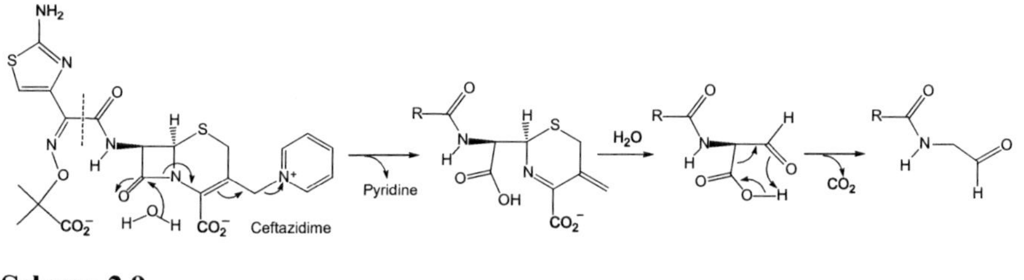

Scheme 2.9

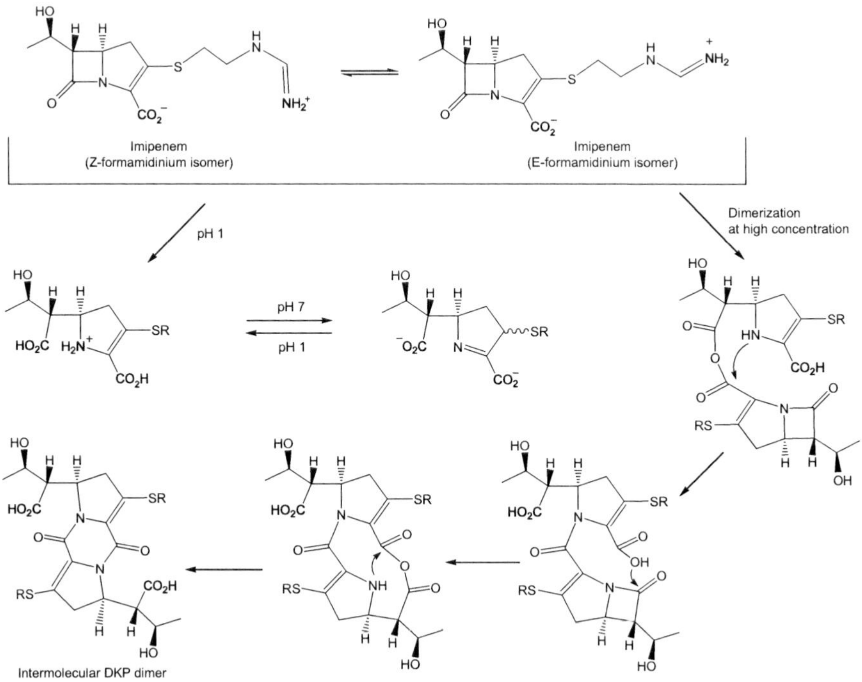

Scheme 2.10

Biapenem, another antibiotic in the carbapenem family, has an almost identical bicyclic lactam core structure. As such, it undergoes the same degradation pathways as illustrated in Scheme 2.10 except for side chain Z/E isomerization,[48] owing to a different substituent attached to the sulfur moiety. The formation of a DKP degradant is a common degradation pathway in peptide and protein-based drugs, which will be discussed in both Chapter 4 and Chapter 7.

2.2.5 Carbamates

Estramustine, a nitrogen mustard derivative of estradiol, and its water soluble pro-drug, estramustine 17-phosphate, are used to treat prostate cancer. Loftsson *et al.* determined the kinetics and thermodynamic parameters of the hydrolysis of the N,N-bis(2-chloroethyl)carbamyl moiety.[49] In pH range ~ 1 to 9 where the rate of hydrolysis remained constant, the enthalpy of activation was determined as $89.3\,\text{kJ}\,\text{mol}^{-1}$ ($21.3\,\text{kcal}\,\text{mol}^{-1}$) and the entropy of activation $-62.0\,\text{J}\,\text{mol}^{-1}$ ($-14.8\,\text{cal}\,\text{mol}^{-1}$). According to the authors, the relatively high enthalpy and small negative entropy values suggest that the hydrolytic degradation is a unimolecular reaction, since a bimolecular mechanism involving attack of the carbamate moiety by water should result in a much larger negative entropy, typically in the range -170 to $-200\,\text{J}\,\text{mol}^{-1}$.[50,51] Hence, a spontaneous

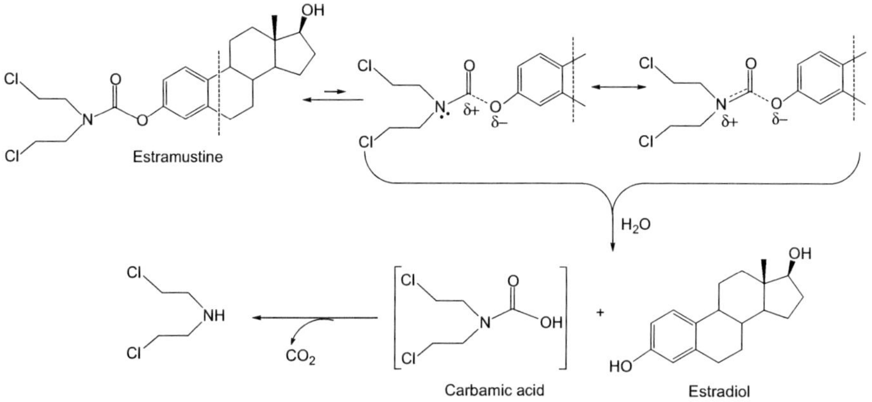

Scheme 2.11

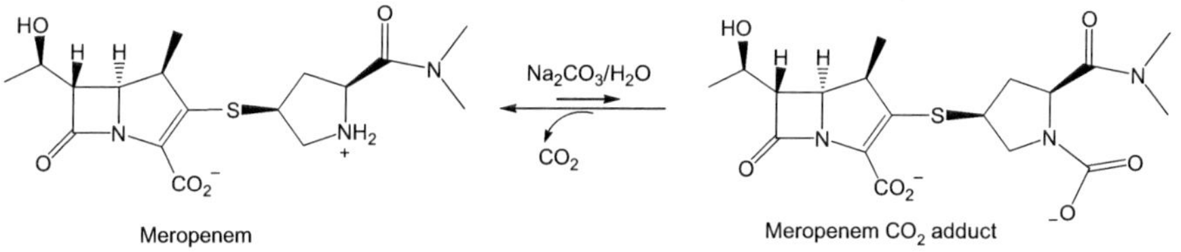

Scheme 2.12

hydrolytic degradation mechanism, rather than a tetrahedron intermediate mechanism, was proposed (Scheme 2.11).

The carbamic acid formed upon the cleavage of the carbamate bond is not stable and it spontaneously degrades to di-(2-chloroethyl)amine and carbon dioxide. This instability is also true of other carbamic acids; only in solutions or lyophilized formulations where carbonate or bicarbonate buffers are present, can a carbamic acid survive. For example, when the drug product Merrem, a powder blend of crystalline meropenem active pharmaceutical ingredient (API) and sodium carbonate, was dissolved in aqueous solution, approximately one-third of the API molecules were present in the carboxylated form with a CO_2 molecule covalently attached to the secondary amino group of meropenem (Scheme 2.12).[52]

When an aqueous solution of Merrem was lyophilized, the covalent linkage between the secondary amine (of a small fraction of the API) and CO_2 could be detected, although such a linkage is absent in the product itself. The only known case in which a carbamic acid moiety is stable stoichiometrically in the solid state appears to be in the active site of a metalloenzyme, urease, where a carboxylated lysine residue is stabilized through complexation with nickel ions.[53]

Carzelesin is an investigational anticancer pro-drug that incorporates an carbamate moiety that can be activated. Upon hydrolytic cleavage of the

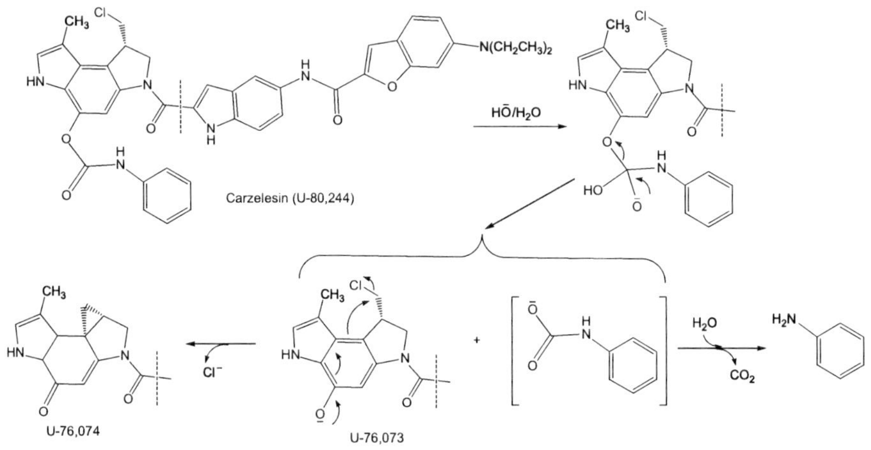

Scheme 2.13

carbamyl linkage under basic conditions, an intermediate, U-76,073, formed, which further decomposed to the active drug, U-76,074.[54] Under acidic conditions, the distribution of degradation products seemed to be more complicated, attributed to the fact that the intermediate, U-76,073, degraded to a number of further degradants under such conditions. The activation energies of the hydrolytic degradation of carzelesin were found to be $111.8\,\mathrm{kJ\,mol^{-1}}$ ($26.7\,\mathrm{kcal\,mol^{-1}}$) and $130.7\,\mathrm{kJ\,mol^{-1}}$ ($31.2\,\mathrm{kcal\,mol^{-1}}$) at pH 1.5 and 7.2, respectively. The proposed mechanism for the hydrolytic degradation under basic conditions is shown in Scheme 2.13.

2.2.6 Phosphates and Phosphoramides

Phosphorylation of drugs containing hydroxyl and amine groups is a common way of converting water-insoluble drugs into water-soluble pro-drugs. The resulting phosphoryl ester and amide bonds are expected to be labile *in vivo* and they are also susceptible to hydrolytic degradation *in vitro*. For example, prednisolone sodium phosphate is the pro-drug of prednisolone, obtained through phosphorylation of the 21-hydroxyl group of the corticosteroid. It is believed that the hydrolysis of the phosphate ester group is its primary degradation pathway. According to a study by Stroud *et al.*, prednisolone sodium phosphate (Figure 2.5) has an activation energy of $126.2\ \mathrm{kJ\ mol^{-1}}$ ($30.16\ \mathrm{kcal\ mol^{-1}}$) for its hydrolysis at pH 8.[55] Structurally similar 6α-methylprednisolone sodium phosphate (Figure 2.5) has an activation energy of $113\ \mathrm{kJ\ mol^{-1}}$ ($27\ \mathrm{kcal\ mol^{-1}}$) at pH 7.5.[56] These values are comparable to or higher than the activation energy for hydrolysis of a typical amide in a similar pH range, suggesting the hydrolytic stability of these steroidal phosphoryl esters is similar to, or somewhat higher than a typical carbonyl amide. A recent study of the degradation chemistry of betamethasone sodium phosphate (Figure 2.5), another structural

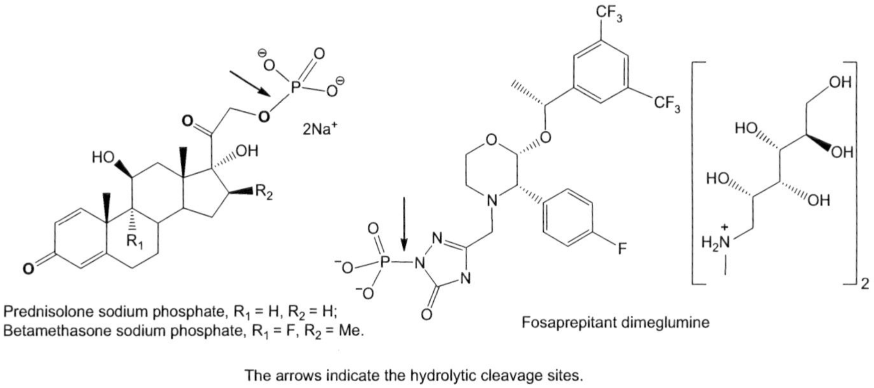

Prednisolone sodium phosphate, R_1 = H, R_2 = H;
Betamethasone sodium phosphate, R_1 = F, R_2 = Me.

Fosaprepitant dimeglumine

The arrows indicate the hydrolytic cleavage sites.

Figure 2.5 Hydrolytic degradation of prednisolone sodium phosphate and betamethasone sodium phosphate.

analog of prednisolone sodium phosphate, shows that betamethasone sodium
phosphate also undergoes other significant degradation pathways such as
elimination of the phosphate and isomeric D-homoannular ring expansion
(refer to Section 4.5.6),[57] in addition to hydrolytic degradation.

The phosphoryl ester bond is a critical linkage in RNA and DNA molecules. In
recent years, RNA and DNA-based drugs have emerged as a novel category of
drug candidates. The degradation chemistry of the phosphoryl ester bond in these
drugs will be discussed in Chapter 7, Chemical Degradation of Biological Drugs.

Phosphoramides are less common than phosphate pro-drugs and they
are generally less hydrolytically stable than the phosphates. For example,
fosaprepitant dimeglumine (Figure 2.5) is the phosphoramide pro-drug of
fosaprepitant. The activation energy for the hydrolysis of the pro-drug in an
acetonitrile–0.1% phosphoric acid (50:50, v/v) mixture was determined as
91 kJ mol^{-1} (22 kcal mol^{-1}).[58] This activation energy is higher than that of a
typical carbonyl ester drug, but lower than that of a phosphoryl ester drug.

Cyclophosphamide is a phosphoramide mustard used for the treatment of a
relatively wide variety of cancers. It contains both phosphoramide and phosphoryl ester linkages in a single cyclized phosphorus-containing moiety, which
can be considered to be a phosphoryl analog of a cyclic carbamate. Hydrolytic
degradation via various pathways constitutes the major degradation chemistry
of the drug. Friedman *et al.* postulated that the degradation is triggered by a
rate-limiting intramolecular displacement of the chlorine in the mustard moiety
by the phosphoramide amine.[59] This mechanism was confirmed by Zon *et al.*
through a comprehensive degradation study of the drug.[60] Subsequent
hydrolysis of the unstable bicyclic intermediate produced in the initiation
step gives a linear diamine degradant. Zon *et al.* also demonstrated that further
hydrolytic degradation of the latter degradant proceeds through the intermediacy of a symmetrical aziridinium ion. Under drastic hydrolytic conditions,
the phosphoryl ester functionality will ultimately be hydrolyzed. The degradation chemistry of cyclophosphamide is summarized in Scheme 2.14.

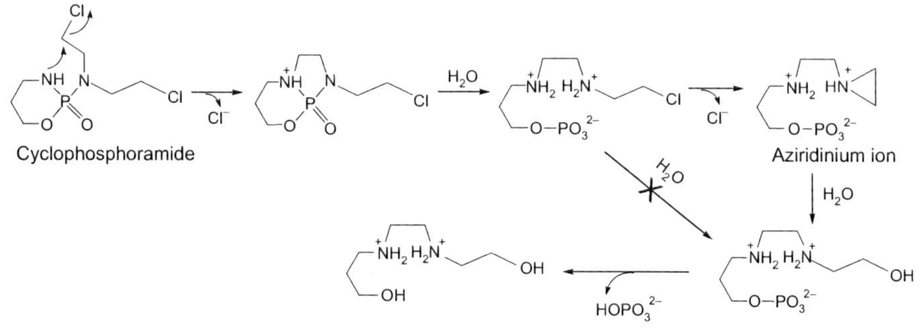

Cyclophosphoramide

Aziridinium ion

Scheme 2.14

2.2.7 Sulfonamide Drugs

The sulfonamide functionality provides a moiety that has been widely used in drug design ever since the introduction of the first of the so-called sulfa drugs, Prontosil, in the 1930s. The sulfa drugs are antibacterial antibiotics. Although their use has been largely replaced by β-lactam antibiotics, a few sulfa drugs are still in clinical use today including sulfanilamide, the active metabolite of Prontosil, and sulfamethoxazole (Figure 2.6). The latter is usually used in combination with trimethoprim.

The sulfonamide bond is extremely stable toward acidic and basic hydrolysis,[61] and because it is so stable, accurate determination of the sulfonamide hydrolysis rate constant turns out to be difficult. For hydroxide ion-catalyzed hydrolysis of *N,N*-dimethylmethanesulfonamide, the second-order rate constant is estimated to be less than 2×10^{-9} $M^{-1}s^{-1}$, approximately four orders of magnitude lower than that of the corresponding carboxamide.[62] Nevertheless, the cyclic β-sultams are enormously more reactive toward hydrolysis than their acyclic counterparts and the ratio of the hydrolysis enhancement is estimated to be at least 10^7 fold. Owing to their structural similarity to β-lactams, β-sultams were evaluated as inhibitors of serine proteases.[63]

In addition to the sulfa drugs, the sulfonamide moiety is also utilized in other drugs, most notably in a number of thiazide and hydrothiazide diuretic drugs, for example, benzthiazide, chlorothiazide, hydrochlorothiazide, hydroflumethiazide, cyclopenthiazide, methyclothiazide, and polythiazide. Because of their high stability, the two sulfonamide bonds remain intact during the hydrolytic cleavage of the thiazide ring (Scheme 2.15).

Yamana *et al.* studied the hydrolytic degradation mechanism of chlorothiazide under both basic and acidic conditions.[64,65] They found that the hydrolysis proceeded via two different intermediates, namely *N*-(2-amino-4-chloro-5-sulfamoylphenylsulfonyl) formamide (Pathway a) and 5-chloro-2,4-disulfamoylformanilide (Pathway b), under basic and acidic conditions, respectively. Owing to the structural similarity, benzthiazide would be expected to follow the same hydrolytic degradation pathways. In the hydrolytic degradation of the hydrothiazide drugs, it is not clear if hydrolysis would go through

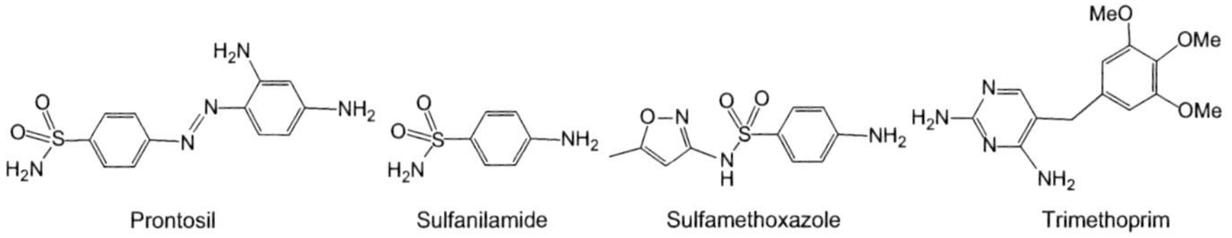

Figure 2.6 Three sulfa drugs and trimethoprim.

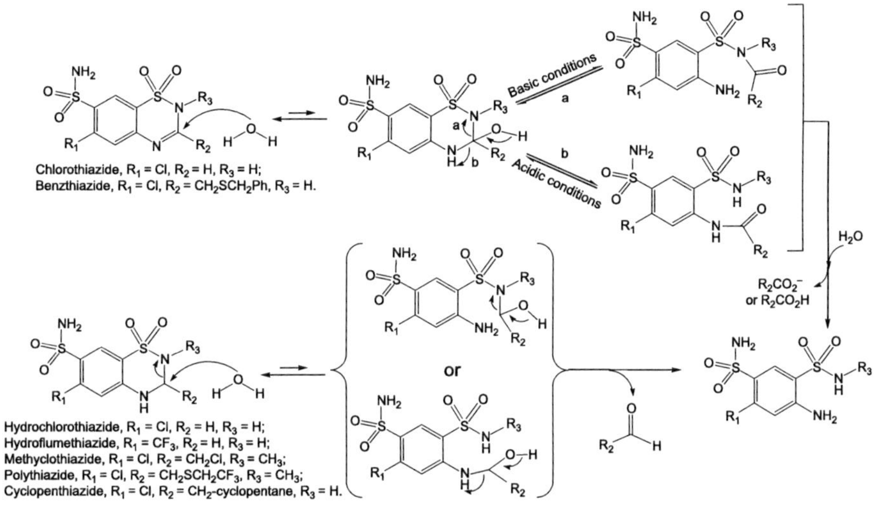

Scheme 2.15

two analogous intermediate,s as illustrated in Scheme 2.15, although it is known that they produce the same type of hydrolytic degradants as the thiazide drugs.

2.2.8 Imides and Sulfonylureas

Phenobarbital is the oldest and used to be the most widely used anticonvulsant. Its core six-membered ring is made by condensation between urea and α-ethyl-α-phenylmalonate or its equivalent. In humans, its major metabolic pathway is the N-glycosidation of one imide nitrogen, leading to the formation of two diastereomers of 1-(1-β-D-glucopyranosyl)phenobarbital. The metabolites are susceptible to hydrolytic degradation at their 1,6- or 3,4-imide linkages followed by decarboxylation (Scheme 2.16).[66] Other barbiturates, including phenobarbital itself, also undergo this degradation pathway.[67,68]

The activation energies for the hydrolysis of the two diastereomers are 19.0 and 19.1 kcal mol^{-1} at pH 7.0 and 16.1 and 16.3 kcal mol^{-1} at pH 9.95, respectively. The latter values at pH 9.95 are lower than that for the hydrolysis

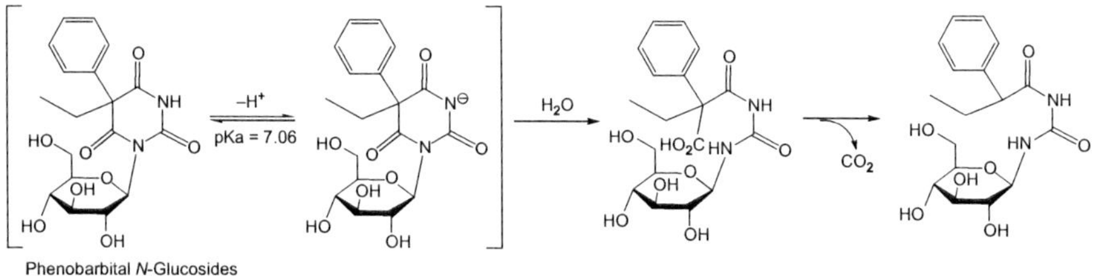

Phenobarbital *N*-Glucosides

Scheme 2.16

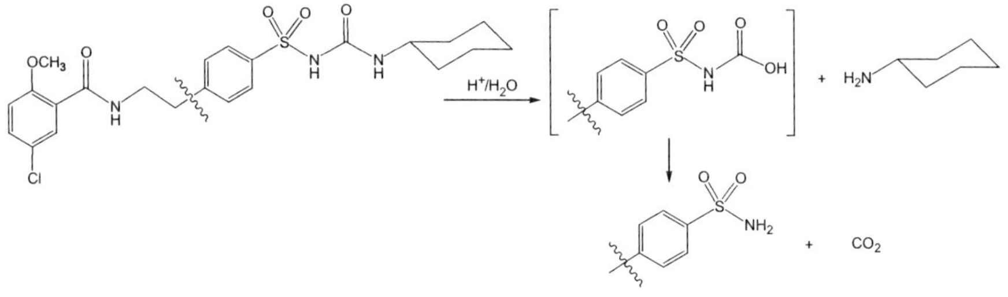

Scheme 2.17

of phenobarbital, which is $78.96 \text{ kJ mol}^{-1}$ ($18.87 \text{ kcal mol}^{-1}$) at pH 10.12.[69,70] In a study by Vest *et al.*,[66] phenobarbital was never observed as a degradant, indicating that the *N*-glycosidic bond is much more resistant to hydrolysis than the two imide bonds at the 1,6- and 3,4-positions. The authors also did not observe any evidence for the cleavage of the two urea bonds at the 1,2- and 2,3-positions, which is consistent with the fact that urea is remarkably resistant toward hydrolytic degradation.[71]

Glibenclamide is an antidiabetic drug in the sulfonylurea family. In addition to the sulfonylurea functionality, it also contains an amide linkage. According to a degradation study by Wiseman *et al.*,[72] glibenclamide hydrolyzes under acidic conditions via the cleavage of one of the urea bonds (Scheme 2.17). This study suggests that the urea carbonyl is greatly activated by the strongly electron-withdrawing sulfonyl group. As such, the substituted urea moiety is now weaker than the carboxyl amide bond that is also present in the drug molecule.

2.2.9 Imines (Schiff Bases) and Deamination

An imine, also called Schiff base, is formed by condensation between a ketone/ aldehyde and a primary amine (Scheme 2.18). A secondary amine is capable of undergoing the same condensation, but the product is an iminium salt. This condensation reaction is readily reversible and hence, a typical imine is quite labile towards hydrolytic degradation. On the other hand, the imine formed is

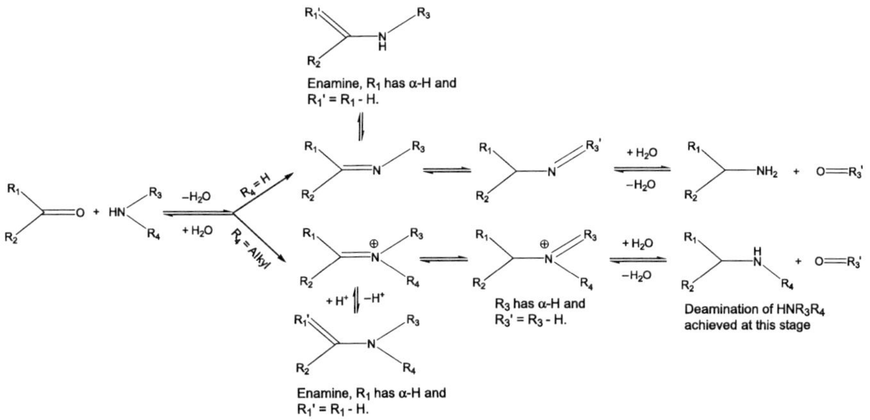

Scheme 2.18

also capable of tautomerization to produce enamine and additional imine isomers.

This combination of reversibility and tautomerization is the culprit that causes a great number of degradation reactions such as racemization and epimerization of chiral drugs that contain amino and/or ketone/aldehyde groups, dimerization via formaldehyde, and deamination of amino-containing drugs. While there will be a relevant discussion of racemization, epimerization, and dimerization in Chapter 4, Sections 4.5.2, 4.5.3, and 4.7, respectively, in this section we discuss a couple of cases of deamination where hydrolysis of an imine or related moiety is the key step.

L-367,073, a potent fibrinogen receptor antagonist, was developed as a potential treatment for a variety of cardiovascular diseases. It is a cyclic heptapeptide analog containing an unnatural amino acid residue, 4-aminomethylphenylanaline. In accelerated stability studies of the drug candidate in a lyophilized formulation containing mannitol, the aminomethyl group of the 4-aminomethylphenylanaline residue was found to be converted to an aldehyde (or formyl) group.[73] The deamination was attributed to a process through the intermediacy of two imines (Schiff bases), followed by hydrolysis of the second imine, as illustrated in Scheme 2.19.

The tautomerization from imine 1 to imine 2 should be favored owing to the fact that the latter should be more stable because of the conjugation of its imine bond to the phenyl group. The aldehyde responsible for the degradation is mostly likely to be a reducing sugar impurity in mannitol. Sometimes, deamination of this type is also categorized as oxidative degradation. In such a case, although the oxygen in the final degradant (the formyl moiety) is from water, the ultimate oxidizing reagent is the reducing aldehyde impurity: the oxygen of the aldehyde functionality is ultimately transferred (via water) to the deamination substrate and in return, the aldehyde impurity is reduced to an amine.

Another example of drug degradation via deamination can be found in the anticancer drug, gemcitabine, a β-difluoronucleoside. Jansen *et al.* studied the

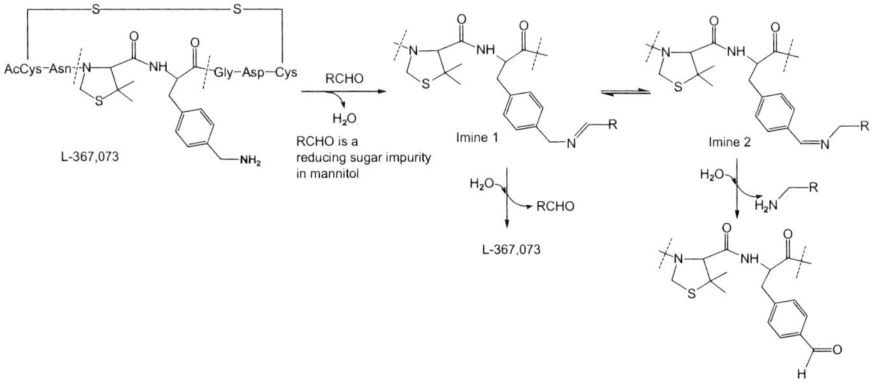

Scheme 2.19

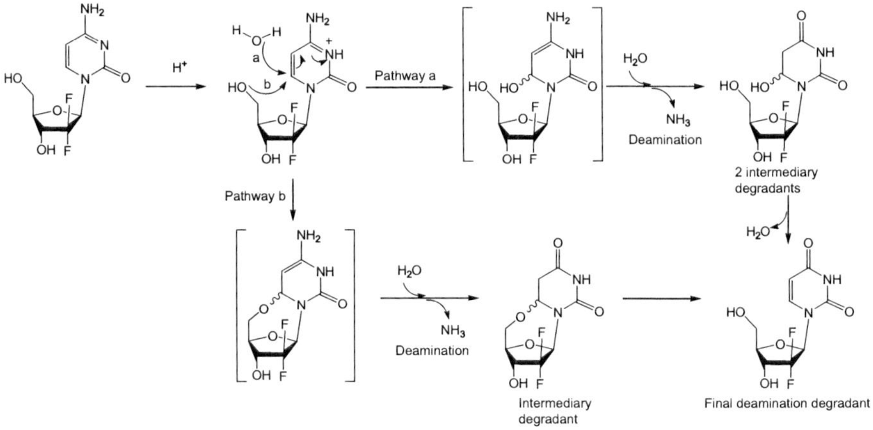

Scheme 2.20

solution stability of the drug at pH 3.2.[74] A major deamination degradant was observed along with three other degradants. The study indicated that the latter are intermediary degradants for the deamination degradant (Scheme 2.20).

In the proposed degradation mechanism, attack by water and the 5'-hydroxyl of the sugar moiety on the C6-position of the cytosine ring is the key activation step. This attack is a nucleophilic addition and it was presumed that the resulting intermediates readily undergo deamination to form the final degradant. This mechanism is consistent with the mechanistic study performed by Shapiro *et al.* on the deamination of cytosine derivatives catalyzed by bisulfite.[75] In the mechanism proposed by Shapiro *et al.*, the bisulfite forms a similar adduct, which readily deaminates, and the optimal pH for the deamination is 5. In both cases,[76,77] details of the deamination step were not

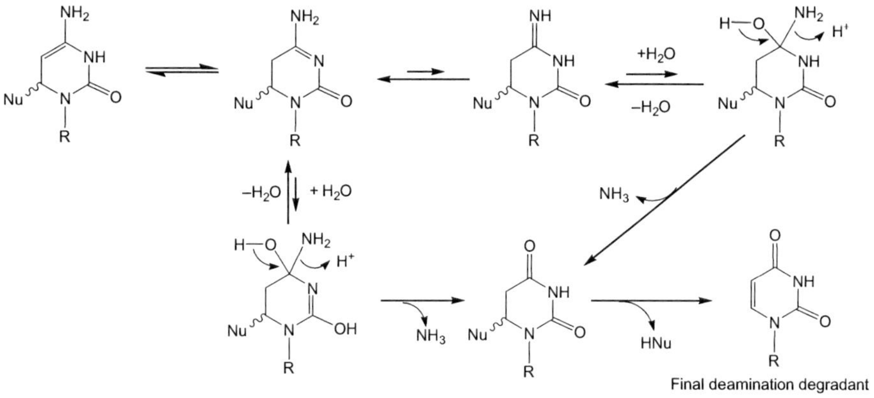

Scheme 2.21

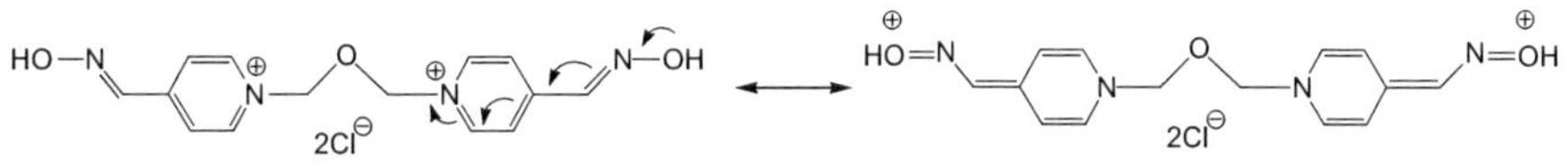

Scheme 2.22

elaborated. It appears that the activated intermediate (either by a hydroxyl or bisulfite) can readily tautomerize to produce two tautomers including the one with an exo-imine bond (Scheme 2.21). Both tautomers should be susceptible to attack by water to form the two tetrahedral intermediates, respectively. The latter undergo rapid deamination and then elimination, giving rise to the final deamination degradant.

Owing to the lability of the imine linkage, it is rare, for a drug to contain a simple imine moiety. Nevertheless, the imine moiety can be stabilized by various neighboring functional groups and/or structural moieties. In the example of gemcitabine above, it can be considered that an imine moiety is embedded in the cytosine ring. Another example is the oxime or oxime ether moiety that is formed by condensing a hydroxylamine or alkoxylamine with a ketone/aldehyde. The resulting oxime or oxime ether is much more stable hydrolytically than its imine counterpart; for example, the activation energy of hydrolysis of the oxime linkage in obidoxime, an antidote for treating poisoning caused by organophosphates, was reported to be 26.2 kcal mol^{-1}.[76] The enhanced stability of the oxime or oxime ether may be explained by the presence of the resonance forms caused by the neighboring oxygen, which can be illustrated by using obidoxime as an example (Scheme 2.22). For this reason, oxime ether functionality is widely utilized in drug design and structure modification. Ample examples can be found in the case of third generation of β-lactam antibiotics (see Section 2.2.4).

2.2.10   Acetal and Hemiacetal Groups

Many drugs, in particular those that originate from fermentation processes, contain a unit or units of saccharides and aminoglycosides. The critical linkage connecting the units of saccharides or aminoglycosides is a glycosidic bond, which belongs to the category of acetal functionality. Acetal and closely related ketal are geminal diethers, which are generally more reactive than regular ethers. Tobramycin is a broad spectrum aminoglycoside antibiotic consisting of three amino-sugar units: nebrosamine, deoxystreptamine, and kanosamine. According to a degradation study performed by Brandl and Gu,[77] under acidic conditions with 1N HCl, only the glycosidic bond between the deoxy-streptamine and kanosamine units was hydrolyzed. The activation energy of the acid-catalyzed hydrolysis was found to be 32 kcal mol^{-1}, indicating that the glycosidic linkage is very stable under acidic conditions. When stressed in 1N NaOH at 80 °C, cleavage in both glycosidic bonds was observed, leading to the formation of component mono-aminosugars and di-aminosugars. The activation energy for the basic hydrolysis was found to be only 15 kcal mol^{-1}. The hydrolytic degradation pathways are illustrated in Scheme 2.23.

In the original paper, the detailed mechanism of the acid-catalyzed hydrolysis was not elaborated.[77] Nevertheless, based on two comprehensive reviews on the hydrolysis mechanism for acetals and ketals by Fife[78] and Cordes and Bull,[79] respectively, the acid-catalyzed formation of the kanosamine carbonium ion would be the rate-limiting step, which is the key step in the so-called *A*-1 mechanism. The carbonium ion should be stabilized by resonance from the neighboring oxygen.

Doxorubicin (adriamycin, Figure 2.7) is an anthracycline antibiotic used in cancer treatment. It consists of a tetracyclic doxorubicinone ring and an amino sugar unit. The glycosidic bond linking the two moieties is susceptible to hydrolytic degradation. The activation energy for hydrolysis of doxorubicin in 0.5 M HCl solution was found to be 92.0 kJ mol^{-1} (22.0 kcal mol^{-1}).[80]

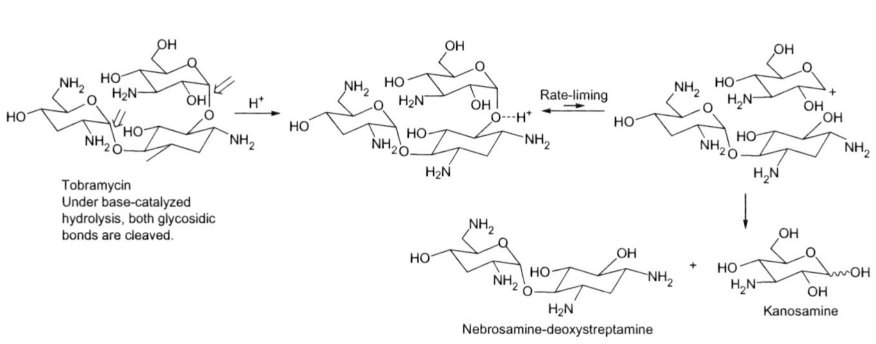

Tobramycin
Under base-catalyzed
hydrolysis, both glycosidic
bonds are cleaved.

Nebrosamine-deoxystreptamine

Kanosamine

Scheme 2.23

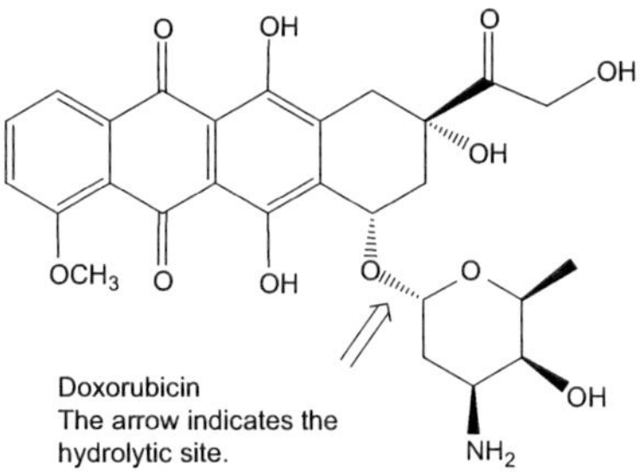

Figure 2.7 Doxorubicin and its hydrolytic site.

2.2.11 Ethers and Epoxides

Ether functionality particularly in alkyl ether is usually stable toward hydrolysis unless structurally activated. The latter case includes an ether moiety that is linked to an allylic or benzylic position. Aromatic ethers are usually more reactive owing to the fact that they are better leaving groups. For example, O^6-benzylguanine, which was investigated as part of a combination adjuvant chemotherapy in cancer treatment, has an ether linkage between the benzyl group and the guanine moiety. During formulation studies, it was found that O^6-benzylguanine is quite susceptible to hydrolytic degradation under acidic conditions.[81] A kinetic study of the degradation showed a very small activation entropy ($\Delta S = -2.4$ e.u. or cal/(mol K)), suggesting that it undergoes hydrolysis *via* an *A*-1 mechanism. The finding is consistent with the $H_2^{18}O$ experiment conducted by the same workers, which showed the ^{18}O ended up in the degradant, benzyl alcohol, rather than guanine. The hydrolytic degradation of O^6-benzylguanine is illustrated in Scheme 2.24.

Another ether hydrolysis in drug degradation involves (*S*)-duloxetine, a serotonin–norepinephrine reuptake inhibitor for the treatment of depression. It was found that at pH less than 2.5, it is not stable and the ether linkage is hydrolyzed. Based on the degradants identified, it is apparent that the hydrolysis proceeds through an *A*-1 mechanism (Scheme 2.25),[82] which is consistent with the fact that 1-nathphol is a good leaving group and the initially formed carbocation can be stabilized by the thiophene group. For this reason, the carbocation is capable of alkylating 1-nathphol, as the latter degradant builds up particularly in a concentrated solution of the drug. Owing to its instability in the acidic pH, the drug is formulated in enteric coated tablets.[83]

Epoxides can be considered to be three-membered cyclic ethers. Owing to the tension caused by the small oxirane ring, epoxides are more reactive toward hydrolysis (or hydration) as well as reaction with nuleophiles. As such, epoxide moiety and its nitrogen analog, aziridine, are frequently employed in chemotherapeutic agents with DNA alkylation capability.[84] The reactivity of an epoxide can vary dramatically depending upon its neighboring substituents.[85] The activation energy of the spontaneous hydrolysis of propylene oxide, a simple epoxide, is approximately 19.0 to 19.5 kcal mol^{-1}.[86,87] This value is

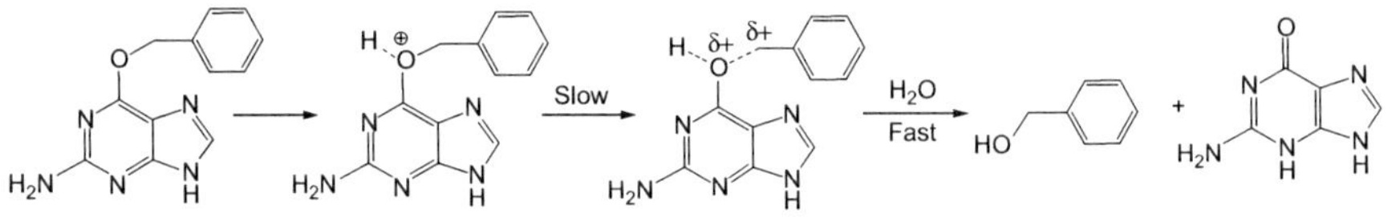

Scheme 2.24

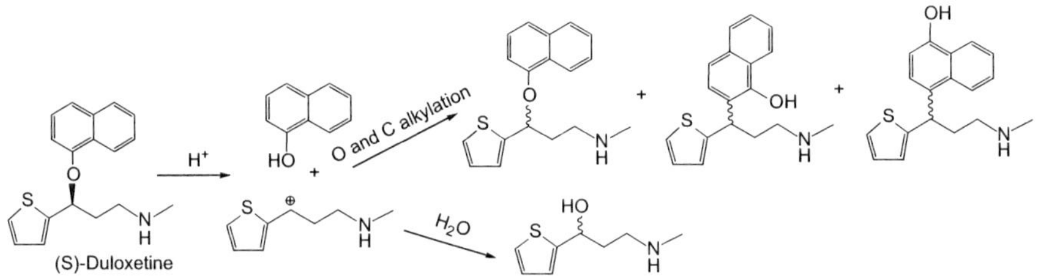

(S)-Duloxetine

O and C alkylation

Scheme 2.25

comparable to the activation energy of hydrolysis of acetamide under neutral conditions (see Section 2.2.3), suggesting simple alkylepoxides are reasonably stable hydrolytically under non-catalytic conditions. When a vinyl group attaches to an epoxide, its hydrolytic reactivity increases:[88] the pseudo first-order rate constant for the spontaneous hydrolysis of butadiene oxide, the simplest vinyl epoxide, is $1.4 \times 10^{-5}\,s^{-1}$, which is approximately 20 times greater than that of propylene oxide $(6.9 \times 10^{-7}\,s^{-1})$. If the vinyl epoxide moiety is contained in a ring, its hydrolytic reactivity depends upon the size of the ring; cyclized vinyl epoxides with a ring size smaller than 7 are more reactive than their larger ring counterpart. The pseudo first-order rate constants for the spontaneous hydrolysis of cycloheptadiene oxide, cyclohexadiene oxide, and cyclopentadiene oxide are 2.7×10^{-5}, 2.6×10^{-4}, and $5.2 \times 10^{-3}\,s^{-1}$, respectively, as the ring size decreases from 7 to 5.

On the extreme side, the exo-8,9-epoxide of mycotoxin aflatoxin B1 is instantaneously hydrolyzed in an aqueous environment; its half-life was esti-mated to be about 1 s at $25\,^{\circ}C$.[89] Aflatoxin B1 is among the most potent car-cinogens, attributed to its metabolic activation in which the exo-8,9-epoxide is formed. This epoxide is a very reactive electrophile, which alkylates the N7 position of guanine residues in DNA. The exo-8,9-epoxide is structurally fused to a five-membered cyclic ether with the two oxygen atoms connected to the same carbon. The latter cyclic ether is fused to another five-membered cyclic ether. Hydrolysis of the epoxide ring triggers the opening of the two con-secutively fused five-membered cyclic ethers as shown in Scheme 2.26.

Hydrolysis of epoxides can be catalyzed by both acid and base. Under acid-catalyzed conditions, the hydrolysis proceeds via an A-1 mechanism that involves the intermediacy of a carbocation; the point of hydrolytic attack

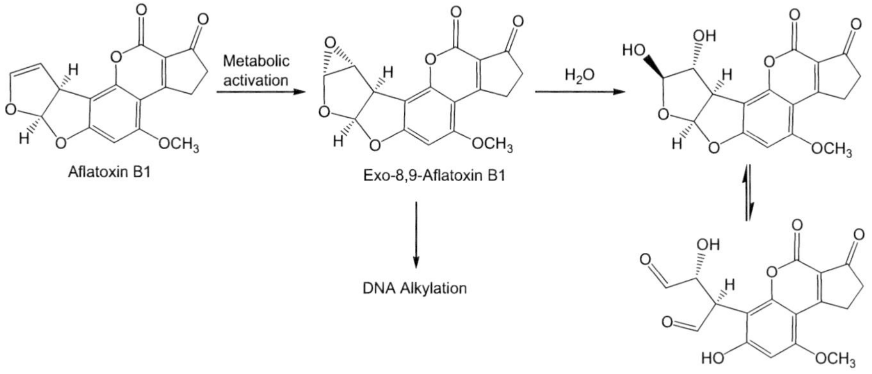

Scheme 2.26

occurs preferentially on the carbon that is usually either more branched or the carbocation formed on such a carbon is stabilized by other mechanisms.[90] Under base-catalyzed conditions, the hydrolysis proceeds via an *A*-2 mechanism; the point of hydrolytic attack occurs preferentially on the carbon that is less sterically hindered.[91]

2.3 Esterification, Transesterification and Formation of an Amide Linkage

Formation of an ester (esterification) and an amide linkage is the reverse reaction of the hydrolysis of ester and amide, respectively. As we have discussed before in this chapter, the reverse reaction proceeds through the same transition state as the hydrolysis and as such, the formation of ester and amide can also be catalyzed by acid and base. Transesterification is very similar to hydrolysis, except that the reagent water is replaced by an alcohol. Generally, degradation reactions by these mechanisms mostly occur between a drug molecule and the excipients or their impurities. Hence, these degradation reactions will be discussed in Chapter 5, Drug-Excipient Interaction and Adduct Formation. Previously in this chapter (Section 2.2.2), we have mentioned that the hydrolysis of the lactone ring in both lovastatin and simvastatin is reversible, indicating a very efficient lactonization (formation of cyclic esters) of the hydrolyzed products. In this section, we will discuss case studies involving intramolecular amide linkage formation and intramolecular transesterification.

Diclofenac, a non-steroidal anti-inflammatory drug, can be considered to be a substituted *N*-(2,6-dichlorophenyl)aniline which contains a methylenecarboxyl group ortho to the amine group. In an accelerated stability study of an aqueous formulation, diclofenac was found to undergo cyclization between the carboxyl and amine groups, resulting in the formation of a five-membered lactam degradant (Scheme 2.27).[92]

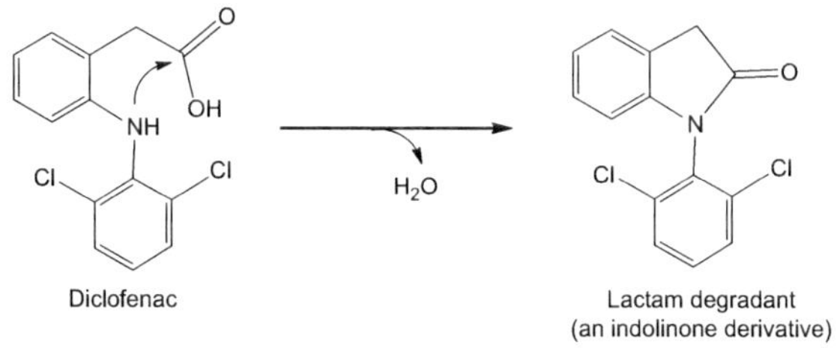

Scheme 2.27

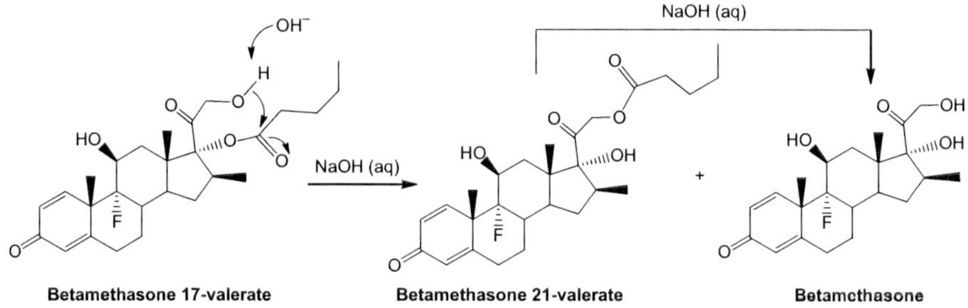

Scheme 2.28

This lactam is an indolinone derivative, which is also a synthetic intermediate of the API. It was also observed as a degradant in a topical formulation.[93]

As mentioned above, transesterification usually occurs between a drug substance and excipients or excipient-related impurities. Nevertheless, it can also occur within the same drug molecule. For example, betamethasone 17-valerate is an anti-inflammatory drug substance, which undergoes intra-molecular transesterification to produce an isomeric degradant, betamethasone 21-valerate, especially under basic conditions (Scheme 2.28).[94]

References

1. K. C. Waterman, R. C. Adami, K. M. Alsante, A. S. Antipas, D. R. Arenson, R. Carrier, J. Hong, M. S. Landis, F. Lombardo, J. C. Shah, E. Shalaev, S. W. Smith and H. Wang, *Pharm. Dev. Technol.*, 2002, **7**, 113.
2. J. Fletcher, A. Wirz, J. Young, R. Vallance and K. E. L. McColl, *Gastroenterology*, 2001, **121**, 775.
3. W. P. Jencks and J. Carriuolo, *J. Am. Chem. Soc.*, 1961, **83**, 1743.
4. T. H. Fife and T. J. Przystas, *J. Am. Chem. Soc.*, 1982, **104**, 2251.
5. T. H. Fife and T. J. Przystas, *J. Am. Chem. Soc.*, 1985, **107**, 1041.
6. T. J. Przystas and T. H. Fife, *J. Chem. Soc., Perkin Trans.*, 1990, **2**, 393.
7. T. H. Fife and T. J. Przystas, *J. Chem. Soc., Perkin Trans.*, 1987, **2**, 143.
8. D. W. Newton and K. W. Miller, *Am. J. Hosp. Pharm.*, 1987, **44**, 1633.

9. B. A. Robinson and J. W. Tester, *Int. J. Chem. Kinet.*, 1990, **22**, 431.
10. K. A. Connors, G. L. Amidon and V. J. Stella, *Chemical Stability of Pharmaceuticals: a Handbook for Pharmacists*, John Wiley & Sons, New York, 2nd edn, 1986.
11. E. A. Gadkariem, F. Belal, M. A. Abounassif, H. A. El-Obeid and K. E. E. Ibrahim, *Il Farmaco*, 2004, **59**, 323.
12. M. G. Bartlett, J. C. Spell, P. S. Mathis, M. F. A. Elgany, B. E. El Zeany, M. A. Elkawy and J. T. Stewart, *J. Pharm. Biomed. Anal.*, 1998, **18**, 335.
13. L. Bajerski, R. C. Rossi, C. L. Dias, A. M. Bergold and P. E. Fröehlich, *Chromatographia*, 2008, **68**, 991.
14. A. D. Marcus and S. Baron, *J. Am. Pharm. Assoc.*, 1959, **48**, 85.
15. E. R. Garrett, *J. Am. Chem. Soc.*, 1957, **79**, 3401.
16. L. J. Edwards, *Trans. Faraday Soc.*, 1950, **46**, 723.
17. L. J. Edwards, *Trans. Faraday Soc.*, 1952, **48**, 696.
18. M. L. Bender, *Chem. Rev.*, 1960, **60**, 53.
19. M. L. Bender, F. Chlouprek and M. C. Neveu, *J. Am. Chem. Soc.*, 1958, **80**, 5384.
20. A. R. Fersht and A. J. Kirby, *J. Am. Chem. Soc.*, 1967, **89**, 4857.
21. M. J. Kaufman, *Int. J. Pharm.*, 1990, **66**, 97.
22. W. Muangsiri and L. E. Kirsch, *J. Pharm. Sci.*, 2001, **90**, 1066.
23. A. Bennet, V. Somayaji, R. Brown and B. D. Santarsiero, *J. Am. Chem. Soc.*, 1991, **113**, 7563.
24. S. L. Ali, *J. Chromatogr.*, 1978, **154**, 103.
25. A. D. Marcus and A. J. Taraszka, *J. Am. Pharm. Assoc.*, 1959, **48**, 77.
26. T. Ojha, M. Bakshi, A. K. Chakraborti and S. Singh, *J. Pharm. Biomed. Anal.*, 2003, **31**, 775.
27. H. Krasowska, *Int. J. Pharm.*, 1979, **4**, 89.
28. M. F. Powell, *Pharm. Res.*, 1987, **4**, 42.
29. N. P. Gensmantel, D. McLellan, J. J. Morris, M. I. Page, P. Proctor and G. S. Randahawa, in *Recent Advances in the Chemistry of β-Lactam Antibiotics*, ed. G. I. Gregory, Royal Society of Chemistry Special Publication No. 38, Royal Society of Chemistry, London, 1981, pp. 227–239.
30. A.M. Davis, P. Proctor and M. I. Page, *J. Chem. Soc., Perkins 2*, 1991, 1213.
31. J. P. Hou and J. W. Poole, *J. Pharm. Sci.*, 1969, **58**, 447.
32. R. Chadha, N. Kashid and D. V. S. Jain, *J. Pharm. Pharmacol.*, 2003, **55**, 1495.
33. R. Oliyai and S. Lindenbaum, *Int. J. Pharm.*, 1991, **73**, 33.
34. T. Yamana and A. Tsuji, *J. Pharm. Sci.*, 1976, **65**, 1563.
35. A. Dimitrovska, K. Stojanoski and K. Dorevski, *Int. J. Pharm.*, 1995, **115**, 175.
36. M. J. Pikal and K. M. Dellerman, *Int. J. Pharm.*, 1989, **50**, 233.
37. W. A. Cressman, E. T. Sugita, J. T. Doluisio and P. J. Niebergall, *J. Pharm. Sci.*, 1969, **58**, 1471.
38. N. P. Gensmantel, P. Proctor and M. I. Page, *J. Chem. Soc., Perkins 2*, 1980, 1725.
39. M. I. Page, *Acc., Chem. Res.*, 1984, **17**, 144.

40. J. O. Fubara and R. E. Notari, *J. Pharm. Sci.*, 1998, **87**, 1572.
41. D. Wang and R. E. Notari, *J. Pharm. Sci.*, 1994, **83**, 577.
42. G. V. Kaiser and S. Kukolja in *Cephalosporins and Penicillins. Chemistry and Biology*, ed. E. H. Flynn, Academic Press, New York, 1972, pp. 125–128.
43. D. B. Boyd and W. H. W. Lunn, *J. Med. Chem.*, 1979, **22**, 778.
44. T. Sugioka, T. Asano, Y. Chikaraishi, E. Suzuki, A. Sano, T. Kuriki, M. Shirotsuka and K. Saito, *Chem. Pharm. Bull.*, 1990, **38**, 1998.
45. K. Itakura, I. Aoki, F. Kasahara, M. Nishikawa and Y. Mizushima, *Chem. Pharm. Bull.*, 1981, **29**, 1655.
46. R. N. Jones, *Am. J. Med.*, 1985, **78**(Suppl. 6A), 22.
47. R. W. Ratcliffe, K. J. Wildonger, L. Di Michele, A. W. Douglas, R. Hajdu, R. T. Goegelman, J. P. Springer and J. Hirshfield, *J. Org. Chem.*, 1989, **54**, 653.
48. M. Xia, T.-J. Hang, F. Zhang, X.-M. Li and X.-Y. Xu, *J. Pharm. Biomed. Anal.*, 2009, **49**, 937.
49. T. Loftsson, B. J. Olafsdottir and J. Baldvinsdottir, *Int. J. Pharm.*, 1992, **79**, 107.
50. A. J. Kirby, in *Comprehensive Chemical Kinetics, Ester Formation and Ester Hydrolysis*, ed. C. H. Bamford and C. F. H. Tipper, Elsevier, Amsterdam, 1972, pp. 156–158.
51. T. Loftsson and N. Bodor, *J. Pharm. Sci.*, 1981, **70**, 750.
52. Ö. Almarsson1, M. J. Kaufman, J. D. Stong, Y. Wu, S. M. Mayr, M. A. Petrich and J. M. Williams, *J. Pharm. Sci.*, 1998, **87**, 663.
53. E. Jabri, M. B. Carr, R. P. Hausinger and P. A. Karplus, *Science*, 1995, **268**, 998.
54. J. D. Jonkman-De Vries, W. G. Doppenberg, R. E. C. Henrar, A. Bult and J. H. Beijnen, *J. Pharm. Sci.*, 1996, **85**, 1227.
55. N. Stroud, N. E. Richardson, D. J. G. Davies and D. A. Norton, *Analyst*, 1980, **105**, 455.
56. G. L. Flynn and D. J. Lamb, *J. Pharm. Sci.*, 1970, **59**, 1433.
57. M. Li, X. Wang, T.-M. Bin Chen, Chan and A. Rustum, *J. Pharm. Sci.*, 2009, **98**, 894.
58. P. J. Skrdla, A. Abrahim and Y. Wu, *J. Pharm. Biomed. Anal.*, 2006, **41**, 883.
59. O. M. Friedman, S. Bien and J. K. Chakrabarti, *J. Am. Chem. Soc.*, 1965, **87**, 4978.
60. G. Zon, S. M. Ludeman and W. Egan, *J. Am. Chem. Soc.*, 1977, **99**, 5786.
61. S. Searles and S. Nukina, *Chem. Rev.*, 1959, **59**, 1077.
62. N. J. Baxter, L. J. M. Rigoreau, A. P. Laws and M. I. Page, *J. Am. Chem. Soc.*, 2000, **122**, 3375.
63. M. I. Page, *Acc. Chem. Res.*, 2004, **37**, 297.
64. T. Yamana, Y. Mizukami and F. Ichimura, *Yakugaku Zasshi*, 1965, **85**, 654.
65. T. Yamana and Y. Mizukami, *Yakugaku Zasshi*, 1967, **87**, 1304.
66. F. B. Vest, W. H. Soine, R. B. Westkaemper and P. J. Soine, *Pharm. Res.*, 1989, **6**, 458.

67. E. R. Garrett, J. T. Bojarski and G. J. Yakatan, *J. Pharm. Sci.*, 1971, **60**, 1145.
68. J. T. Bojarski, J. L. Mokrosz, H. J. Barton and M. H. Paluchowska, *Adv. Heterocycl. Chem.*, 1985, **38**, 229.
69. E. R. Garrett, J. T. Bojarski and G. J. Yakatan, *J. Pharm. Sci.*, 1971, **60**, 1145.
70. M. Tarsa, G. Zuchowski, A. Stasiewicz-Urban and J. Bojarski, *Acta Pol. Pharm. (Drug Res. Warsaw)*, 2009, **66**, 123.
71. B. Zerner, *Bioorg. Chem.*, 1991, **19**, 116.
72. E. H. Wiseman, J. Chiaini and R. Pinson, Jr., *J. Pharm. Sci.*, 1964, **53**, 766.
73. D. C. Dubost, M. J. Kaufman, J. A. Zimmerman, M. J. Bogusky, A. B. Coddington and S. M. Pitzenberger, *Pharm. Res.*, 1996, **13**, 1811.
74. P. J. Jansen, M. J. Akers, R. M. Amos, S. W. Baertschi, G. G. Cooke, D. E. Dorman, C. A. J. Kemp, S. R. Maple and K. A. Mccune, *J. Pharm. Sci.*, 2000, **89**, 885.
75. R. Shapiro, V. DiFate and M. Welcher, *J. Am. Chem. Soc.*, 1974, **96**, 906.
76. S. Rubnov, I. Shats, D. Levy, S. Amisar and H. Schneider, *J. Pharm. Pharmacol.*, 1999, **51**, 9.
77. M. Brandl and L. Gu, *Drug Dev. Ind. Pharm.*, 1992, **18**, 1423.
78. T. H. Fife, *Acc. Chem. Res.*, 1972, **5**, 264.
79. E. H. Cordes and H. G. Bull, *Chem. Rev.*, 1974, **74**, 581.
80. K. Wassermann and H. Bundgaard, *Int. J. Pharm.*, 1983, **14**, 73.
81. M. Safadi, D. S. Bindra, T. Williams, R. C. Moschel and V. J. Stella, *Int. J. Pharm.*, 1993, **90**, 239.
82. S. W. Baertschi and K. M. Alsante, in *Pharmaceutical Stress Testing: Predicting Drug Degradation*, ed. S. W. Baertschi, Informa Healthcare, 2005, pp. 87–88.
83. P. J. Jansen, P. L. Oren, C. A. Kemp, S. R. Maple and S. W. Baertschi, *J. Pharm. Sci.*, 1998, **87**, 81.
84. A. G. Bosanquet, *Cancer Chemother. Pharmacol.*, 1985, **14**, 83.
85. J. G. Pritchard and F. A. Long, *J. Am. Chem. Soc.*, 1956, **78**, 2667.
86. J. Lichtenstein and G. H. Twigg, *Trans. Faraday Soc.*, 1948, **44**, 905.
87. J. Koskikallio and E. Whalley, *Can. J. Chem.*, 1959, **37**, 783.
88. A. M. Ross, T. M. Pohl, K. Piazza, M. Thomas, B. Fox and D. L. Whalen, *J. Am. Chem. Soc.*, 1982, **104**, 1658.
89. W. W. Johnson, T. M. Harris and F. P. Guengerich, *J. Am. Chem. Soc.*, 1996, **118**, 8213.
90. J. G. Pritchard and F. A. Long, *J. Am. Chem. Soc.*, 1956, **78**, 2667.
91. F. A. Long and J. G. Pritchard, *J. Am. Chem. Soc.*, 1956, **78**, 2663.
92. M.-J. Galmier, B. Bouchona, J.-C. Madelmont, F. Mercier, F. Pilotaz and C. Lartigue, *J. Pharm. Biomed. Anal.*, 2005, **38**, 790.
93. R. Hajkova, P. Solich, M. Pospigilovad and J. Sicha, *Anal. Chim. Acta*, 2002, **467**, 91.
94. M. Li, M. Lin and A. Rustum, *J. Pharm. Biomed. Anal.*, 2008, **48**, 1451.

CHAPTER 3
Oxidative Degradation

3.1 Introduction

Oxidative degradation of drugs is one of the most common degradation pathways but perhaps the most complex one. In the vast majority cases, the ultimate source of the oxidizing agent is molecular oxygen (O_2), which accounts for approximately 21% of the atmosphere. Because the oxidation of many organic compounds by molecular oxygen is seemingly "spontaneous and uncatalyzed", this type of oxidation is usually called "autooxidation" or "autoxidation".[1] Other terms such as "aerial oxidation" or "allomerization" are also used. The term "allomerization" was initially used by Willstatter and Stoll between 1911 and 1913 to describe the solution degradation of chlorophylls by exposure to molecular oxygen.[2,3] Hence, the autooxidation of chlorophylls is referred to as "allomerization". Since most organic compounds are in a singlet state, that is, electron-paired, while molecular oxygen in its ground state is a triplet species, the reaction between most organic compounds and molecular oxygen is a kinetically forbidden process owing to violation of the spin conservation rule.[4] Hence, the "spontaneous" autooxidation reaction usually involves activation of ground state molecular oxygen, during which process the latter can be activated into a few species of various reactivity such as superoxide anion radical ($O_2^{-\bullet}$), hydrogen peroxide (H_2O_2), hydroxyl free radical ($HO^{\bullet}$), and singlet oxygen ($^1O_2$). Collectively, these species are usually called "reactive oxygen species" or "ROS".[5] Redox-active transition metal ions, most commonly iron and copper ions, usually play a key catalytic role in the activation process that produces $O_2^{-\bullet}$, H_2O_2, and $HO^{\bullet}$. This process, involving electron transfer and free radicals, is the most significant one in autooxidation of drugs. On the other hand, ozone, typically formed by electric sparking or vacuum UV irradiation, is usually not a concern for the oxidative degradation of drugs. Singlet oxygen, usually generated under photosensitization conditions, plays an important role in photooxidative degradation of drugs, which will be discussed in Chapter 6.

RSC Drug Discovery Series No. 29
Organic Chemistry of Drug Degradation
By Min Li
Published by the Royal Society of Chemistry, www.rsc.org

Certain electron-rich species, like many phenol or polyphenol type compounds, seem to be capable of reacting with molecular oxygen without apparent activation of the latter; one example is tetrachlorohydroquinone (TCHQ), a metabolite of pentachlorophenol (PCP).[6] Nevertheless, whether the autooxidation of these compounds or any singlet organic molecules is truly without the involvement of transition metal catalysis is still debatable. For one thing, it is extremely difficult to completely remove all residual transition metal ions experimentally. Miller *et al.* hypothesized that "true" autooxidation, that is, autooxidation without redox transition metal catalysis, is negligible and the rate constant of such a true autooxidation is estimated to be $\sim 10^{-5}$ $M^{-1} s^{-1}$.[4]

Drug substances that contain somewhat "acidic" carbonated protons (CH_n, *n* is typically 1 to 2) tend to undergo autooxidation *via* carbanion/enolate-type intermediates through deprotonation. The autooxidation of these compounds, which is also referred to as base-catalyzed autooxidation, apparently does not involve radical species and its degradation kinetics is usually much faster than a free radical-mediated autooxidation. This type of non-radical-mediated autooxidation is much less known for its role in drug degradation, although it can be a significant degradation pathway particularly in liquid formulations.[7,8]

3.2 Free Radical-mediated Autooxidation

Free radical-mediated autooxidation of drugs usually involves redox-active transition metal ions and/or exposure to light. The latter will be covered in Chapter 6, Photochemical Degradation. The role of the metal ions in the initiation stage of a free radical-mediated autooxidation is to act as an electron donor from its lower oxidation state (or reduced state) to molecular oxygen. The commonly encountered redox-active transition metal ion pairs are Fe(II)/Fe(III), Cu(I)/Cu(II), Mn(II)/Mn(III), Ni(II)/Ni(IV), Pb(II)/Pb(IV), Ti(III)/Ti(IV), and Co(II)/Co(III). The most relevant redox-active transition metal ions in drug degradation are iron ions, followed perhaps by copper ions. This type of transition metal-catalyzed process which generates reactive oxygen species, in particular HO• radicals, is generally referred to as the Fenton reaction or Fenton-type reaction by a great number of researchers. Nevertheless, a closely related process, called the Udenfriend reaction, is more directly relevant in the autooxidative degradation of drugs.

3.2.1 Origin of Free Radicals: Fenton Reaction and Udenfriend Reaction

While still a London college student in 1894, H.J.H. Fenton described the oxidation of tartaric acid in aqueous solution using a mixture of H_2O_2 and Fe(II) salt.[9] The reaction did not receive much attention until 40 years later when Haber and Weiss suggested that HO• might be produced in the Fenton reaction as the oxidizing agent (Scheme 3.1).[10]

$$HO–OH + Fe(II) \rightarrow HO^{\bullet} + HO^{-} + Fe(III)$$

Scheme 3.1

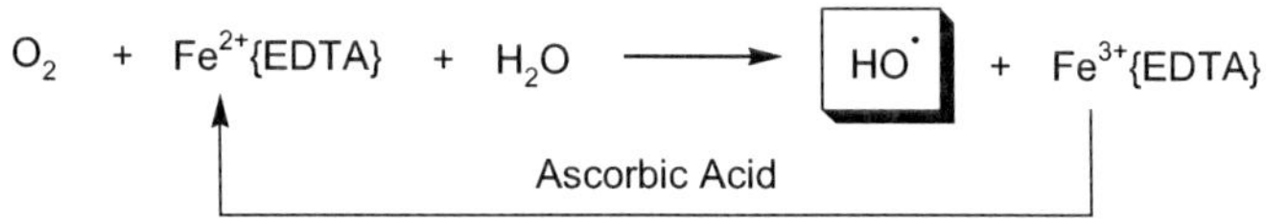

$$O_2 \;+\; Fe^{2+}\{EDTA\} \;+\; H_2O \longrightarrow \boxed{HO^{\bullet}} \;+\; Fe^{3+}\{EDTA\}$$

Ascorbic Acid

Scheme 3.2

In 1954, Sydney Udenfriend and co-workers at National Heart Institute, Bethesda, Maryland, published a study which showed that aromatic compounds could be effectively hydroxylated in an aqueous solution containing Fe(II), ascorbic acid, and ethylenediamine tetraacetic acid (EDTA) when the resulting mixture was exposed to air.[11] Udenfriend *et al.* also demonstrated that H_2O_2 is a critical intermediate in the reaction. This process (the Udenfriend reaction) is illustrated in Scheme 3.2.

Multiple intermediary steps are involved in the Udenfriend reaction and one of them is likely to be the Fenton reaction which turns H_2O_2 into $HO^{\bullet}$. It appears that Fe(II){EDTA} activates molecular oxygen by transferring an electron to it. Consequently, molecular oxygen is reduced, becoming a superoxide anion radical, while Fe(II){EDTA} is oxidized into Fe(III){EDTA}. The superoxide anion radical can then transform to hydrogen peroxide by three possible routes: (1) by abstracting an $H^{\bullet}$ radical, (2) *via* reduction by Fe(II){EDTA}, and (3) by disproportionation. The hydrogen peroxide formed can be dissociated into hydroxyl radicals upon further reaction with Fe(II){EDTA}, (the Fenton reaction). On the other hand, Fe(III){EDTA} can be recycled back to the catalytically active Fe(II){EDTA} through reduction by ascorbic acid. All the plausible steps of the Udenfriend reaction are shown in Scheme 3.3.

As shown in Scheme 3.3, the Fenton reaction can be considered an important step in the multiple-step Udenfriend reaction. In order for both the Fenton reaction and Udenfriend reaction to be fully operative under near neutral pH conditions, a good iron chelating agent such as EDTA is needed to prevent iron ions, in particular Fe(III), from precipitating out of solution, especially at pH approaching neutral. A consequence of the use of a chelating agent is that it could lower the reduction potential (E°) of Fe(III)/Fe(II), depending upon the nature of the chelator. For example, $E^{\circ\prime}$ for Fe(III)/Fe(II) at pH 7.0 is 0.11 V, while $E^{\circ\prime}$ for Fe(III){ferrioxamine}/Fe(II){ferrooxamine} is -0.45 V.[12] On the other hand, $E^{\circ\prime}$ for Fe(III){EDTA}/Fe(II){EDTA} at pH 7.0 is 0.12 V, which is essentially the same as $E^{\circ\prime}$ for the non-chelated Fe(III)/Fe(II).[13] Please note that the frequently quoted standard reduction potential (E°) value of 0.77 V for Fe(III)/Fe(II) is obtained under the "standard" condition in which the pH is 0 (the standard concentration of H^{+} is 1 M).[14] Owing to the low reduction

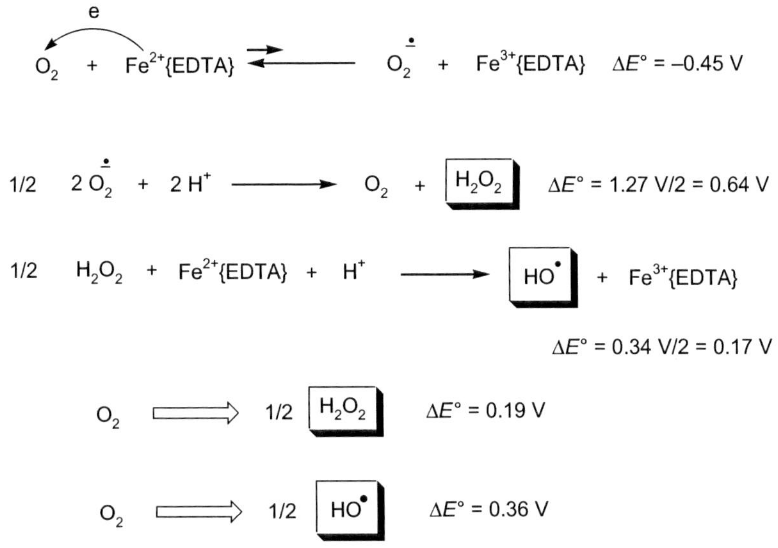

$$O_2 \ + \ Fe^{2+}\{EDTA\} \ \overset{e}{\underset{}{\rightleftharpoons}} \ O_2^{\bullet} \ + \ Fe^{3+}\{EDTA\} \quad \Delta E° = -0.45 \ V$$

$$1/2 \quad 2\,O_2^{\bullet} \ + \ 2\,H^+ \ \longrightarrow \ O_2 \ + \ \boxed{H_2O_2} \quad \Delta E° = 1.27 \ V/2 = 0.64 \ V$$

$$1/2 \quad H_2O_2 \ + \ Fe^{2+}\{EDTA\} \ + \ H^+ \ \longrightarrow \ \boxed{HO^{\bullet}} \ + \ Fe^{3+}\{EDTA\}$$

$$\Delta E° = 0.34 \ V/2 = 0.17 \ V$$

$$O_2 \ \Longrightarrow \ 1/2 \ \boxed{H_2O_2} \quad \Delta E° = 0.19 \ V$$

$$O_2 \ \Longrightarrow \ 1/2 \ \boxed{HO^{\bullet}} \quad \Delta E° = 0.36 \ V$$

Scheme 3.3

potential ($E°'$) and much improved solubility at pH 7.0 for EDTA chelated Fe(III)/Fe(II), EDTA greatly facilitates the Fenton and Udenfriend reactions because the soluble Fe(II)EDTA should be a much more efficient electron donor to molecular oxygen at neutral pH. In the experiments carried out by Udenfriend *et al.*, the use of EDTA markedly enhanced the rate of the autooxidation reactions.

Note that the steps shown in Scheme 3.3 are probably simplified working models for both the Fenton and Udenfriend reactions and the thermodynamic feasibility of the reaction sequence is demonstrated by the use of the standard reduction potential, $E°$, rather than the reduction potential at neutral pH, $E°'$. A great deal of effort has been put into studying the detailed mechanism of the Fenton reaction over the past few decades.[15–17] One of the key questions, which is still debatable today, has been whether a HO$^{\bullet}$ free radical is really produced in the Fenton reaction.[18] The alternate hypothesis for the oxidation inter- mediate is a ferryl species either in the form of a Fe(IV)O^{2+} ion or a Fe(IV)O$^{+\bullet}$ radical cation. The Fe(IV)O^{2+} ion has been generated and characterized by a number of techniques.[19,20] Based on the discovery of some chemistry that is unique to the Fe(IV)O^{2+} ion, such as oxygen atom transfer to sulfoxides, its involvement in the Fenton reaction was ruled out.[21] With regard to the Fe(IV)O$^{+\bullet}$ radical cation, although its hypothesized presence as the critical oxidizing intermediate in oxidative enzymes such as cytochrome P450[22] has been recently verified experimentally,[23] the possibility of its being the inter- mediate in the regular Fenton reaction seems still very low. In order to stabilize a high valence iron species like the Fe(IV)O$^{+\bullet}$ radical cation, strong, electron- rich ligands such as porphyrins are required. Hence, in the regular Fenton and related Udenfriend reactions, where the ligands are typically not as strong or electron-rich as porphyrins, HO$^•$ radical is most likely to be the oxidizing intermediate. Obviously, nobody believes that this HO$^•$ radical would behave like one that is generated by γ-radiation of water. The HO$^•$ radical intermediate in the regular Fenton and related Udenfriend reactions is most probably formed in a site-specific manner.[24–26] Such a HO$^•$ radical would not diffuse or react too far away from the point of its formation. In addition, the site-specific HO$^•$ radical displays muted reactivity compared to that generated by γ-radiation.

As illustrated above, the Udenfriend reaction consists of three key components, that is, a transition redox metal ion (Fe^{2+}), a good chelating agent (EDTA), and a reducing agent (ascorbic acid). In other words, the combination of these three types of components would effectively convert molecular oxygen into a few reactive oxygen species (ROS) including H$_2$O$_2$ and HO$^•$ radicals. Indeed, studies have shown that other transition metal ions, chelating agents and reducing agents/antioxidants could replace Fe(II), EDTA, and ascorbic acid, respectively, in the Udenfriend reaction. For example, in several mechanistic studies where hydroxylation of aromatic compounds was used as the indicator for HO$^•$ formation or DNA damage under oxidative stress, it has been demonstrated that a number of transition redox metal ions, such as Cu(I), can replace Fe(II).[27,28] On the other hand, several metal chelators, such as citrate[29] and diethylenetriamine pentaacetic acid (DTPA) (also called DETA-PAC),[30] can substitute for EDTA. This is consistent with the fact that the reduction potentials of the DTPA chelated and citrate chelated iron pairs are 0.165 V[31,32] and $\sim$0.1 V,[33] respectively, which are similar to the reduction potential (0.12 V) of an EDTA chelated iron pair. These studies, along with that of Kasai and Nishimura,[28] also implied that several reducing agents, like derivatives of phenol (*e.g.* trolox, a vitamin E analog)[34] and catechol,[35] are capable of recycling Fe(III) back to Fe(II), suggesting that they can take the role of ascorbic acid in the Udenfriend reaction. Among all of the species implicated in the above studies as replacements for the three key components of the Udenfriend reaction, those that are pharmaceutically and/or physiologically relevant are summarized in Table 3.1.

Since chelating agents and antioxidants are frequently used in the formulation of drug products for the purpose of product preservation and stability, the Udenfriend reaction has a direct impact on the stability of a drug product formulated with a combination of a chelating agent and an antioxidant (not

Table 3.1 Chemical species that may substitute in Udenfriend reaction.

Original components in Udenfriend reaction	Species that may substitute
Redox metal ion (Fe^{2+})	Cu^{1+},[a,b] Sn^{2+},[b] Co^{2+},[b] Ti^{2+}.[b]
Metal chelator (EDTA)	Citrate,[c] diethylenetriaminepentaacetate (DTPA),[d] pyrophosphate,[e] triphosphate,[e] tetraphosphate,[e] lactate,[e] desferrioxamine B.[f]
Reducing agent (ascorbic acid)	Phenols (trolox),[g] catechols,[h] gallate,[h] bisulfite,[b,i] hydroxylamine,[b] hydrazine,[b] dihydroxymaleic acid.[b]

[a]Ref. 4; [b]Ref. 28; [c]Ref. 29; [d]Ref. 30; [e]Ref. 36; [f]Ref. 37; [g]Ref. 34; [h]Ref. 35; [i]Ref. 38.

necessarily limited to those listed in Table 3.1). Such a drug product could potentially be intrinsically vulnerable to autooxidation, because a slight increase of a transition redox metal ion, either from the primary packaging, raw materials or during manufacturing, into the formulation could trigger the Udenfriend process, causing decreased stability of the finished drug product. Nevertheless, this does not mean that any combination of the three components from each of the three categories in Table 3.1 above would automatically constitute a Udenfriend reaction system, because the thermodynamics and/or kinetics of such a combination may not always be favorable for the reaction to proceed.

Sometimes, the drug molecule itself can be the chelating agent for redox transition metal ions. As a result, the drug may be oxidized at a particular site by the ROS formed nearby. In such cases, use of additional chelating agent such as EDTA can inhibit the oxidation occurring at that particular site. Nevertheless, the drug may be oxidized at yet another site or sites by a new Udenfriend system that now consists of EDTA that replaces the drug molecule.

3.2.2 Origin of Free Radicals: Homolytic Cleavage of Peroxides by Thermolysis and Heterolytic Cleavage of Peroxides by Metal Ion Oxidation

As shown in Section 3.2.1, hydrogen peroxide is generated during the activation of molecular oxygen in the Udenfriend reaction. Certain polymeric excipients are prone to autooxidation leading to the formation of peroxides. For example, it was reported that pharmaceutical grade polyethylene glycol (PEG) and povidone contain various levels of peroxides including hydrogen peroxide.[39–41] The O–O bond of the peroxide is weak and susceptible to thermal decomposition, in addition to the transition metal ion-catalyzed cleavage (*e.g.* the Fenton reaction). Of the various degradation pathways of organic peroxides under thermolysis, which was reviewed by Antonovskii and Khursan,[42] homolytic cleavage of the O–O bond is a main pathway (Scheme 3.4).

On the other hand, the oxidative state of certain metal ions such as Fe(III) and Mn(III) are capable of oxidizing hydroperoxides (ROOH) into peroxy radicals (ROO$^{\bullet}$) (Scheme 3.5)[43] owing to their strong oxidation capability as

$$R_1O–OR_2 \rightarrow R_1O^{\bullet} + {}^{\bullet}OR_2,$$

$$R_1 = \text{Alkyl group}, R_2 = \text{Alkyl group or H}$$

Scheme 3.4

$$ROOH + Fe(III) \rightarrow ROO^{\bullet} + Fe(II) + H^+$$

Scheme 3.5

ROOH + Fe(II){EDTA} → RO˙ + Fe(III){EDTA} + HO⁻

Scheme 3.6

evidenced by the relatively high reduction potentials ($E°$) for the two metal ion pairs: Fe(III)/Fe(II), 0.77 V; Mn(III)/Mn(II) 1.5 V.[44]

Since the EDTA-complexed Fe(III)/Fe(II) pair has a much lower $E°$ (0.12 V), Fe(III){EDTA} would not be expected to oxidize hydroperoxides effectively into the corresponding alkylperoxyl radicals anymore near neutral pH, because the $E°$ of a typical alkylperoxyl radical is in the range 0.77–1.44 V,[45] resulting in a thermodynamically unfavorable positive ΔG value. On the other hand, Fe(II){EDTA} is capable of decomposing ROOH to RO˙ (Scheme 3.6) in a way similar to the Fenton reaction.

3.2.3 Autooxidative Radical Chain Reactions and Their Kinetic Behavior

As discussed above, various oxygen-based free radicals can be formed during the Fenton reaction, Udenfriend reaction, and decomposition of peroxides and hydroperoxides. Once these radicals are formed, they can trigger a chain reaction which consists of the following three stages: initiation, propagation, and termination. The following scheme (Scheme 3.7) uses peroxyl radicals (XOO˙, X = alkyl, H) as representative oxygen-based free radical initiators.

In Scheme 3.7, RH represents any species that can donate an H˙ which includes the oxidation substrate. In autooxidation, the initiation stage is usually a slow process, which can be impacted by a number of factors such as temperature, pH, moisture level (in solid state autooxidation), and low levels of impurities, in particular trace levels of transition metal ions. In pharmaceutical products, some components or impurities of the components can inhibit (or slow down) the autooxidation process. Because of these factors, radical-mediated autooxidation displays a variable induction period, during which time no significant oxidation is observed. During the propagation stage, the chain reaction is sustained by continuous generation of ROO˙ and R˙ radicals at the expense of consuming the oxidation substrates (RH) and molecular oxygen. The reaction between R˙ and O_2 (Step 2 in Scheme 3.7) is diffusion-controlled (*i.e.* the rate constant k is $\sim 10^9 \, M^{-1} \, s^{-1}$),[46] while the rate constant of an allylic H abstract reaction by ROO˙ is typically in the range of ~ 0.1–60 $M^{-1} \, s^{-1}$.[47] In the final termination stage, when enough radical species are present, combination of any two radicals can contribute to the termination of the chain reaction. The last step shown in the chain reaction illustrated in Scheme 3.7 is known as the Russell mechanism,[48] giving rise to ketone/aldehyde, alcohol, and singlet oxygen. It should be noted that the Russell mechanism is not the only pathway to generate the ketone/aldehyde and alcohol. These degradants may also be formed from further degradation of the hydroperoxide (ROOH) produced in the propagation stage. For an inhibited autooxidation reaction, the rate of the oxidation can be described as follows.[49]

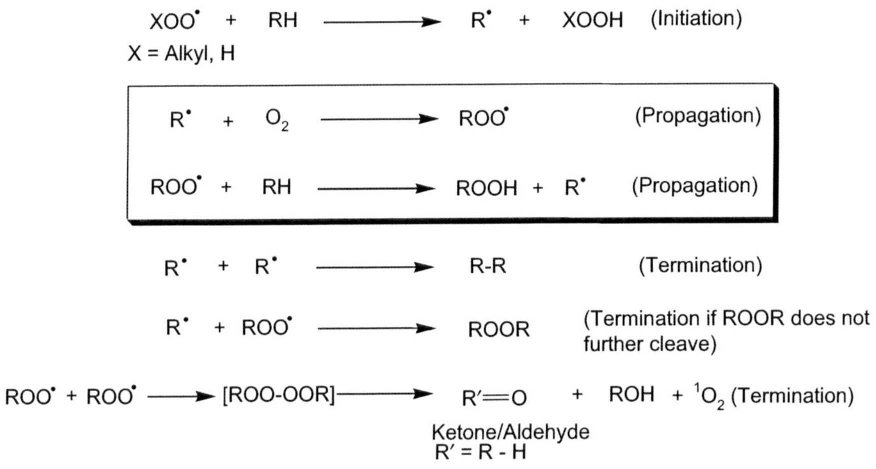

$$XOO^\bullet \ + \ RH \longrightarrow R^\bullet \ + \ XOOH \quad \text{(Initiation)}$$
$$X = \text{Alkyl, H}$$

$$R^\bullet \ + \ O_2 \longrightarrow ROO^\bullet \qquad \text{(Propagation)}$$

$$ROO^\bullet \ + \ RH \longrightarrow ROOH \ + \ R^\bullet \quad \text{(Propagation)}$$

$$R^\bullet \ + \ R^\bullet \longrightarrow R\text{-}R \qquad \text{(Termination)}$$

$$R^\bullet \ + \ ROO^\bullet \longrightarrow ROOR \qquad \text{(Termination if ROOR does not further cleave)}$$

$$ROO^\bullet + ROO^\bullet \longrightarrow [ROO\text{-}OOR] \longrightarrow R'{=}O \ + \ ROH \ + \ {}^1O_2 \text{ (Termination)}$$
$$\text{Ketone/Aldehyde}$$
$$R' = R\text{-}H$$

Scheme 3.7

$$-d[O_2]/dt = k_3\,[RH]Ri/(nk_5[\text{Inhibitor}])$$

where k_3 is the rate constant of the propagation stage, k_5 is the rate constant of the inhibition reaction ($R^\bullet +$ Inhibitor $\rightarrow$ RH + relatively stable inhibitor radical), Ri is the rate of chain initiation, and n is the stoichiometric factor of the inhibitor (antioxidant).

According to this equation, the rate of an inhibited autooxidation is proportional to the concentration of the oxidation substrate, [RH], and inversely proportional to the concentration of the inhibitor, [inhibitor]. The inhibitor is usually an antioxidant in a pharmaceutical formulation. This equation also indicates that the oxygen partial pressure has no impact on the rate of auto-oxidation. A practical implication of this conclusion is that reducing the oxygen concentration in a pharmaceutical formulation will not usually slow down the rate of the free radical-mediated autooxidation, unless oxygen can be almost completely removed from the formulation. The autooxidation kinetic study performed by Burton and Ingold with styrene as the oxidation substrate and vitamin E as the antioxidant showed a clear induction period, followed by a rapid surge of oxidation when the antioxidant was consumed;[49] an illustrative plot is shown in Figure 3.1 based on their work, which resembles the kinetic behavior of oxidizable drugs in autooxidation where generation of radicals is rate limiting.[50]

In addition to the pathways involved in the three stages of a typical chain reaction shown in Scheme 3.7, the two additional pathways that were discussed in Section 3.2.2 can also occur in an autooxidative chain reaction. These two pathways relate to further decomposition of the organic peroxide and hydroperoxide, which are described below.

Both the peroxide and hydroperoxide can undergo homolytic cleavage, as discussed in the previous section, to give alkoxyl and hydroxyl radicals. The alkoxyl radical can then abstract an H to produce alcohol (ROH).

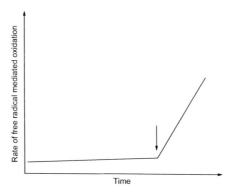

Figure 3.1   Kinetic behavior of an inhibited free radical mediated autooxidation. The arrow indicates the time point when the inhibitor was consumed.

The hydroperoxide can also be oxidized by certain metal ions to produce peroxyl radical. In a typical case of autooxidation, various oxygen-based free radicals are formed: $O_2^{-\bullet}$, $HO^\bullet$, $ROO^\bullet$, and $RO^\bullet$, where $ROO^\bullet$ is predominant. The reactivity of these radicals is in the following order: $HO^\bullet > RO^\bullet > ROO^\bullet \sim O_2^{-\bullet}/HO_2^\bullet$ based on the O–H bond cleavage energies listed in Table 3.2.

Table 3.2 also indicates that the allylic and benzylic methylene moieties should be quite susceptible to H abstraction by the oxygen-centered radicals, compared to secondary and tertiary alkyl CH moieties, because the bond dissociation energies of the former are significantly lower than the latter.

Owing to the characteristics and complexity of the chain reaction discussed above, autooxidative degradation kinetics is usually difficult to replicate and predict accurately. In addition, it usually cannot be sped up by increasing temperature, because under higher temperature, homolytic cleavage of the various peroxides produced in the chain reaction, illustrated in Scheme 3.8, would become significant. This could alter the kinetic behavior, complicate the degradation pathways, and ultimately result in different degradation profiles.

3.2.4   Additional Reactions of Free Radicals

In the previous section, two important radical reactions were mentioned: H abstraction and a radical termination reaction. Additional radical reactions that are significant in drug degradation include radical addition to a molecule containing unsaturated bonds. This addition reaction can lead to the formation of polyperoxides,[58] oligomerization and polymerization of the oxidation substrate,[59] while cleavage of the initially formed radical adduct can lead to the formation of epoxides and alkoxy radicals[60] (Scheme 3.9).

For a drug substance containing a benzylic moiety, its reaction with free radicals can be either on the benzylic CH through H abstraction or *via* addition onto the aromatic ring as shown in Scheme 3.10. The rates of both reactions are comparable based on a study using carbon-centered radicals as the attacking free radicals.[61]

Table 3.2 Homolytic bond dissociation energy of relevant and model species in free radical-mediated autooxidation.

Bond	Dissociation energy D^o_{298} $(kJ\ mol^{-1})^a$	D^o_{298} $(kcal\ mol^{-1})^a$
HO–H	498	119
CH$_3$O–H	436.8	104.5
CH$_3$OO–H	365;[b] 359.7[c]	87.3;[b] 86.1[c]
HOO–H	369.0; 367.4[c]	88.3; 87.89[c]
CH$_3$C(O)OO–H	386[b]	92.3[b]
PhO–H	361.9	86.6
HO–OH	213	51.0
CH$_3$O–OCH$_3$	157.3	37.6
CH$_3$CH(OH)–H	389	93[d]
(CH$_3$)$_2$C(OH)–H	380	91[d]
CH$_3$OCH$_2$–H	390	93.3[e]
CH$_3$CH$_2$–H	422.8	101.1
(CH$_3$)C–H	403.5	96.5
PhCH$_2$–H	368.2	88.1
CH$_2$=CHCH$_2$–H	362.0[f]	86.6[f]

[a]All data, except noted otherwise, are from Ref. 51. The underlined values are converted between kJ mol^{-1} and kcal mol^{-1} by a factor of 4.18.
[b]Calculated value taken from Ref. 52.
[c]Calculated values from Ref. 53.
[d]Ref. 54.
[e]Ref. 55 and 56.
[f]Ref. 57.

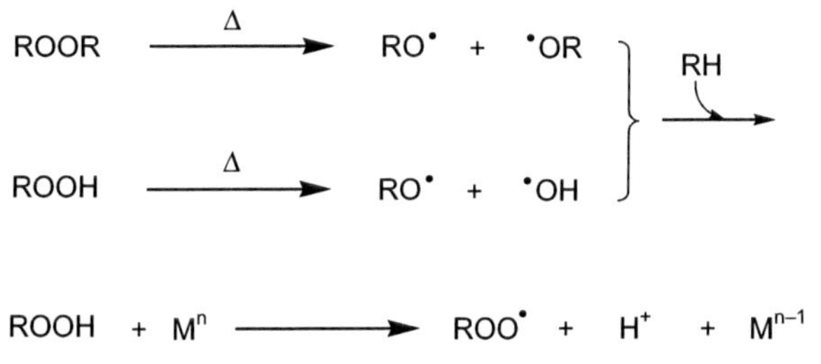

Scheme 3.8

3.3 Non-radical Reactions of Peroxides

3.3.1 Heterolytic Cleavage of Peroxides and Oxidation of Amines, Sulfides, and Related Species

Hydrogen peroxide is a key intermediate formed during the activation of molecular oxygen *via* the Udenfriend reaction. In addition to its role as the

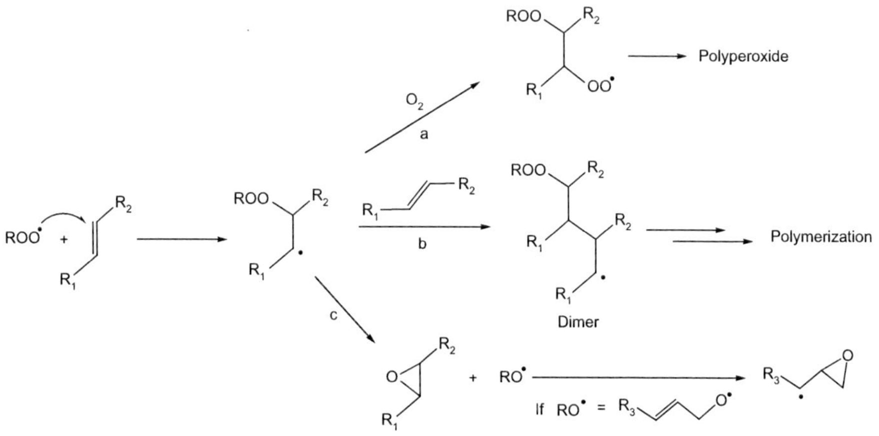

Scheme 3.9

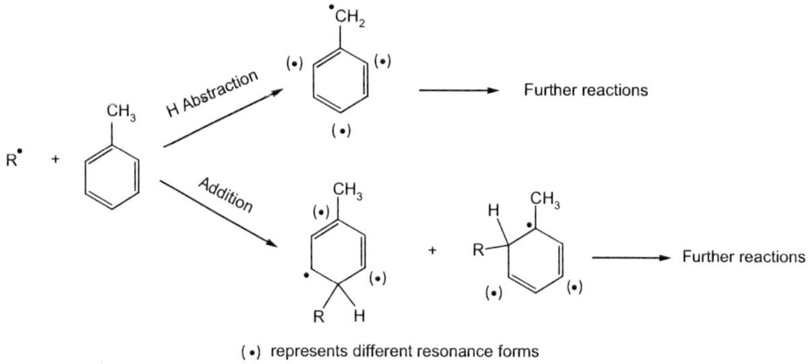

(•) represents different resonance forms

Scheme 3.10

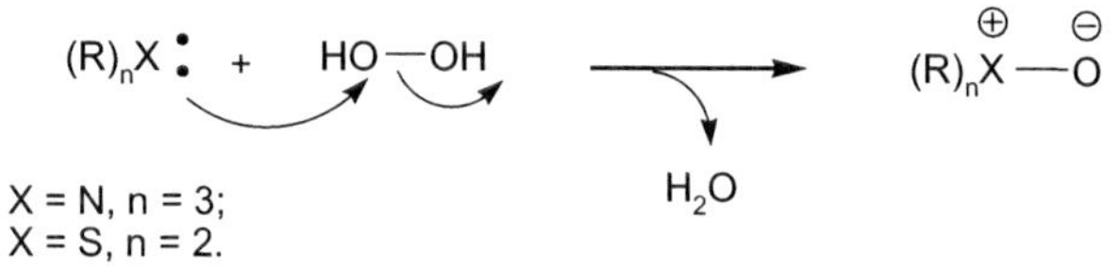

X = N, n = 3;
X = S, n = 2.

Scheme 3.11

precursor for HO• free radical *via* the Fenton reaction, hydrogen peroxide can also undergo non-radical-mediated oxidation. The most pharmaceutically significant oxidation pathway in this category is the oxidation of amines, sulfides, and similar functional groups containing nucleophilic lone electron pairs. The key step in this oxidation is an S_N2 nucleophilic attack by the N and S moieties of the oxidation substrates (Scheme 3.11).

Hence, the rate of this S_N2 reaction depends upon electronic and steric factors of the oxidation substrates. Among all alkyl amines, tertiary amines would most easily undergo this pathway,[62] because the lone pair of electrons on the tertiary amines are the most nucleophilic owing to the presence of three electron-pushing alkyl groups. Toney *et al.* measured the oxidation rates of two tertiary amines (*N*-lauryl morpholine and dimethyllaurylamine) and one secondary amine (piperidine) by hydrogen peroxide.[63] It was found that *N*-lauryl morpholine oxidizes the fastest, while piperidine oxidizes the slowest. The results are consistent with the S_N2 reaction mechanism: *N*-lauryl morpholine has both the electronic and steric advantage in that it is a cyclized tertiary amine which should be less sterically hindered than the acyclic tertiary amine, dimethyllaurylamine. If at least one of the three alkyl groups is replaced by an aryl group, the lone pair electrons of the amino group in the resulting aromatic amines can conjugate with the aryl group and thus become much less nucleophilic. Consequently, aromatic amines usually could not be oxidized by hydrogen peroxide *via* the nucleophilic route without the use of catalysts.[64]

In a kinetic study of *N*-oxidation of 4-substituted *N*,*N*-dimethylanilines by hydrogen peroxide in the presence of rhenium trioxide as the catalyst, Zhu and Esperson found that the oxidation is inhibited by electron-withdrawing groups in the 4-position.[65] This finding is again consistent with the S_N2 reaction mechanism. For the same reason, pyridine and pyridine-like moieties are not oxidized by hydrogen peroxide alone; for example, imatinib contains a piperizine ring, a pyridine, and a pyrimidine moiety and only the two nitrogens in the piperizine ring are oxidized under excessive stress with 10% hydrogen peroxide.[66] When amines are protonated at low pH, their nucleophilicity is greatly reduced, resulting in a much muted reactivity toward autooxidation.[67] On the other hand, the ability of non-ionizable thioethers (sulfides) and related compounds, such as disulfides, to undergo nucleophilic oxidation by hydrogen peroxide is not negatively impacted by low pH.

Hydrogen peroxide formed during autooxidation may react or interact with certain components of a drug formulation or stress system to yield stronger oxidants. Specifically, hydrogen peroxide can react with carboxylic acids, bicarbonate, and organonitriles (particularly acetonitrile, owing to its widespread use) to yield carboxyl peracids,[68] peroxymonocarbonate,[69–71] and peroxycarboximidic acids,[72] respectively. The resulting species owe their increased oxidation capability to the formation of a better leaving group than hydroxide/water in each case, facilitating the nucleophilic oxidative degradation (Scheme 3.12).

These "activated" forms of hydrogen peroxide have been used in synthetic organic chemistry to generate *N*-oxides, epoxides, sulfoxides, and sulfones, and so on. This can also explain why different levels of oxidation or oxidative degradation profiles may be obtained, dependent upon the choice of organic solvent (*e.g.* acetonitrile *versus* methanol) and other reagents, during forced degradation of a drug substance using hydrogen peroxide.

3.3.2 Heterolytic Cleavage of Peroxides and Formation of Epoxides

Although epoxides can be formed *via* a free radical pathway as discussed in Section 3.2.4, they can also be formed through non-radical pathways.

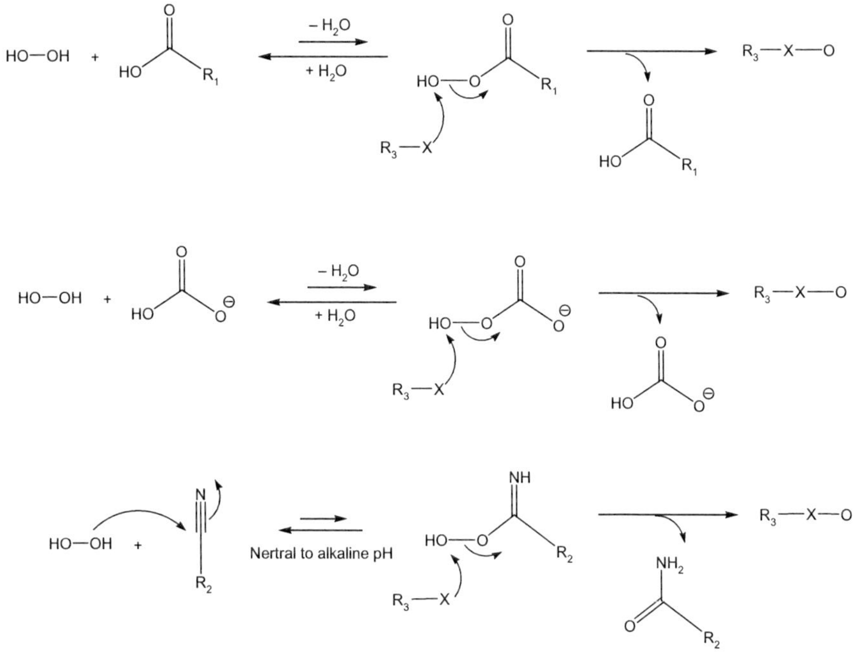

Scheme 3.12

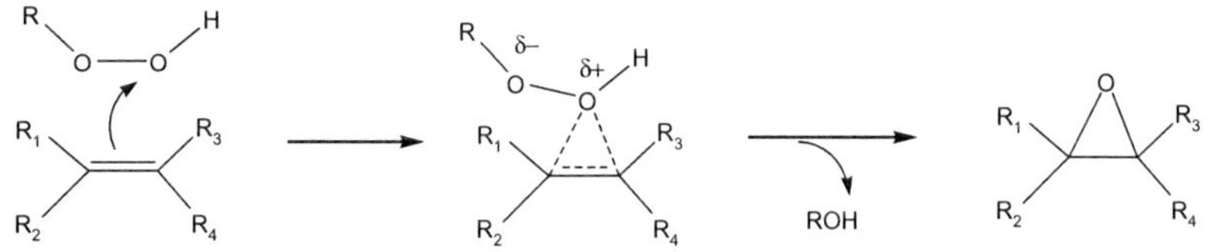

Scheme 3.13

Hydroperoxides, in particular the activated forms of hydrogen peroxide, *e.g.* peracids[73] and related compounds,[74] can react directly with electron-rich double bonds *via* an electrophilic oxygen transfer process, leading to the formation of epoxides, as shown in Scheme 3.13.

The double bonds in olefins that contain electron-withdrawing groups are electron-deficient. In such cases, the epoxidation can proceed through a nucleophilic oxygen transfer process. The epoxidation of α,β-unsaturated carbonyl compounds with hydroperoxides under basic conditions, that is, the Weitz–Scheffer reaction[75] is such an example (Scheme 3.14). According to the mechanism proposed by Bunton and Minkoff,[76] this oxidation proceeds *via* a two-step, addition and ring-closure mechanism shown in Scheme 3.14.

The epoxide formed may be isolatable but frequently would decompose further. Oxidative degradation involving heterolytic peroxide cleavage will be

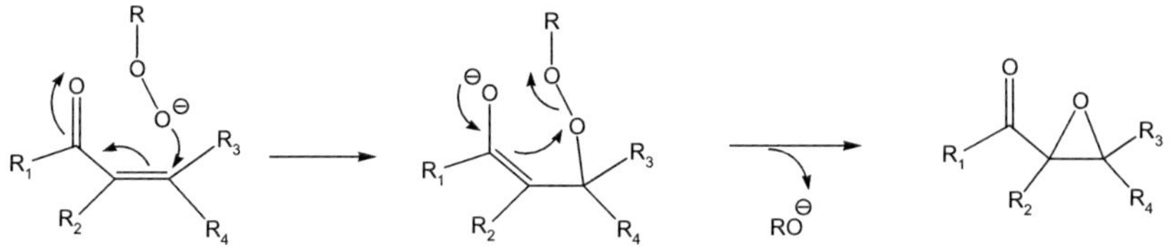

Scheme 3.14

discussed further in Sections 3.5.2 with representative drugs containing oxidizable carbon–carbon double bonds.

3.4 Carbanion/enolate-mediated Autooxidation (Base-catalyzed Autooxidation)

In contrast to free radical-mediated autooxidation, carbanion/enolate-mediated autooxidation is much less known for its role in the autooxidation of drugs. Interestingly, oxidation of carbanion/enolate by molecular oxygen (*i.e.* autooxidation) has been known since 1930s with its main utility being in synthetic organic chemistry.[77] More recently, some of the syntheses based on carbanion/enolate autooxidation are promoted as "green chemistry" because the oxidizing agent is molecular oxygen rather than hazardous metal-based oxidizing agents.[78] Since the vast majority of the carbanion/enolate species are generated using strong bases, the process is also called "base-catalyzed autooxidation". For drug substances containing somewhat "acidic" carbonated protons (CH_n, n is typically 1 to 2), the acidic CH_n can be slightly deprotonated by a relatively weak base or a general base in the drug formulation, which could result in the formation of impurities exceeding the International Conference on Harmonisation (ICH) thresholds for impurity identification or qualification (typically between 0.1% and 0.5%, dependent upon its daily maximum dose and potency for non-genotoxic impurities). Carbanion/enolate-mediated autooxidation (base-catalyzed autooxidation) can be described by Scheme 3.15.

Once the carbanion/enolate is formed by deprotonatation of an "acidic" carbon-bonded proton, it can quickly react with molecular oxygen to give the organic peroxide. The latter usually further decomposes to yield a number of final degradation products, including alcohol, ketone, anhydride, carboxylic acid, and rearrangement products depending upon the structure of the organic peroxide intermediate and other factors such as pH and solvent. Since a carbanion/enolate is usually a singlet species, its apparent swift reaction with the triplet ground state molecular oxygen appears to contradict the spin conservation rule.[4,79–81] To overcome this controversy, it was proposed that the carbanion/enolate would transfer an electron to molecular oxygen to form a pair of a carbon-based free radical and superoxide anion radical in a "caged" complex, which is then followed by spin inversion and subsequent combination to produce the peroxide or peroxide anion. On the other hand, whether or not

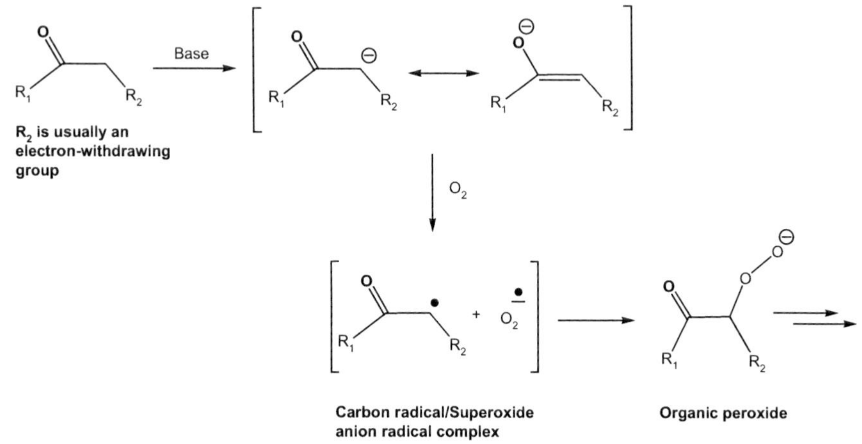

Scheme 3.15

the caged process exists, the overall rate of the carbanion/enolate-mediated autooxidation appears much faster than a usual free radical-mediated process and apparently displays no characteristics typical of a free radical-mediated reaction.[7,82,83]

3.5 Oxidation Pathways of Drugs with Various Structures

The origins and mechanisms of oxidative degradation of drugs of several major types of autooxidation mechanisms have been described in Sections 3.2 to 3.4. In this section, specific oxidation pathways of drugs with various functional groups and functional structures will be discussed in relation to each type of the oxidation mechanism. Note that the same type of functional group may undergo different pathways under different conditions including different dosages. For example, the carbon–carbon double bonds can undergo allylic oxidation as well as epoxidation. On the other hand, formation of the same degradant may derive from multiple degradation pathways. Epoxide, for example, can be formed from both radical and non-radical pathways.

3.5.1 Allylic- and Benzylic-type Positions Susceptible to Hydrogen Abstraction by Free Radicals

Drugs containing allylic- and benzylic-type moieties are susceptible to free radical attack, because the resulting carbon-centered radicals are stabilized by conjugation with the nearby double bond or aromatic ring. The carbon-centered radicals usually react with O_2 at an extremely fast rate (approaching the diffusion controlled rate). Nonetheless, carbon-centered radicals that are

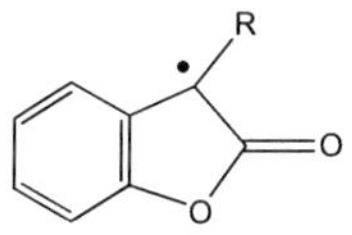

Figure 3.2 Structure of the benzylic radicals derived from 2-coumaranone derivatives.

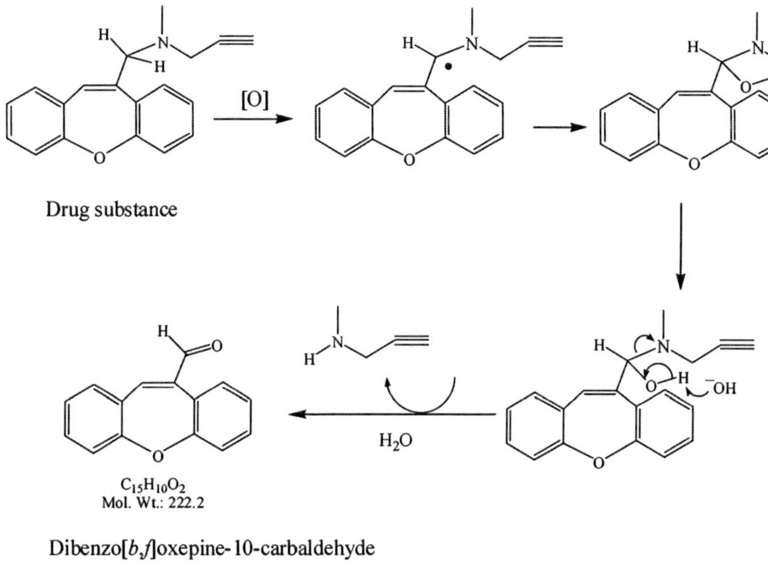

Drug substance

$C_{15}H_{10}O_2$
Mol. Wt.: 222.2

Dibenzo[*b,f*]oxepine-10-carbaldehyde

Scheme 3.16 Reproduction from Reference 85 with permission.

stabilized by extensive resonance, such as triphenylmethyl and 9-phenyl-fluorenyl, have greatly reduced reactivity with O_2. Other factors can also have an impact on the reactivity with O_2. For example, Bejan *et al.* reported that the benzylic radical derived from 2-coumaranone, which has a lactone function-ality next to the radical, completely lacks any reactivity with O_2 (Figure 3.2).[84]

During pharmaceutical development of a novel drug candidate, TCH346, for neurodegenerative disorders, a degradant devoid of the original amine moiety was found during long term and accelerated stability studies of a tablet for-mulation.[85] The authors proposed the degradation mechanism shown in Scheme 3.16.

In this drug substance, the allylic position is quite susceptible to H abstraction by a free radical generated in autooxidation, since the radical formed can be stabilized by an extended conjugated system. The allylic radical should readily react with O_2 to give a peroxide which would in turn produce the alcohol intermediate. It is possible that the peroxide could directly decompose to give the final aldehyde degradant, while producing a hydroxylamine as the

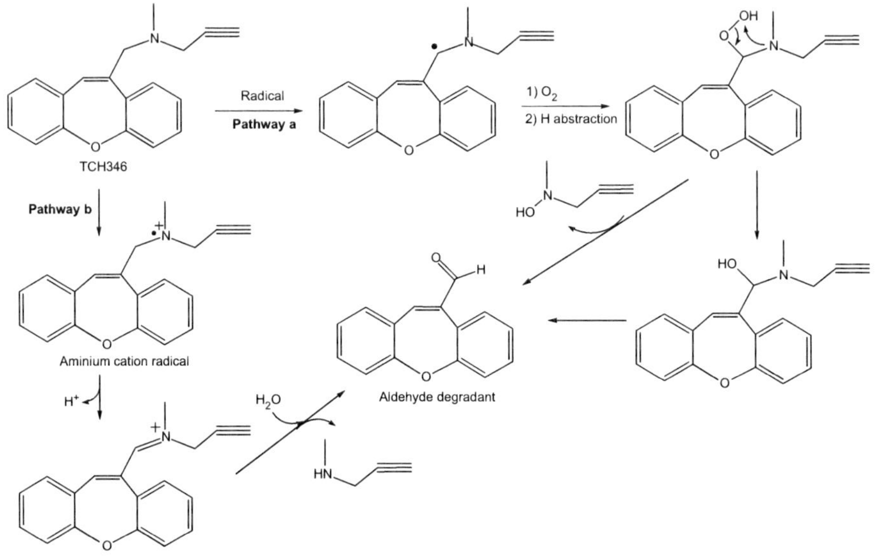

Scheme 3.17

leaving group. Hence, an alternate degradation mechanism (Scheme 3.17, pathway a) can be proposed.

In this particular case, the benzylic position is also α to the tertiary amine moiety. This type of structural moiety can also form an aminium radical cation intermediate through a one electron transfer oxidation of the nitrogen atom, which will give the same aldehyde degradant while producing an amine as the leaving group (Scheme 3.17, pathway b). The degradant distribution in the latter case is the same as that proposed by the original researchers. Since the structure of the leaving group was not determined in the original study, it is not possible to tell which degradation pathway is more likely. There will be a more detailed discussion of the autooxidation mechanism *via* the aminium radical cation intermediate in Section 3.5.3.3.

Clopidogrel bisulfate, the active pharmaceutical ingredient (API) of the second best selling drug Plavix, contains a moiety that is similar to that in TCH346 above: a benzylic-type 3-thiothenylmethyl position that is also α to a tertiary amine functionality. Recently, a significant new oxidative degradant was observed in the clopidogrel bisulfate drug substance and drug product.[86] This new degradant elutes close to the void volume in the clopidogrel bisulfate USP method[87] due to the polar iminium cation moiety. Despite the thorough structure characterization carried out by the original workers, no formation mechanism was proposed for this new degradant. Based on the similarity of the key functionalities between clopidogrel bisulfate and that of TCH346, formation of the new oxidative degradant can proceed through either of the two radical pathways (Scheme 3.18) *via* intermediates that are analogous to those in Scheme 3.17.

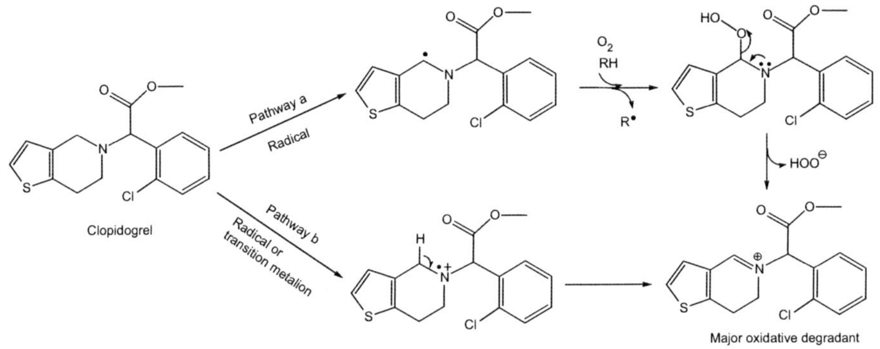

Scheme 3.18

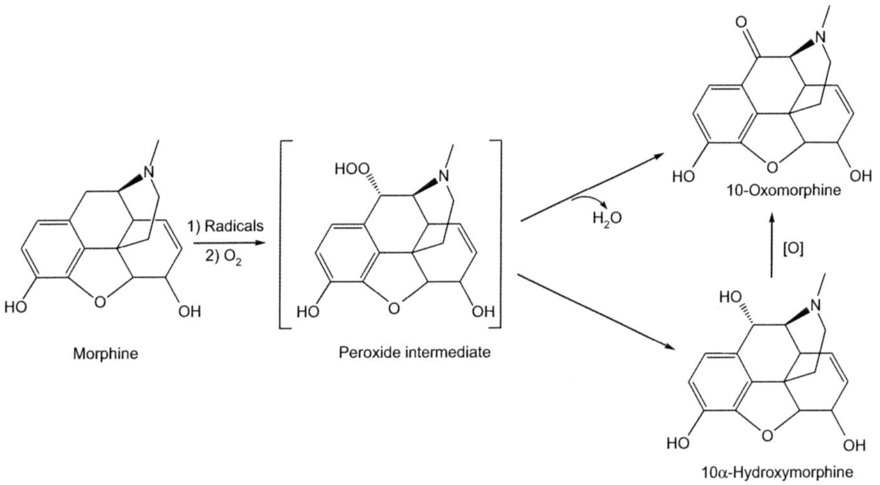

Scheme 3.19

The molecule of morphine contains both a benzylic (C10) and an allylic (C14) position. Two degradants resulting from C10 oxidation were observed in morphine sulfate drug substance as well as in several pharmaceutical preparations.[88–90] The formation of these two degradants, 10α-hydroxy and 10-oxomorphine, is probably *via* the pathway shown in Scheme 3.19.

The stability results showed that 10-oxomorphine continuously increased over time, while 10α-hydroxymorphine remained relatively unchanged. This phenomenon is consistent with the above mechanism where 10α-hydroxy-morphine is an intermediary degradant leading to the terminal degradant, 10-oxomorphine. Under chemical transformation conditions, no oxidation was seen on C14 position,[90] indicating the C14 allylic position is less reactive than the benzylic C10 position towards what is presumed to be radical-mediated autooxidation. On the other hand, the phenolic moiety of morphine (and related drugs, *e.g.*, naloxone, nalbuphine, and oxymorphone) can undergo oxidative 2,2′-dimerization in solutions to yield primarily 2,2′-morphine dimer (pseudomorphine). Refer to Section 3.5.9 for a more detailed discussion on the oxidation mechanism of drugs containing a phenolic moiety.

Ezlopitant, a non-peptide substance P receptor antagonist, possesses a 4-methoxybenzylic moiety as well as a diphenylmethyl moiety (Figure 3.3). Although it is quite stable in solution, ezlopitant was found to autooxidize relatively quickly at the 4-methoxybenzylic position to give the benzylic peroxide as the major degradant when stored in solid.[91] Based on consideration of the electronic factor, the diphenylmethyl position may be more susceptible to the radical-mediated autooxidation. Nevertheless, steric hindrance may inhibit the autooxidation at this position.

Avermectins and related compounds are macrolides with broad anti-parasite activities. Their core structures contain a number of allylic and allylic-like sites, among which the 8α-site is the most reactive one owing to its linkage to the neighboring butadiene and ether functionalities. The resulting autooxidative degradant is 8α-oxoavermectin (Scheme 3.20).[92,93]

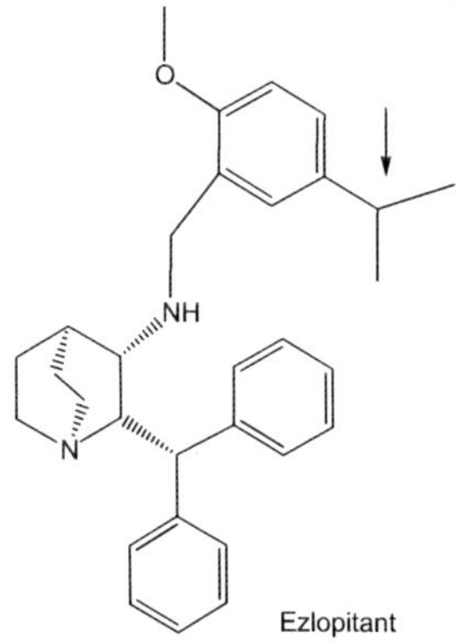

Ezlopitant

Figure 3.3   The structure of ezlopitant. The arrow indicates where peroxidation occurs.

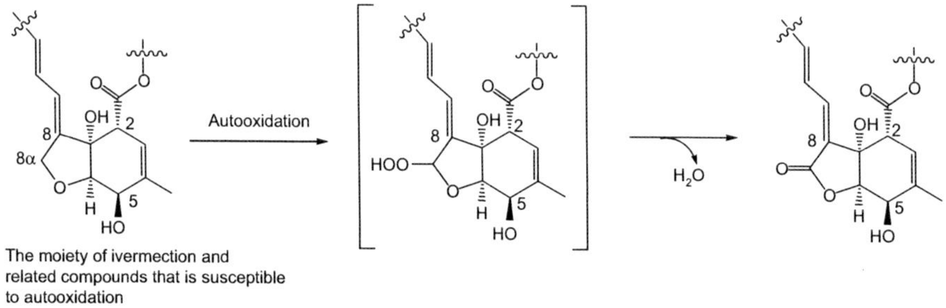

The moiety of ivermection and
related compounds that is susceptible
to autooxidation

Scheme 3.20

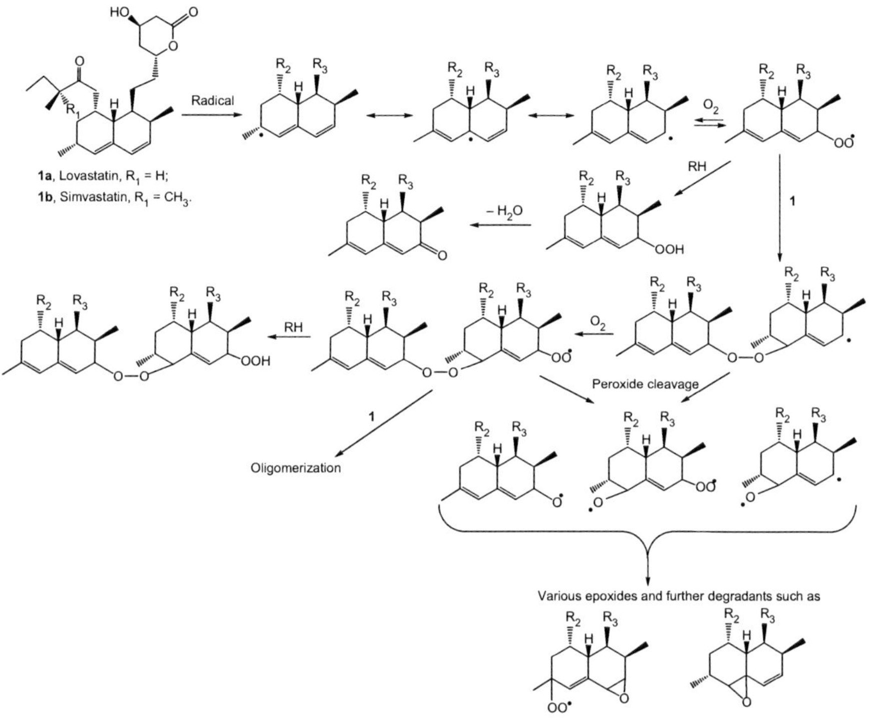

1a, Lovastatin, R_1 = H;
1b, Simvastatin, R_1 = CH_3.

Radical

O_2

RH

$-H_2O$

OOH

O_2

RH

OOH

O—O

1

Oligomerization

Peroxide cleavage

Various epoxides and further degradants such as

OO·

Scheme 3.21

Other examples of autooxidation involving a diene functionality include lovastatin and simvastatin, which are the first and second generation HMG-CoA reductase inhibitors used for the treatment of hypercholesterolemia. Both drug substances have a diene functionality embedded in their ten-membered fused ring core structures. This diene functionality is particularly reactive towards free radical-mediated autooxidation in both solid and solution states.[94,95] Owing to the presence of various resonance forms of the initial radical generated and their reactions with molecular oxygen and/or among themselves, a great number of oxidative degradants can be produced (Scheme 3.21).

Structural variations of the benzylic and allylic moieties, such as the CH positions that are α to a carbon–heteroatom double bond or α to a heterocyclic aromatic ring, are also susceptible to the same free radical-mediated auto-oxidation. For example, the anti-psychotic drug risperidone (Figure 3.4) contains a fused pyrimidin-4-one ring (ring A). The α-position (9-position) next to the heterocyclic pyrimidin-4-one ring was found to undergo autooxidation in bulk drug as well as in a tablet formulation, resulting in the formation of 9-hydroxyrisperidone as the main degradant.[96] This degradant is also a metabolite of the drug.[97] On the other hand, *N*-oxidation on the middle piperidine ring also occurred; the *N*-oxide was the second most significant degradant.

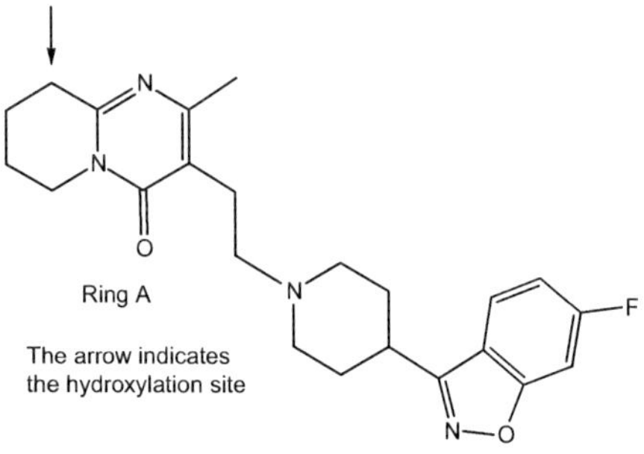

Figure 3.4 Structure of risperidone.

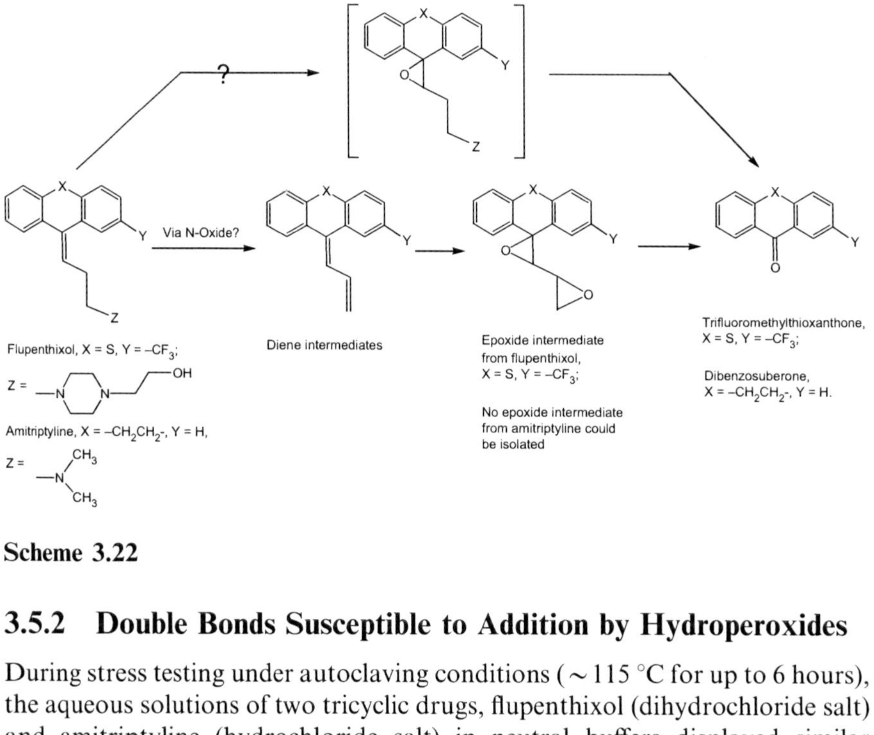

Scheme 3.22

3.5.2 Double Bonds Susceptible to Addition by Hydroperoxides

During stress testing under autoclaving conditions ($\sim 115\ ^{\circ}$C for up to 6 hours), the aqueous solutions of two tricyclic drugs, flupenthixol (dihydrochloride salt) and amitriptyline (hydrochloride salt) in neutral buffers displayed similar degradation patterns.[98,99] Analogous tricyclic ketones, namely trifluoromethylthioxanthone and dibenzosuberone, were formed respectively (Scheme 3.22). It would be intuitive to postulate that the formation of the tricyclic ketones may be mediated through an epoxide intermediate of the parent drugs *via* the electrophilic oxygen transfer process as shown in the upper pathway of Scheme 3.22, the mechanism of which has been discussed above in Section 3.3.2. Nevertheless, such an intermediate could not be isolated in either case, although an epoxide formed from an intermediary degradant of flupenthixol was isolated. The latter epoxide quickly decomposed to the corresponding tricyclic ketone, trifluoromethylthioxanthone upon exposure to air. This observation, along with the fact that a few other intermediary degradants were also seen during the stress test, led the authors to propose a stepwise degradation pathway leading to the final formation of the tricyclic ketones (Scheme 3.22).

In the course of a liquid and tablet formulation study of tiagabine, a potent inhibitor of gamma-aminobutyric acid (GABA) uptake for the treatment of epilepsy, two major degradants, dihydroxytiagabine and ketotiagabine, were observed.[100] The formation of dihydroxytiagabine is most likely to take place *via* a transient epoxide intermediate, while ketotiagabine appears to be a further dehydration degradant of dihydroxytiagabine (Scheme 3.23).

The indole ring is an important functional moiety that is present in the amino acid, tryptophan; its UV absorption property is mostly responsible for the absorbance of proteins at 280 nm, a wavelength widely used for protein detection and assays. The indole ring also exists in many natural products, fragrances, as well as drugs. Part of the fused pyrrole ring in indole can be considered to be an embedded enamine moiety. As such, the double bond in the fused pyrrole ring is quite electron-rich and hence susceptible to oxidation by hydroperoxides. The oxidation proceeds *via* nucleophilic attack of the double bond with hydroperoxide (or electrophilic oxygen transfer from the perspective of hydroperoxide). The resulting epoxide usually further degrades into various final degradants depending upon the structures connecting to the indole ring. For example, epoxides of simple alkyl-substituted indoles decompose to 2-oxindoles[101] (Scheme 3.24).

Nevertheless, during a hydrogen peroxide stress study of indomethacin, a non-steroidal anti-inflammatory drug containing a 5-methoxyindole ring, two major degradants were produced, which can be rationalized as being formed

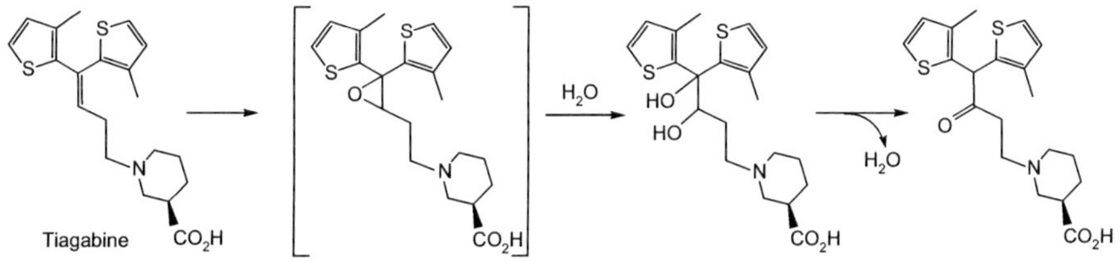

Scheme 3.23

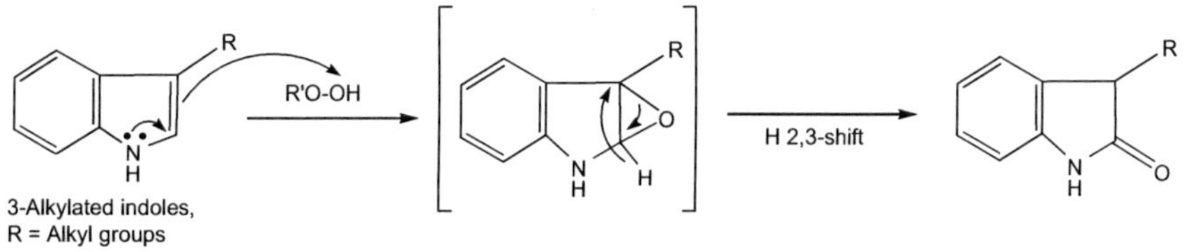

Scheme 3.24

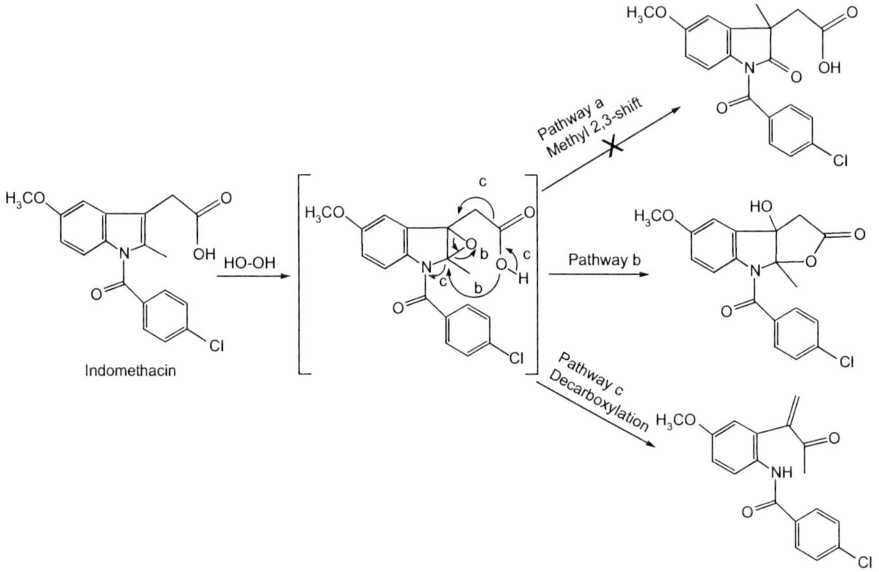

Scheme 3.25

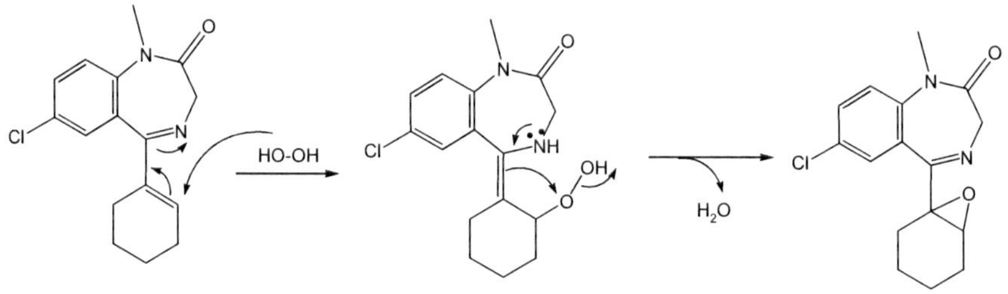

Scheme 3.26

from the epoxide intermediate through pathways that differ from Scheme 3.24.[102] The expected methyl 2,3-shift (pathway a, Scheme 3.25) did not occur.

As mentioned above, drugs containing electron-deficient double bonds can undergo epoxidation through nucleophilic oxygen transfer from hydroperoxides. One example can probably be found in the hydrogen peroxide stress of tetrazepam, a benzodiazepine used clinically as a myorelaxant. When the stress was conducted at 40 °C in dark, the epoxide was formed as the only degradant.[103] Owing to the presence of the conjugated imine moiety, the nucleophilic oxygen transfer in Scheme 3.26 can be proposed.

In the degradation of a tablet formulation of tetrazepam, however, the epoxide was observed only as a minor degradant, while the main degradation that occurred was oxidation at the 3′-allylic position. This suggests that the degradation pathway to the epoxide degradant in a tablet formulation may proceed through a radical-mediated process. Indeed, stress on a tetrazepam solution with a radical initiator, azobisisobutyronitrile (AIBN), generated a degradation profile that is quite similar to that of the tablet formulation in an accelerated stability study.[103]

3.5.3 Tertiary Amines

3.5.3.1 *Formation of N-oxides* via *Nucleophilic Attack on Hydrogen Peroxide*

In the case of amine oxidation by hydrogen peroxide (or hydroperoxides in general) *via* a nucleophilic pathway (Scheme 3.27), tertiary alkyl amines are most prone to producing *N*-oxides by oxidation, as discussed above in Section 3.3.1. The *N*-oxide of a tertiary alkyl amine is reasonably stable and in most cases, can be isolated.

A great number of drugs have alkyl tertiary amine functionality and hence are susceptible to autooxidation *via* the nucleophilic pathway shown in Scheme 3.27 and subsequent degradation pathways. Tertiary amine drugs are also susceptible to free radical-mediated autooxidation *via* the aminium cation intermediate, which will be discussed in Section 3.5.3.3. A good class example of the nucleophilic oxidative degradation can be found in phenothiazine-derived drugs; there are more than two dozen such drugs according to a search on the web site, http://drugbank.wishartlab.com.[104] These drugs all contain a tricylic phenothiazine ring with different substituents on the N atom of the ring. The majority of the *N*-substituents contain either *N,N*-disubstituted piperazine ring or acyclic tertiary amine moieties, most of which can be illustrated in Figure 3.5 where the sites for *N*-oxide formation are indicated by arrows.

In this class of drugs, the sulfur in the tricyclic ring can compete with the side chain tertiary amines for oxidation during the early stage of autooxidation. Recently, Wang *et al.* reported a stress study of perphenazine solution in methanol with hydrogen peroxide, which showed the formation of all the three mono-oxidized degradants, that is, perphenazine 17*N*-oxide, perphenazine 14*N*-oxide, and perphenazine sulfoxide (Figure 3.6), in addition to lower levels of dioxidized degradants.[105]

Among these degradants, 17*N*-oxide is most abundant in the early stage of the stress; interestingly, 17*N*-oxide is also the most abundant oxidative

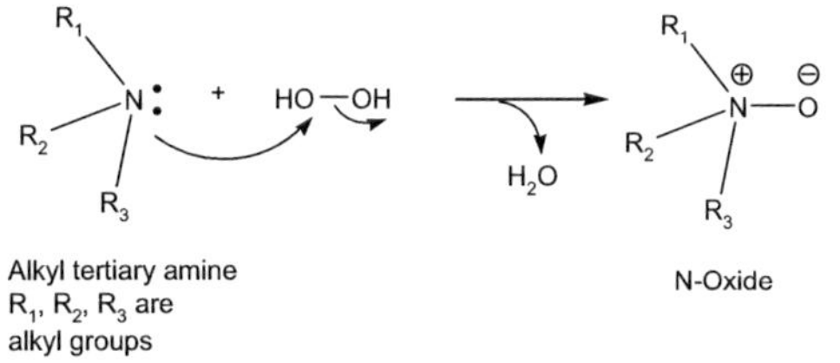

Alkyl tertiary amine
R$_1$, R$_2$, R$_3$ are
alkyl groups

N-Oxide

Scheme 3.27

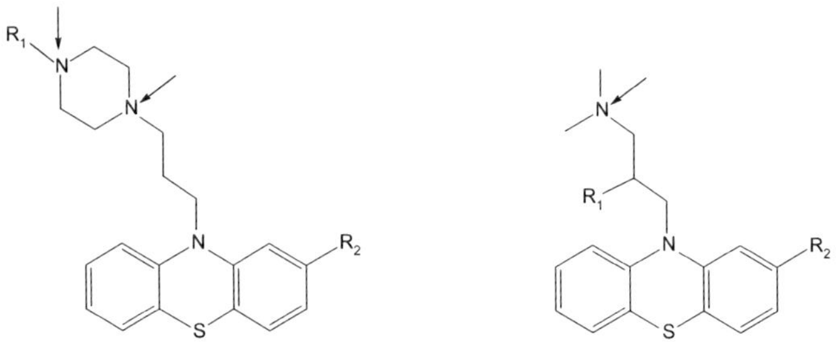

Perphenazine, R_1 = 2-hydroxyethyl, R_2 = chloro;
Prochlorperazine, R_1 = methyl, R_2 = chloro;
Fluphenazine, R_1 = 2-hydroxyethyl, R_2 = trifluoromethyl;
Triethylperazine, R_1 = methyl, R_2 = ethylmercapto;
Carphenazine, R_1 = 2-hydroxyethyl, R_2 = propionyl;
Trifluoperazine, R_1 = methyl, R_2 = trifluoromethyl;
Thioproperazine, R_1 = methyl, R_2 = dimethylsulfamyl.

Chlorpromazine, R_1 = hydrogen, R_2 = chloro;
Trimeprazine, R_1 = methyl, R_2 = hydrogen;
Promazine, R_1 = hydrogen, R_2 = hydrogen;
Methotrimeparzine, R_1 = (S)-methyl, R_2 = methoxyl;
Triflupromazine, R_1 = hydrogen, R_2 = trifluoromethyl;
Acepromazine, R_1 = hydrogen, R_2 = acetyl;

The arrows indicate the N-oxidation sites.

Figure 3.5   Structures of phenothiazine-derived drugs.

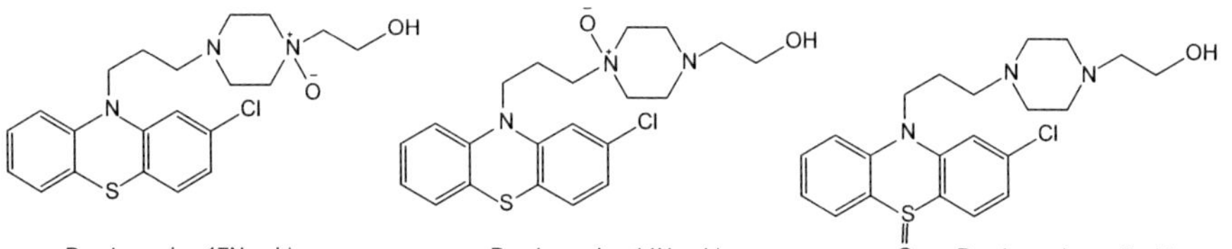

Perphenazine 17N-oxide   Perphenazine 14N-oxide   Perphenazine sulfoxide

Figure 3.6   Structures of three mono-oxidized degradants of perphenazine.

degradant in a solid formulation containing perphenazine. No oxidation occurred on the aromatic nitrogen, as expected. This result is not in a complete agreement with the stress study of perphenazine reported by Li *et al.*[106] According to this study, perphenazine sulfoxide was observed as the only mono-oxidized degradant.[106] The discrepancy may be caused by one of the following two factors: (1) the stress solution used by Li *et al.* is mostly aqueous, which is quite different from the mostly methanolic solution employed by Wang *et al.* (2) The two N-oxides might be formed in the stress by Li *et al.* but not separated by the method they used.[106]

   In a formulation study assessing pH effect on the control of N-oxidative degradants, Freed *et al.* found that oxidation of alkyl tertiary amines by hydroperoxides (including hydrogen peroxide in a few stress studies) can be inhibited by lowering the pH of stress solutions or by acidifying solid dosage formulations with citric acid.[67] In one of their solution stress studies, it is apparent that the pH of the solution needs to be controlled well below the pK_a of the tertiary amines in order to suppress the N-oxidation effectively.

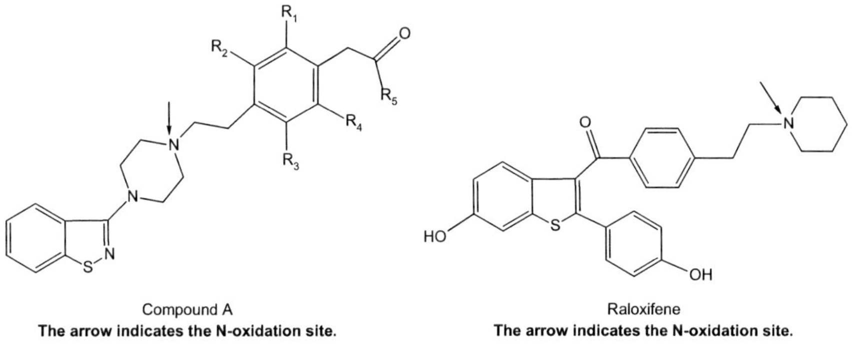

Compound A
The arrow indicates the N-oxidation site.

Raloxifene
The arrow indicates the N-oxidation site.

Figure 3.7 Compound A and raloxifene.

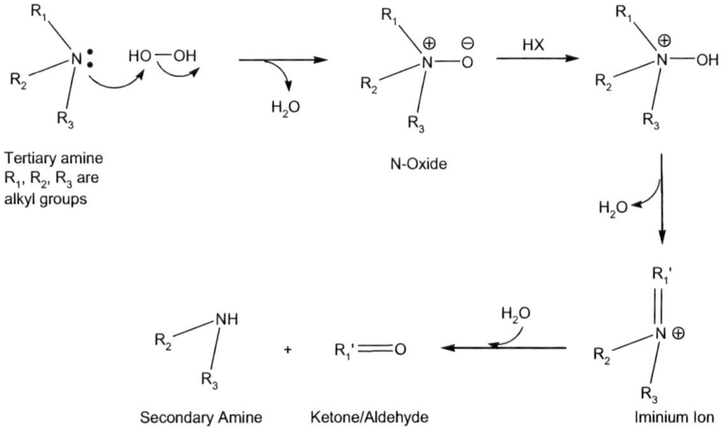

Scheme 3.28

For example, the two tertiary amines used in the study, compound A (a proprietary drug candidate with partial structure revealed) and raloxifene, displayed significantly different rates of N-oxidation at pH 6: Compound A with a pK_a of 6.45 oxidizes approximately twice as fast as raloxifene which has a calculated pK_a of 8.67 (Figure 3.7).

3.5.3.2 Decomposition of N-oxides: Secondary and Tertiary Degradation Pathways

As discussed previously, alkyl tertiary amine oxides are reasonably stable and can be isolated in most cases. Therefore, further degradation of N-oxides is usually not significant for drug substances and products stored under ICH long term and accelerated stability conditions. Nevertheless, under somewhat excessive reaction conditions such as various stress conditions, N-oxides are capable of degrading further into an array of secondary and tertiary degradants (Scheme 3.28). These further degradation pathways include dealkylation through the iminium ion intermediate, if at least one alkyl group contains an α-H.

This pathway, in particular demethylation and de-ethylation (R_1 = methyl or ethyl), is significant in the metabolism of alkyl tertiary amine drugs, which is catalyzed by metabolic enzymes,[107,108] but is generally insignificant for drugs (particularly in solid dosage forms) under commercial storage conditions which typically follow the ICH long term stability storage conditions. During the course of writing this book, despite repeated searches of the literature, the author has not yet been able to find a meaningful dealkylation case as a result of the nucleophilic oxidation of tertiary amines except for those under somewhat excessive stress conditions.[109,110] On the other hand, dealkylation of tertiary amines is possible under free radical-mediated autooxidation conditions, which will be discussed in the next section. Nevertheless, the alkyl groups that are cleaved from the tertiary amines under the latter conditions are usually more complicated than straightforward alkyl groups like methyl and ethyl groups.[85,111]

Other degradation pathways of N-oxides include several thermolytic reactions such as deoxygenation, in which N-oxides are reduced back to the original tertiary amines,[112,113] Cope elimination, where an olefin and a hydroxylamine are produced if the N-oxides contain a β-hydrogen,[114,115] and Meisenheimer rearrangement.[116] These reactions occur only at very high temperature and their only relevance to drug degradation study may be limited to two areas. The examples in the first area include special studies such as the autoclaving studies, discussed in Section 3.5.2, where decomposition of flupenthixol and amitriptyline into the corresponding diene intermediates is very likely *via* Cope elimination of the two N-oxides as illustrated in Scheme 3.29.

The examples in the second area are the thermal degradation pathways that occur during gas chromatography (GC) and atmospheric pressure chemical ionization-mass spectrometry (APCI-MS) analyses in which processes the temperatures inside the MS detectors can be as high as several hundred degrees

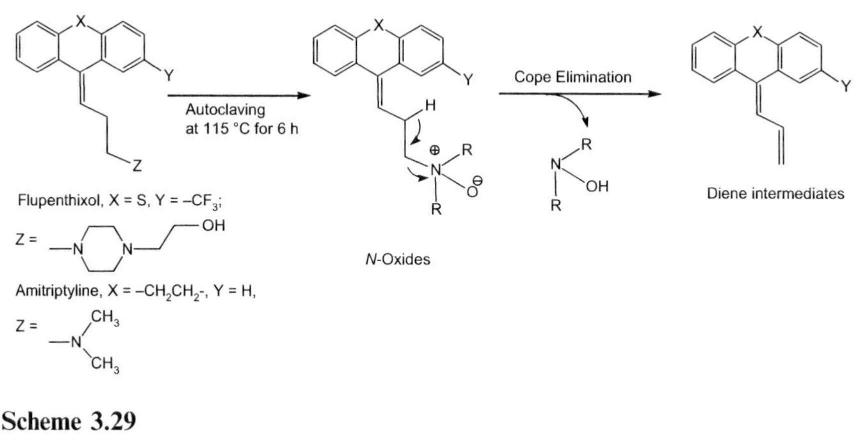

Scheme 3.29

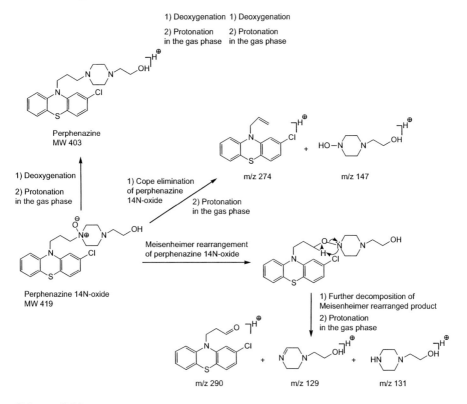

Scheme 3.30

centigrade. An example is a recent liquid chromatography-mass spectrometry/ mass spectrometry (LC-MS/MS) study of perphenazine 14*N*-oxide, in which the analyte undergoes all these degradation pathways leading to the formation of various species in the gas phase (Scheme 3.30).[105]

3.5.3.3 Free Radical-mediated Autooxidation of Tertiary Amines *via the Aminium Radical Cation Intermediate*

In the presence of free radical sources, such as those produced by Udenfriend chemistry, tertiary amine drugs can also undergo autooxidation through the aminium radical cation intermediate.[117,118] During a formulation study of a potential therapeutic agent (X) for the treatment of stroke and severe head trauma, the drug candidate was found to autooxidize in an intravenous dosage form.[38] The liquid formulation contains a 10 mM, pH 4.5 lactate buffer. Various antioxidants, such as ascorbic acid, thioglycerol and sodium bisulfate, were tried to stabilize the formulation. Nevertheless, all these antioxidants were found actually to promote the oxidative degradation of the drug candidate. Based on these results as well as the structure of the drug candidate, it appears that the autooxidation is caused by the Udenfriend reaction. Although the

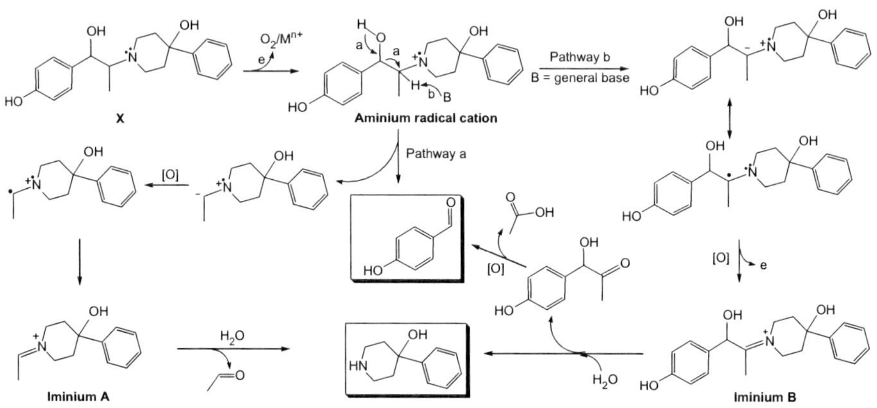

Scheme 3.31

original authors did not specify that the Udenfriend chemistry may be responsible for the observed instability of the formulated drug, the mechanism they proposed, in which a transition metal ion catalyzes the formation of reactive oxygen species from molecular oxygen and the oxidized metal ion is then recycled back to its reduced state by the antioxidants, largely resembles a typical Udenfriend pathway (see Section 3.2.1).

The proposed mechanism is consistent with a ferrous ion spiking experiment they performed, in which the oxidative degradation of the drug candidate was accelerated. In this particular case, Udenfriend chemistry can operate in the following two ways. In the first scenario, the role of transition metal chelator can be played by the lactate buffer, which is a reasonably good iron chelator.[36] On the other hand, since the drug candidate itself contains a 2-hydroxyamine moiety, which is a good chelator for transition metal ions,[119] the authors surmised that it can directly chelate with the ferrous ion and the bound metal ion can therefore activate molecular oxygen. The reactive oxygen species formed can then extract an electron from the tertiary nitrogen of the drug candidate to produce the critical aminium radical cation as shown in Scheme 3.31.

In the above degradation mechanism, "pathway a" is proposed by the original authors, while "pathway b" is proposed by the current author. Both pathways involve the initial formation of the aminium radical cation intermediate. Both De La Mare[117] and Hong *et al.*[38] described this type of aminium radical cation and its conversion to the subsequent carbon-centered radical.

3.5.4   Primary and Secondary Amines

In the case of primary and secondary amine oxidation, direct nucleophilic oxidation by hydrogen peroxide does not appear to be a significant degradation pathway, presumably because the nitrogen atoms in primary and secondary amines are less nucleophilic compared to tertiary amines. Aromatic primary

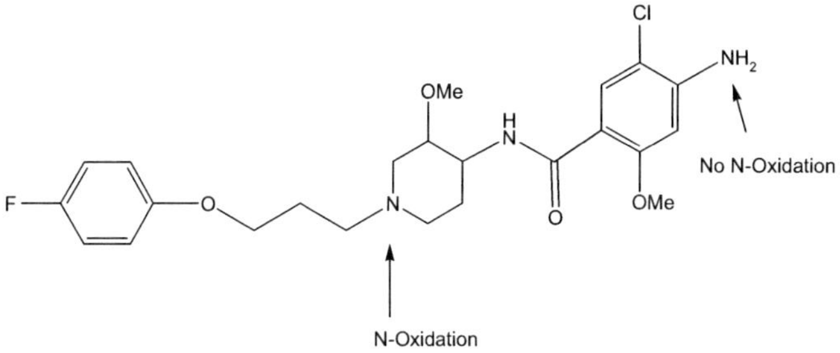

Figure 3.8 *N*-Oxidation site of cisapride under stress by hydrogen peroxide.

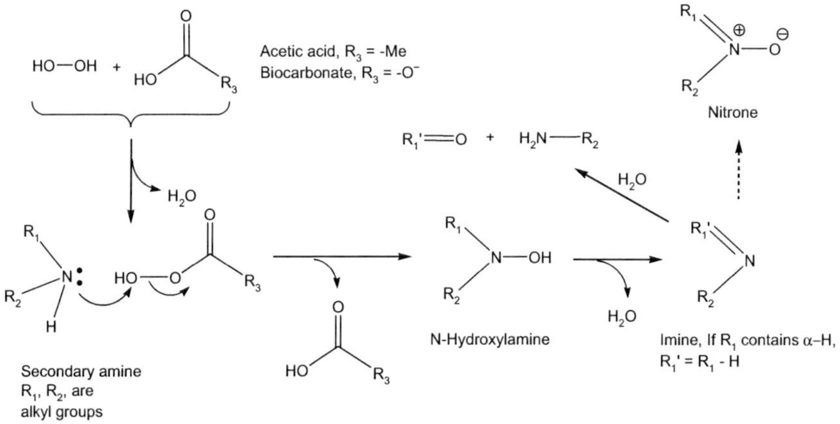

Scheme 3.32

and secondary amines are even less reactive. For example, excessive stress of cisapride by hydrogen peroxide at elevated temperature produced ~50% cisapride *N*-oxide, while no oxidation occurred on the aromatic primary amine moiety (Figure 3.8).[120]

On the other hand, hydrogen peroxide can be "activated" by interacting with certain components in a formulation or in a stress system as discussed above in Section 3.3.1 to form peracids, acetonitrile adducts, and peroxymono-carbonates, respectively. The resulting species are more reactive than hydrogen peroxide. For example, peracids are capable of reacting with primary and secondary amines to yield initially hydroxylamines[121,122] which are usually subject to further degradation depending upon the structure of the amines (Scheme 3.32).[123,124]

Scheme 3.32 shows that the initial degradation product formed is *N*-oxide, which immediately isomerizes to *N*-hydroxylamine.[125] This isomerization can be viewed as a special case of the Meisenheimer rearrangement, in which the proton rearranges from the nitrogen to the oxygen. The *N*-hydroxylamine formed can further degrade to imine if one of the alkyl groups contains an α-H.

The imine can further oxidize to become a nitrone. All the synthetically useful procedures for preparing a nitrone from secondary amine use catalysts to assist the oxidation by hydrogen peroxide.[126] Hence, how meaningful the pathway is for secondary amines in autooxidation is questionable. On the other hand, the imine formed is an electrophile which is susceptible to nucleophilic attack. When the nucleophile is water or hydroxide, hydrolysis of the imine can occur.

In the presence of radical species, the initial oxidation product of a secondary amine would be a hydroxylamine radical according to a recent study using electron spin resonance (ESR) and theoretical calculation (Scheme 3.33).[127]

The secondary hydroxylamine radical formed can obviously abstract a hydrogen to yield the corresponding hydroxylamine. In general, secondary hydroxylamines are usually not very stable and hence difficult to isolate. There are a few exceptions, for example, N-hydroxydesloratadine, which is an oxidative degradant of desloratadine and is stable enough to be isolated.[128] In the case of a stable free radical species, TEMPO (2,2,6,6-Tetramethylpiperidin-1-yl)oxyl) (Figure 3.9), the presence of the surrounding 2,2,6,6-tetramethyl groups stabilizes the free radical on the secondary amine.

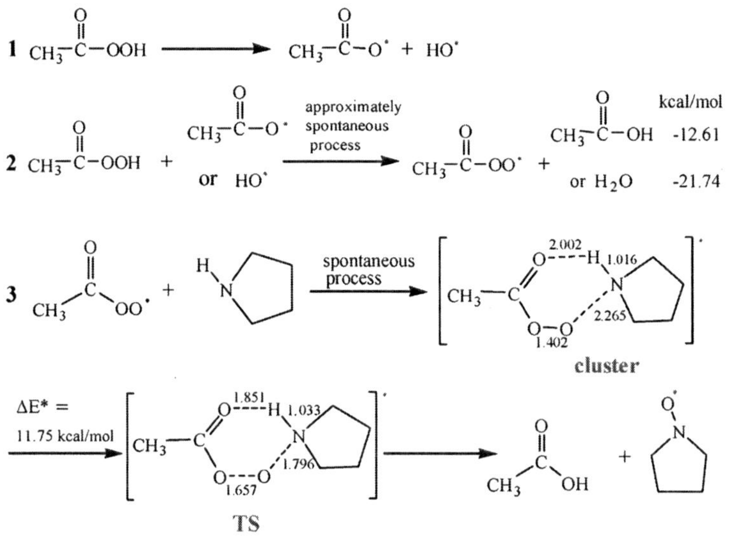

Scheme 3.33 Reproduction from Reference 127 with permission.

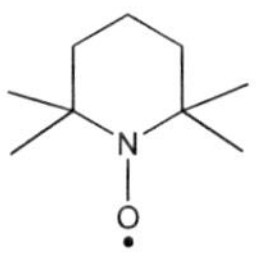

Figure 3.9 Structure of TEMPO.

TEMPO has been extensively used in ESR and nuclear magnetic resonance (NMR) studies as a stable free radical label[129,130] and a catalyst for oxidation of alcohols.[131,132]

3.5.5 Enamines and Imines (Schiff Bases)

Enamines and imines are interconvertible species through tautomerization. Nevertheless, they have quite different reactivities: enamines are nucleophiles while imines are electrophiles. Imines are formed by condensation between primary amines and aldehydes; replacement of the primary amines by secondary amines will produce iminium salts. The condensation reaction is usually reversible except for a few cases where the resulting imines may be stabilized by additional structural features such as conjugation and/or cyclization. Both imines and iminium salts can tautomerize to enamines. The above processes can be summarized in Scheme 3.34.

According to the work by Malhotra *et al.*, enamines and imines of α,β-unsaturated ketones are susceptible to autooxidation at the γ-position, leading to the formation of 1,4-diones.[133] Malhotra *et al.* proposed, using the pyrrolidine enamine of 10-methyl-$\Delta^{1(9)}$-octalone-2 as the model compound, that the enamine of an α,β-unsaturated ketone can directly transfer an electron from the nitrogen to molecular oxygen in the initiation step of a free radical-mediated oxidation. The aminium radical cation intermediate formed, which very much resembles the one generated from tertiary amines during the free radical-mediated autooxidation of the latter (Section 3.5.3.3), then reacts with molecular oxygen at the γ-position to give the γ-peroxy radical. The latter can abstract a H to form the γ-peroxide, which ultimately yields the corresponding 1,4-dione (Scheme 3.35).

The authors also observed a striking catalytic effect by transition metal ions such as Cu^{2+} and Fe^{3+}, and attributed it to their ability to accept an electron from the enamine in the initiation step. Such results may imply that, even in cases where no metal ions were purposely added, the initiation step may be catalyzed by trace levels of metal ions, rather than proceed by direct reaction between the enamine and molecular oxygen as shown in Scheme 3.35. Imines

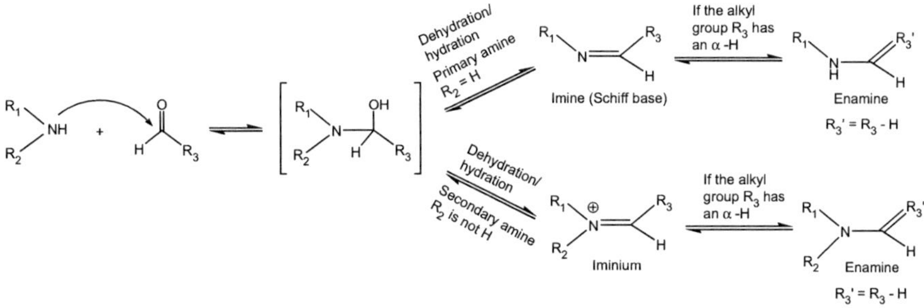

Scheme 3.34

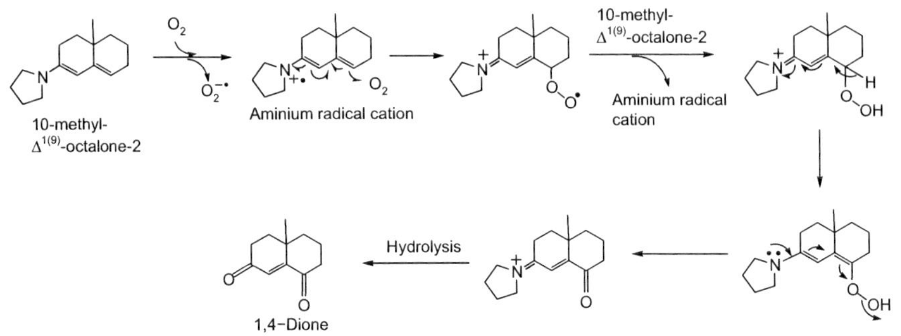

Scheme 3.35

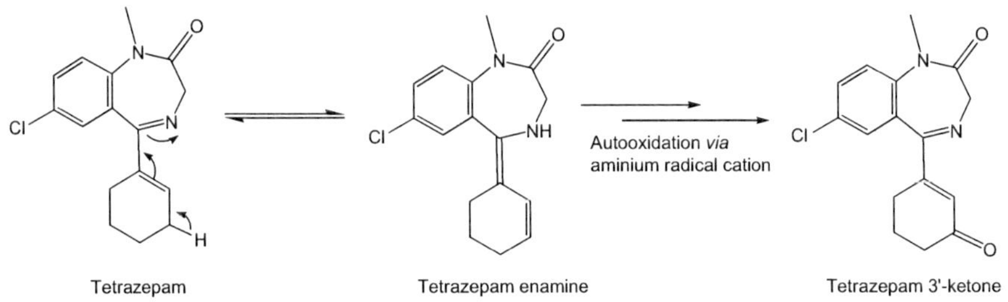

Scheme 3.36

are susceptible to the same autooxidation once they tautomerize to the corresponding enamine forms.

In a stability study of tetrazepam, it was proposed that one of the major degradants, tetrazepam 3′-ketone, formed *via* the pathway outlined in Scheme 3.35, after the embedded imine of α,β-unsaturated ketone tautamerizes to the enamine form.[103] The imine moiety remained intact during this particular degradation pathway, apparently due to the stabilization provided by the seven-membered ring of benzodiazepan and conjugation to the phenyl moiety (Scheme 3.36).

3.5.6   Thioethers (Organic Sulfides), Sulfoxides, Thiols and Related Species

One of the most common degradation pathways of the thioether moiety in drug molecules is probably the oxidation caused by hydroperoxides present in certain excipients such as PEG.[41] This oxidation can take place *via* nucleophilic attack by the sulfur on the hydroperoxide.[134,135] From the perspective of the hydroperoxide, one of its two oxygens is electrophilically transferred to the nucleophilic sulfur. Hence, this process (nucleophilic oxidation of sulfur) is also called "electrophilic oxygen transfer", which is similar to the epoxidation of electron-rich olefins by hydroperoxides (see Scheme 3.13). On the other hand, the hydroperoxide oxygen can also transfer to electrophilic oxidation substrates *via* a "nucleophilic oxygen transfer" process, which is represented by the well-known Baeyer–Villiger oxidation of ketones and aldehydes.[136] Since sulfoxides are both nucleophilic (due to the lone electron pair on the sulfur) and electrophilic, they can be oxidized *via* both the "electrophilic oxygen transfer" and "nucleophilic oxygen transfer" processes.[137]

In other cases, the thioether moiety can undergo autooxidation through one of the following two mechanisms: photochemical oxidation by singlet molecular oxygen (discussed in Chapter 6) and free radical-mediated autooxidation initiated by peroxyl radicals. The latter mechanism was proposed to proceed through a unique $2\sigma/1\sigma^*$ three-electron-bonded disulfide radical cation, which can then react with either superoxide anion radical or water and molecular oxygen (Scheme 3.37).[138–141]

Alternately, other species containing nucleophilic hetero atoms (X) such as oxygen and nitrogen can replace the second molecule of the thioether (in the second step of Scheme 3.37) in the formation of a similar $2\sigma/1\sigma^*$ three-electron-bonded S-X radical cation.[141]

This mechanism seems to be consistent with several observations made during the study of the free radical-mediated formation of sulfoxides:[128] first, the yield from sulfoxide formation increases with increasing pH, probably due to the facilitation of the deprotonation step at higher pH. Second, it has been demonstrated that the oxygen on the sulfoxide originates from water rather than molecular oxygen, although the presence of the latter significantly enhances the sulfoxide yield. This observation is different from the formation of sulfoxide *via* nucleophilic oxidation by hydroperoxide (see also Section 3.3.1), where the sulfoxide oxygen comes from the hydroperoxide. Third, the yield of sulfoxide correlates with the concentration of the oxidation substrate, thioether. The sulfoxide formed *via* either the radical or non-radical-mediated pathways is reasonably stable and can be isolated, although it can be further oxidized to form sulfone upon excessive oxidation.

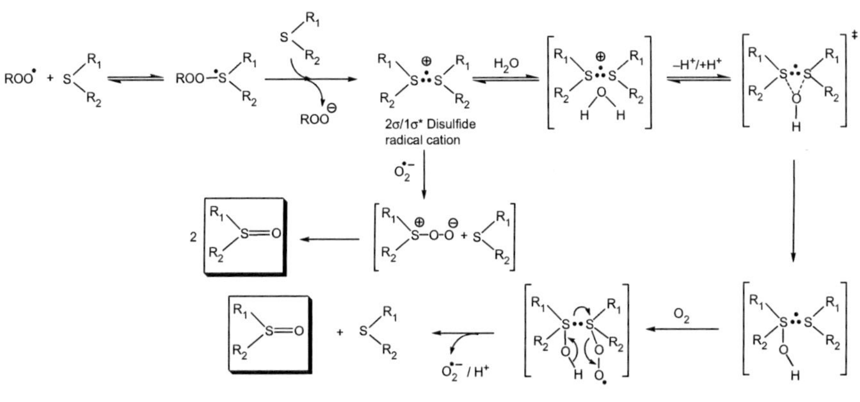

Scheme 3.37

As we have discussed so far, pH can have an impact on the oxidation of thioethers and sulfoxides under both the radical and non-radical-mediated conditions. Under the former condition, we have just shown that oxidation is facilitated at higher pH. Under non-radical conditions (electrophilic and nucleophilic oxygen transfers), electrophilic oxygen transfer is favored at lower pH, while nucleophilic oxygen transfer is favored at higher pH. The oxidation of amines and related molecules is enhanced at higher pH, regardless of radical or non-radical-mediated conditions (see Section 3.5.3). One may take advantage of such differences in formulation development by selecting an optimal pH where the degradation of the drug candidate would be minimal.

In a particular oxidative degradation case involving thioethers and sulfoxides, it may be difficult to tell whether a radical or non-radical mechanism is involved. Sometimes, both mechanisms may be operative. The oxidative degradation mechanism of methionine has been studied quite extensively, owing to its relevance to protein degradation as a protein amino acid.[142–149] Oxidation of methionine to the corresponding sulfoxide (Scheme 3.38) can be either *via* the nucleophilic (non-radical) pathway, which would be straightforward, as discussed in Section 3.3.1, or *via* the radical-mediated pathway which should follow the mechanism illustrated in Scheme 3.37.

Several drug molecules containing aliphatic thioether moieties that undergo the same *S*-oxidation include montelukast sodium,[150,151] cimetidine,[152–154] and ranitidine (Figure 3.10).[155,156]

Since one alkyl group that attaches to the sulfur is chiral in montelukast sodium, the resulting two sulfoxides are diastereomers with respect to each other. In the cases of cimetidine and ranitidine where all the alkyl groups are achiral, the resulting sulfoxides are enantiomers in each case.

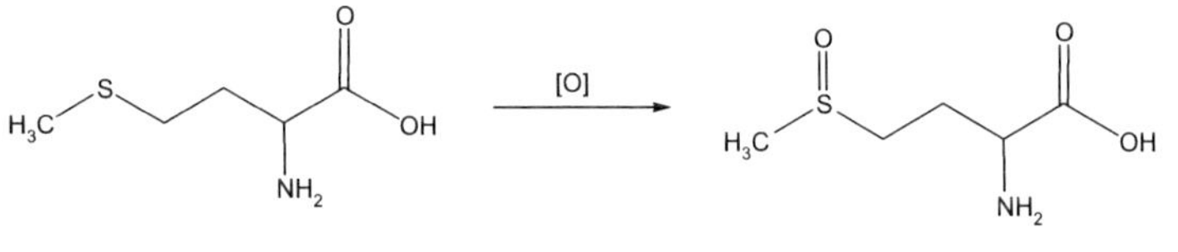

Scheme 3.38

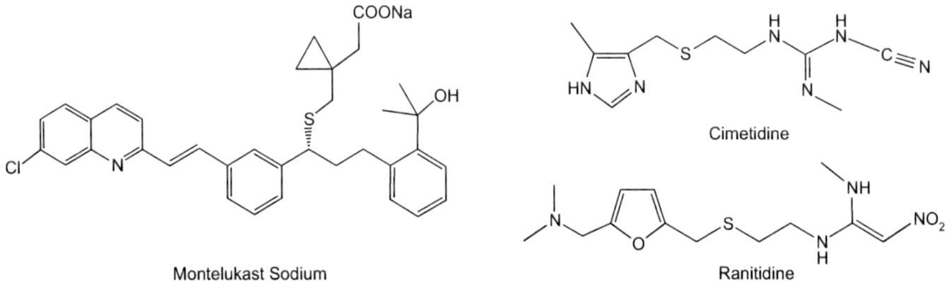

Figure 3.10 Structures of montelukast sodium, cimetidine, and ranitidine.

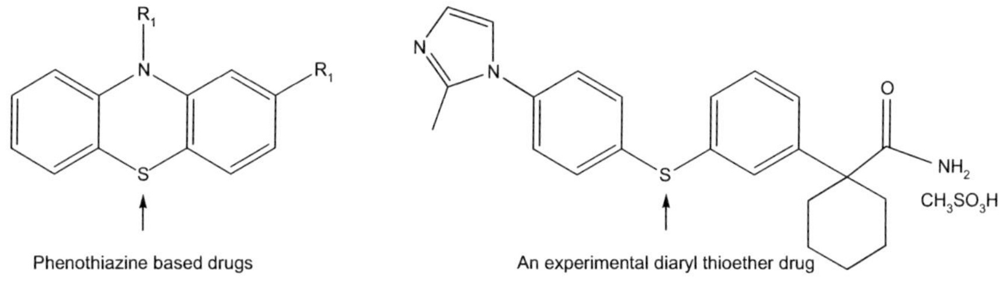

Figure 3.11 Oxidation of the diaryl thioether moiety of phenothiazine-based drugs and an experimental drug. The arrows indicate the *S*-oxidation sites.

The mechanisms discussed above are based on the studies of aliphatic thioethers. For aryl thioethers, it appears that the non-radical, nucleophilic oxidation mechanism can be readily applied. For example, oxidation of the diaryl thioether moiety of phenothiazine-based drugs (Figure 3.11; see also Section 3.5.3.1) by hydroperoxides at room temperature should proceed *via* the nucleophilic oxidation mechanism.[98] In another case where an experimental diaryl thioether drug was formulated using BHT as the antioxidant, Puz *et al.* found that the use of 2% BHT in the tablet coating was able to suppress the majority of the sulfur oxidation in two accelerated stability studies (40 °C/75% RH and 50 °C/20% RH).[157] Since BHT exerts its anti-oxidation effect by inhibiting the free radical propagation step in a radical-mediated autooxidation, it appears that a free radical mechanism is mostly likely to be responsible for the sulfoxide formation in this experimental formula under the above accelerated stability conditions.

Very few small molecule drugs contain the thiol (sulfhydro) functionality. Two notable examples are captopril, a first generation angiotensin-converting enzyme (ACE) inhibitor which is still used clinically for the treatment of hypertension, and *N*-acetylcysteine, a drug mainly used as a mucolytic agent. The main degradants of both compounds are the corresponding dimers (captopril disulfide and *N*-acetylcysteine disulfide) formed through the oxidative coupling of the two thiol groups.[158–160] The degradation pathways of the thiol functional group will be further discussed with the amino acid cysteine in Chapter 7, Chemical Degradation of Biological Drugs.

3.5.7 Examples of Carbanion/enolate-mediated Autooxidation

As discussed in Section 3.4, much less is known about carbanion/enolate-mediated (or base-catalyzed) autooxidation compared to free radical-mediated autooxidation. This is primarily due to the fact that free radicals, in particular peroxy radical, mediate the majority of autooxidative degradation observed in drug substances and drug formulations. Nevertheless, for a group of compounds that contain an acidic CH_n (n is usually 1 to 2), carbanion/enolate-mediated autooxidation may be the predominant process, especially in solutions or liquid formulations where pH is near neutral or alkaline conditions.

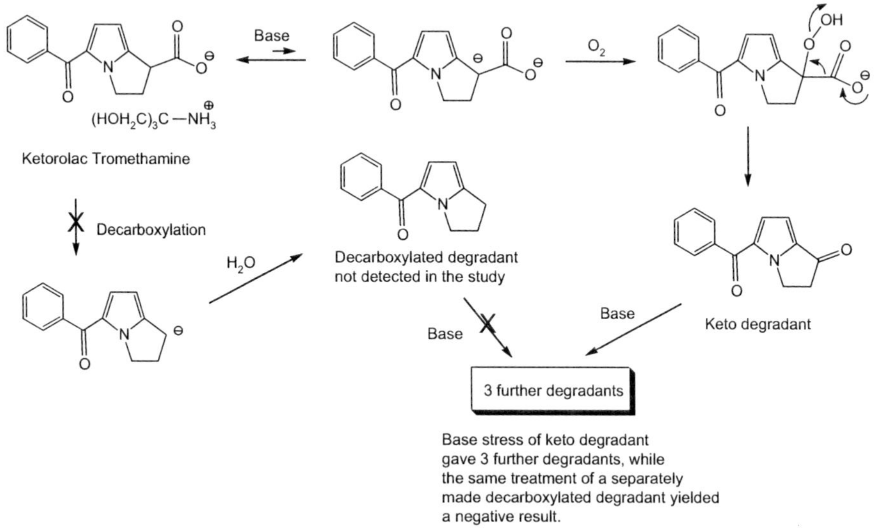

Scheme 3.39

During a study of ketorolac tromethamine stability in aqueous buffers with a wide range of pH under elevated temperatures (60–100 °C), the formation of the initial major degradant at pH ≥ 4.8 was attributed to the carbanion-mediated autooxidation as shown in Scheme 3.39.[161]

In this proposed mechanism, the formation of the carbanion on the position α to the carboxyl group and the subsequent reaction with molecular oxygen precedes the decarboxylation step. The possibility of an alternate mechanism where initial decarboxylation is followed by reaction with molecular oxygen was excluded, based on the absence of a particular degradant (the decarboxylated degradant) which would be formed under the alternate degradation mechanism.

Another good example of carbanion/enolate-mediated autooxidation can be found in the case of rofecoxib. The autooxidation of rofecoxib was originally hypothesized as being mediated by free radicals.[162] Harmon et al. found that this autooxidation proceeds nearly two orders of magnitude faster than a peroxy free radical-mediated autooxidation and the stress condition lacks an obvious source of the free radical.[8] Hence, a systematic mechanistic study was undertaken which provided convincing evidence that the autooxidation of rofecoxib is mediated by the rofecoxib carbanion/enolate which is readily generated under alkaline conditions (Scheme 3.40).

Reddy and Corey performed a similar base-catalyzed autooxidation study of rofecoxib at about the same time;[163] and although their reaction medium was a mixture of tetrahydrofuran (THF) and water with one equivalent LiOH, which is different from the mixture of acetonitrile and phosphate buffer (pH 11 or 12) used by Harmon et al.,[8] the same final degradants were observed. In both cases, refocoxib anhydride was found to exist in appreciable quantities when the

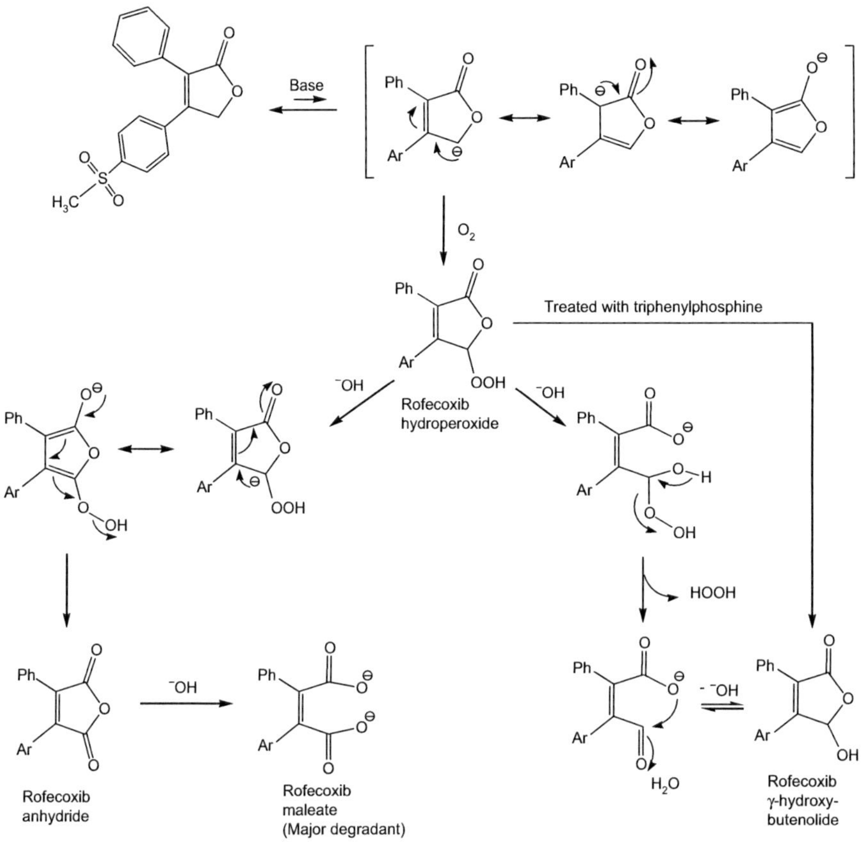

Scheme 3.40

organic co-solvent is present in high percentage. Since Harmon *et al.* used high-performance liquid chromatography (HPLC) to monitor the autooxidation reaction, they were able to observe the transient rofecoxib hydroperoxide intermediate.[8] When treated with triphenylphosphine, the hydroperoxide was converted to rofecoxib γ-hydroxybutenolide.

Most degradation of corticosteroidal drugs containing a 1,3-dihydroxy-acetone side chain such as hydrocortisone, betamethasone and dexamethasone, is oxidative in nature, occurring at the D-ring which contains this side chain.[83,164,165] Edmonds *et al.* studied autooxidative degradation of dexamethasone in neutral to alkaline aqueous buffers.[7] At pH 7.4 and room temperature, autooxidation of dexamethasone proceeded reasonably fast: it decomposed completely in 28 days to give dexamethasone glyoxal (dexamethasone 21-aldehyde). With increased pH, the autooxidation proceeded much faster and further degradants such as dexamethasone etioacid (dexamethasone 17-acid) and dexamethasone glycolic acid were formed. Li *et al.* performed a detailed mechanistic study of the base-catalyzed autooxidation of betamethasone with high resolution LC-MS and $^{18}O_2$, from which the degradation pathway in Scheme 3.41 was proposed.[82]

It appears that this mechanism can explain the majority of the oxidative degradation behaviors of the corticosteroids containing a 1,3-dihydroxyacetone side chain which include not only betamethasone but also hydrocortisone,[164] prednisolone,[165] and dexamethasone.[7,166] In a study of dexamethasone in an aqueous ophthalmic suspension,[166] dexamethasone 17-formyloxy-17-acid was isolated as one of the degradants. In addition, the dexamethasone anhydride intermediate, analogous to constituent (**9**) in Scheme 3.41, was also implicated. These results provide further support for the general applicability of the mechanism proposed in Scheme 3.41 for this class of corticosteroids.

In drug molecules where a more acidic CH_n is present, carbanion/enolate-mediated autooxidation can take place spontaneously. For example, phenylbutazone undergoes very rapid autooxidation on a silica gel thin-layer chromatography (TLC) plate to produce two major degradants, 4-hydroxyphenylbutazone and N-(α-ketocaproyl)hydrozobenzene.[167] This facile autooxidation may be better explained by the mechanism shown in Scheme 3.42, which is different from the one proposed during the original study.

In solid dosage formulations of phenylbutazone containing anti-acid ingredients, phenylbutazone was found to undergo significant oxidative and

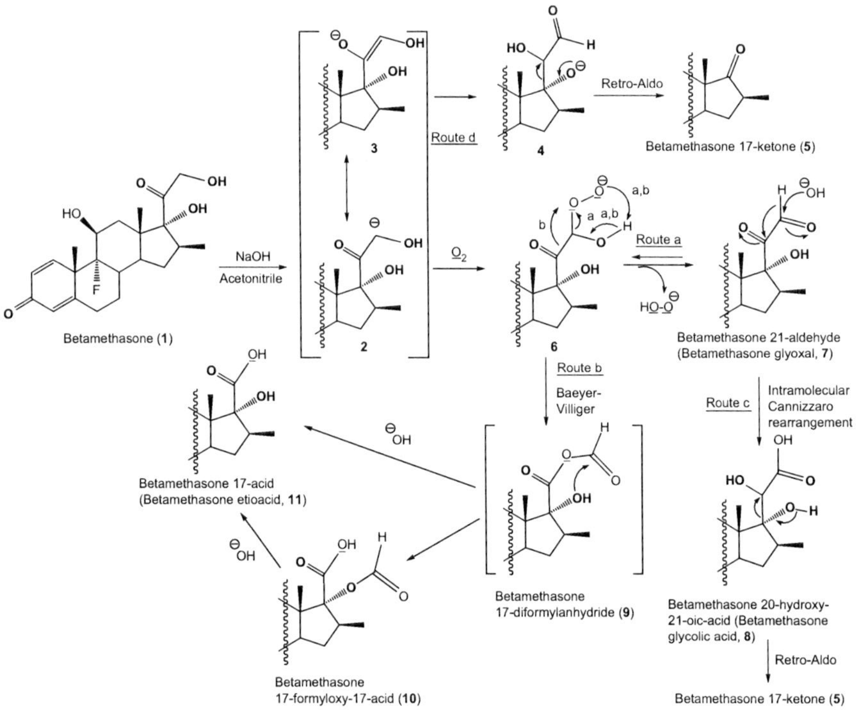

Scheme 3.41

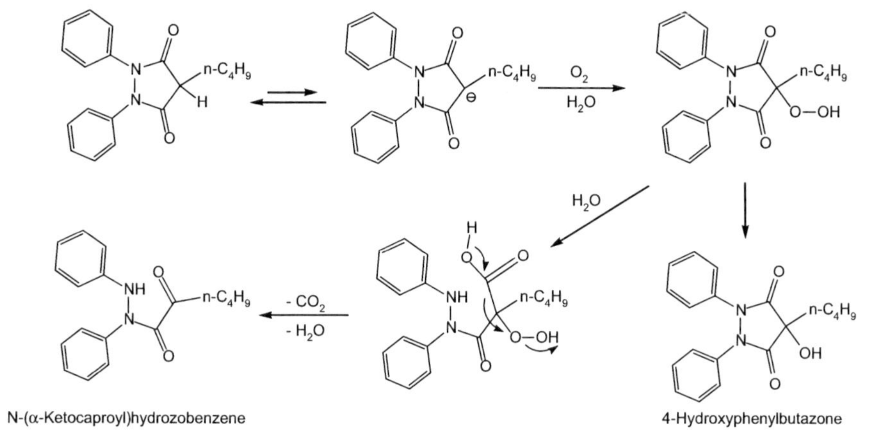

N-(α-Ketocaproyl)hydrozobenzene   4-Hydroxyphenylbutazone

Scheme 3.42

hydrolytic degradation. This degradation behavior is consistent with a base-catalyzed autooxidation mechanism.

3.5.8 Oxidation of Drugs Containing Alcohol, Aldehyde, and Ketone Functionalities

Oxidation of primary alcohols yields aldehydes, while oxidation of secondary alcohols produces ketones. Autooxidation of both primary and secondary alcohols is likely to proceed *via* a free radical-mediated mechanism. Williams *et al.* performed an electron paramagnetic resonance (EPR) study on the auto-oxidation of an amorphous hydroxyl-containing drug substance in which two radical species were detected and can be attributed, respectively, to a carbon-centered radical and a peroxide radical.[168] The bond dissociation energies (BDE) of the C–H bonds α to the hydroxyl group in ethanol and 2-propanol are approximately $93 \, kcal \, mol^{-1}$ ($389 \, kJ \, mol^{-1}$)[54,169] and $91 \, kcal \, mol^{-1}$,[54] respectively. Both values are slightly higher than the BDE of the O–H bond in a typical hydroperoxide (~ 88–$90 \, kcal \, mol^{-1}$).[170,171] This fact indicates that the autooxidation of typical primary and secondary alcohols, which is predominantly mediated by the peroxide radical, would be sluggish, since ethanol and 2-propanol can be viewed as the respective models for the oxidation of typical primary and secondary alcohols. The aldehydes/ketones formed can further oxidize through radical as well as non-radical-mediated oxidation pathways. Aldehydes usually undergo a radical-mediated oxidation pathway ultimately to give the corresponding carboxylic acids *via* the sequential intermediary of acyl radical and peracid.[172] A commonly seen pathway of the non-radical oxidation is the Baeyer–Villiger oxidation, which turns the acyclic ketones into esters, cyclic ketones into lactones, and aldehydes into carboxylic acids or formic esters.[136,173] The above pathways can be summarized in Scheme 3.43. The oxidation agent can be hydrogen peroxide or peracids; the latter can

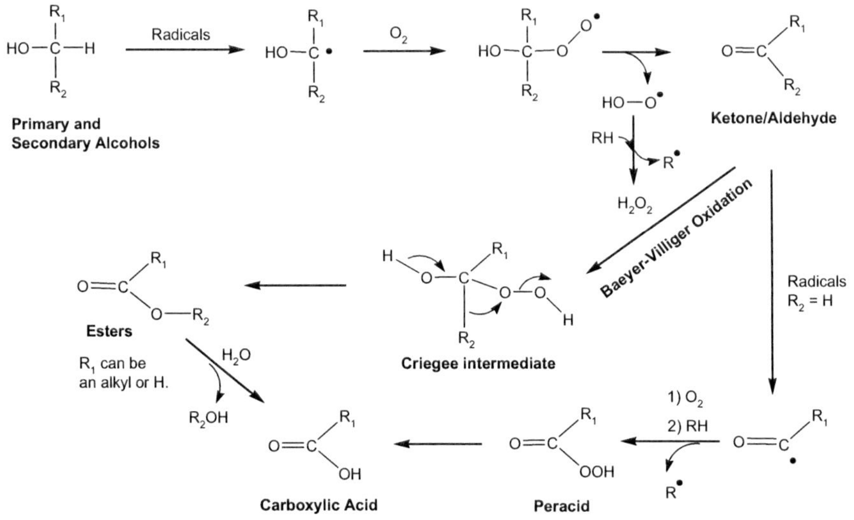

Scheme 3.43

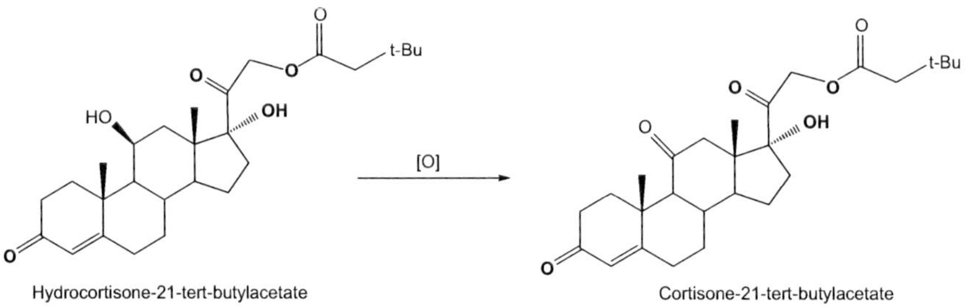

Scheme 3.44

be viewed as activated forms of hydrogen peroxide. Different types of groups (R_1 and R_2) in the Criegee intermediate have different propensities to migrate and, in general, certain types are more likely to migrate than others but the preference for migration can change depending upon the structure of the oxidation substrate and reaction conditions.[136] To simplify the discussion, the R_2 group is assumed to have a higher propensity to migrate in Scheme 3.43.

It appears that there are not many examples of autooxidation of primary and secondary alcohol moieties in drug molecules; this could be attributed to the sluggishness of this type of autooxidation as discussed previously in this section. One notable example is the facile autooxidation of the 11β-hydroxyl group in certain corticosteroids when the latter compounds are present in a few solvated crystal forms.[174] For example, a solvated crystal form of hydrocortisone 21-tert-butylacetate was found to undergo spontaneous oxidation to yield the corresponding 11β-keto degradant (*i.e.* cortisone 21-tert-butylacetate, Scheme 3.44). When stored in ambient conditions for between

1 to 2 years; up to ∼40% of this solvated form was oxidized at the 11β-hydroxyl position.

On the other hand, a non-solvated crystal form of the same compound was found to be completely stable under the same oxidative conditions. According to a study by Lin *et al.* who determined the structures of several crystal polymorphs, the reactivity toward the autooxidation in the solvated form can be attributed to a channel in the crystal through which molecular oxygen is capable of penetrating.[175] No chemical mechanism for the autooxidation was discussed in these studies. It can, nonetheless, be assumed that the autooxidation of the 11β-hydroxyl to 11β-keto proceeds *via* the free radical-mediated pathway as illustrated in Scheme 3.43.

Another example can be found in the autooxidation of dexamethasone in an ophthalmic suspension. In several expired commercial batches, the 21-hydroxyl group of the corticosteroid, which is a primary alcohol, was oxidized to the 21-aldehyde group.[166] The dexamethasone 21-aldehyde formed underwent further oxidation to give a dexamethasone 17-formyloxy degradant, the process of which is most likely to be mediated through a Baeyer–Villiger oxidation of the 21-keto aldehyde moiety as shown in Scheme 3.45.

It is also worthwhile discussing the autooxidation of hydroxyl-bearing pharmaceutical excipients such as benzyl alcohol, polyethylene glycol (PEG), and polysorbate, owing to their wide application. Benzyl alcohol is commonly used as a preservative in pharmaceutical products at levels of 3–5%.[176] It is known that benzyl alcohol is susceptible to autooxidation, producing low level of benzaldehyde.[177–179] Abend *et al.* found that the benzaldehyde formed can react with two moles of benzyl alcohol to give benzaldehyde dibenzyl acetal.[180]

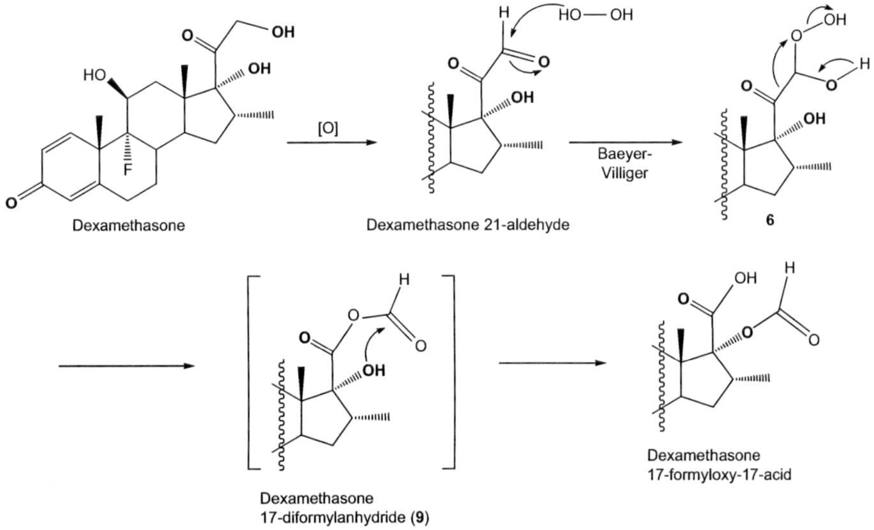

Scheme 3.45

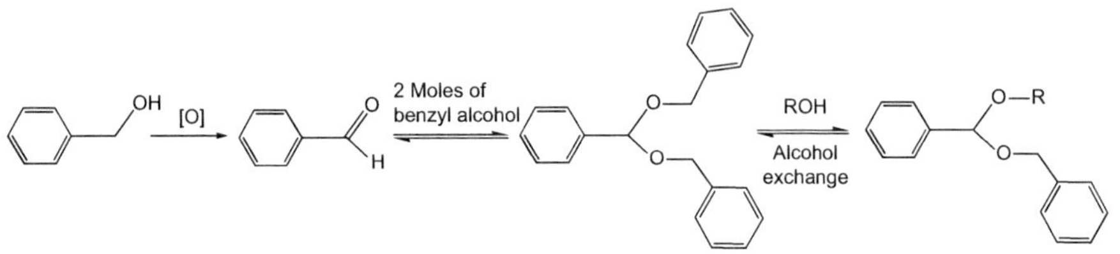

Scheme 3.46

The acetal formed can then readily undergo alcohol exchange with a variety of hydroxyl-bearing excipients such as propylene glycol, resulting in the formation of interfering impurity peaks which could cause considerable analytical challenges during pharmaceutical development. This scenario is summarized in Scheme 3.46.

Polyethylene glycol (PEG) and polysorbate (in particular polysorbate 20 and 80, trade names: Tween 20 and 80) are non-ionic detergents and emulsifiers that are widely used in pharmaceutical formulations. The key structural moiety of both oligomers/polymers is polyethoxylated alcohol which can be represented by the formula R-(OCH$_2$CH$_2$)$_n$-OH. It has long been known that PEG[181,182] and polysorbate[183] are readily susceptible to free radical-mediated autooxidation during storage and handling under ambient conditions. This susceptibility can be attributed to the autooxidative degradation of the polyethoxylated alcohol moiety, during which process the terminal hydroxyl group, as well as the abundant ether functionality, can be readily oxidized to produce a range of aldehydes including formaldehyde, formic acid,[184] alkoxyl formate,[185] as well as numerous hydroperoxides including hydrogen peroxide.[41] Although the degradation mechanism has not been fully elucidated, the overall process can be described in Scheme 3.47, based on the published results.

The above autooxidation is most likely to be catalyzed by redox active transition metal ions as discussed in Section 3.2.1. Generally speaking, two types of carbon-centered radicals would be formed: one that is α to the terminal hydroxyl group (radical A) and the other α to the ether functionality (radical B). Reaction of radicals A and B with molecular oxygen, respectively, would produce the corresponding peroxide radicals which in turn would abstract hydrogen atoms to give the peroxides. The peroxide formed from radical A is a Criegee intermediate[186,187] which can undergo two degradation pathways. The first pathway (a) would eliminate H$_2$O$_2$ yielding an aldehyde, while the second pathway (b) would undergo a Baeyer–Villiger type of rearrangement, resulting in the formation of an α-alkoxyl formate. The peroxide formed from radical B could go through a carbon–carbon bond breakage to produce a formaldehyde and β-alkoxyl formate according to the mechanism proposed by Waterman *et al.*[50,184] Breakage of a similar carbon–carbon bond in analogous β-hydroxyl peroxides to produce α, ω-diketones[188] and dialdehydes[189] has been reported.

In the case of drug degradation caused by the oxidation of a ketone moiety, the autooxidation of tirilazad mesylate in acidic solutions is worth some discussion.[190] This drug substance is a 21-amino substituted steroid that is

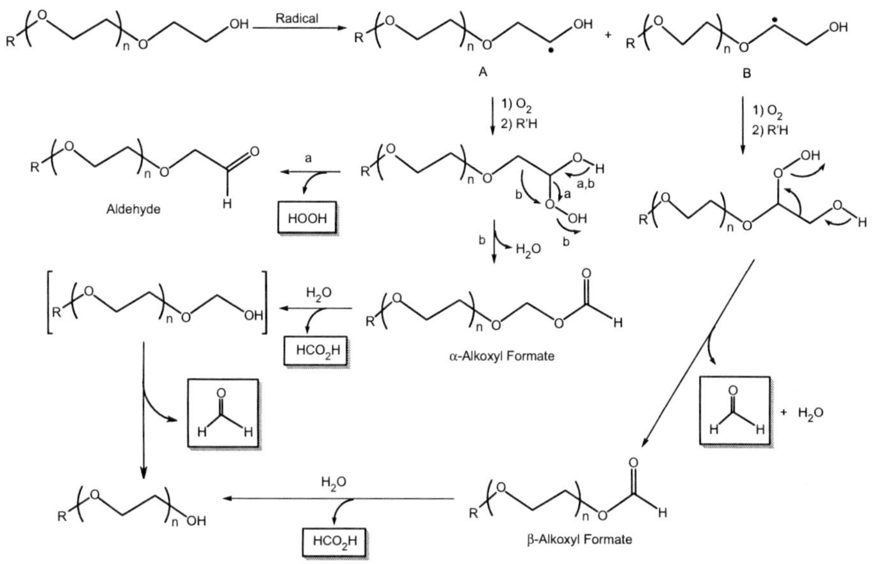

Scheme 3.47

typically formulated in pH 3.0 aqueous solutions for maximaum solubility. Oxidation of the α-aminoketone moiety was found to be one of the two major degradation pathways of the drug substance, the process of which was postulated to begin with the addition of hydrogen peroxide to the ketone to form the Criegee intermediate. The latter species would then undergo a Grob fragmentation process[191] to yield a carboxylic acid and an iminium intermediate, according to a study performed by Wenkert *et al.*[192] Further oxidation of the iminium intermediate by a second molecule of hydrogen peroxide should give the formic amide, which ultimately would hydrolyze to release the amine degradant. Alternately, the iminium intermediate can be directly hydrolyzed to form the amine degradant. Although the latter possibility was not discussed by the original authors, it would not be surprising if the alternate hydrolysis is the major degradation pathway of the iminium intermediate, considering the fact that the study by Wenkert *et al.* was carried out in methanol solution,[192] while the autooxidation of tirilazad mesylate occurred in an entirely aqueous environment.[190] Furthermore, the authors' observation that the formic amide degradant was present in a very low quantity is consistent with the direct hydrolysis pathway. The complete degradation pathways are shown in Scheme 3.48.

Conjugated ketones can also be susceptible to oxidation *via* the Baeyer–Villiger mechanism. For example, steroids containing a cyclohexenone A-ring were found to degrade to the corresponding enol lactones, resulting from the net insertion of an oxygen into the six membered-ring.[193,194] Minor lactone epoxides were also observed (Scheme 3.49).

For such a conjugated system, generally the vinyl group preferentially migrates,[195] but exceptions have been reported.[196,197]

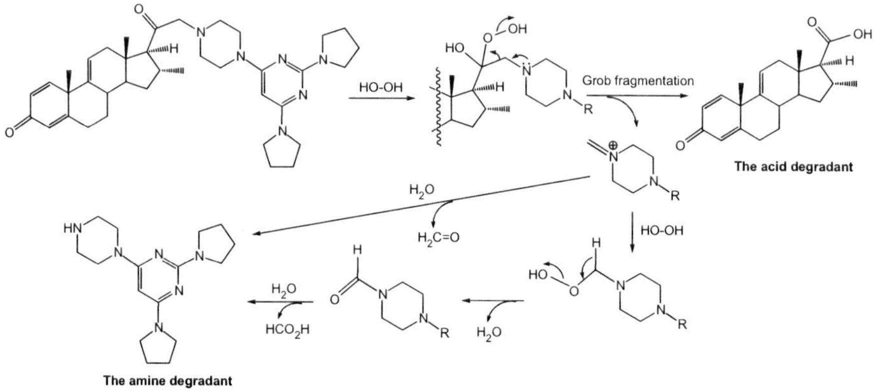

Scheme 3.48

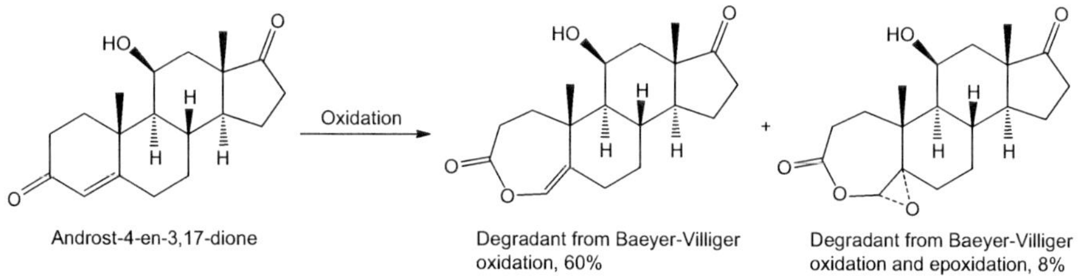

Androst-4-en-3,17-dione

Degradant from Baeyer-Villiger oxidation, 60%

Degradant from Baeyer-Villiger oxidation and epoxidation, 8%

Scheme 3.49

3.5.9 Oxidation of Aromatic Rings: Formation of Phenols, Polyphenols, and Quinones

It has long been known that hydroxylation of unactivated aromatic rings such as monoalkyl-substituted phenyl rings is a very common oxidative degradation pathway in drug metabolism.[198] However, this pathway does not appear to be significant in the autooxidation of drugs, as only a few such cases have been reported in the literature. In principle, HO• radical generated by the Udenfriend and/or Fenton reactions is capable of hydroxylating an unactivated phenyl ring. Frequently, there are "spots" in a drug substance that are more reactive or susceptible to oxidation by HO• or its precursor in the Udenfriend degradation pathway, H_2O_2, which may explain why hydroxylation of unactivated phenyl rings is usually not a significant event in drug degradation.

Wu *et al.* reported the oxidative degradation study of a thrombin inhibitor, L-375,378, in tablet as well as intravenous (i.v.) solution dosage forms.[199]

Various oxidation reactions occurred at the "hot spots" as illustrated in Figure 3.12. Among the oxidative degradants characterized, all the three hydroxylated phenyl degradants, that is, *para-*, *meta-*, and *ortho*-hydroxylated analogs of L-375,378, were observed as relatively minor degradants. The two major degradants observed in the formulations were formed *via* the destruction

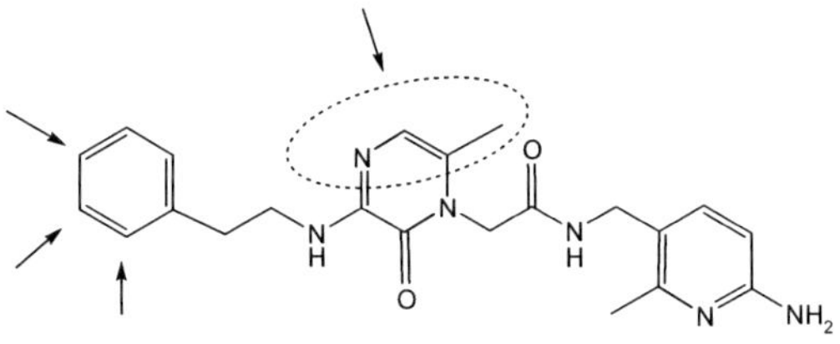

Figure 3.12 Structure of L-375,378 showing various oxidation sites.

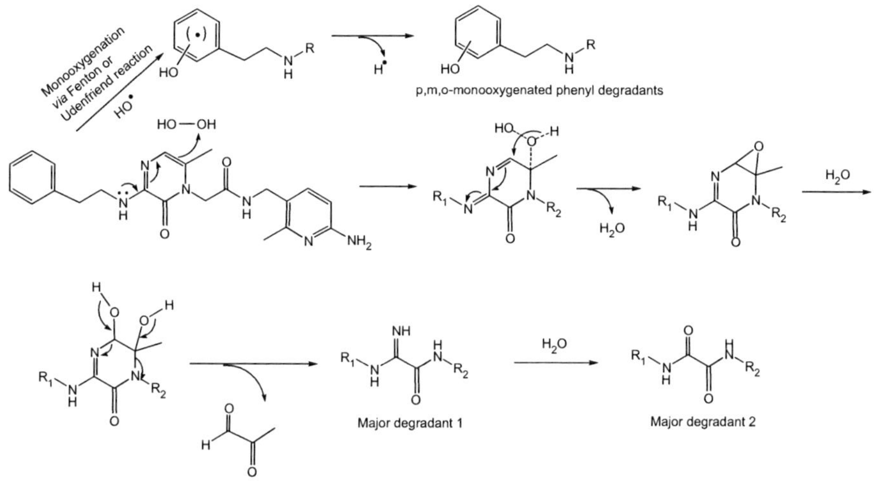

Scheme 3.50

of the central six-membered heterocyclic ring. Although the original authors did not propose a degradation mechanism, it appears that the monooxygenation of the phenyl ring is likely to occur *via* attacks by hydroxyl radical or its equivalent. The oxygen reactive species can be generated either through the Fenton reaction or the Udenfriend reaction; components of the formulation could render either of the oxidative degradation pathways possible as discussed in Section 3.2.1. On the other hand, the destruction of the central six-membered heterocyclic ring would probably start from the formation of an epoxide on the upper double bond of the heterocyclic ring. All the oxidative degradation pathways are summarized in Scheme 3.50.

Major degradants 1 and 2 were also seen as the major degradation products in the forced degradation of L-375,378 with H_2O_2, which is consistent with the degradation pathway proposed above. This pathway can be viewed as another example of autooxidation through an epoxide intermediate as discussed in Section 3.5.2; the presence of the phenylethylamino group could promote the formation of the expoxide intermediate as shown in Scheme 3.50.

In contrast to the non-activated phenyl rings, activated phenyl rings such as hydroxyphenyl (phenol) and alkoxylphenyl derivatives are quite susceptible to

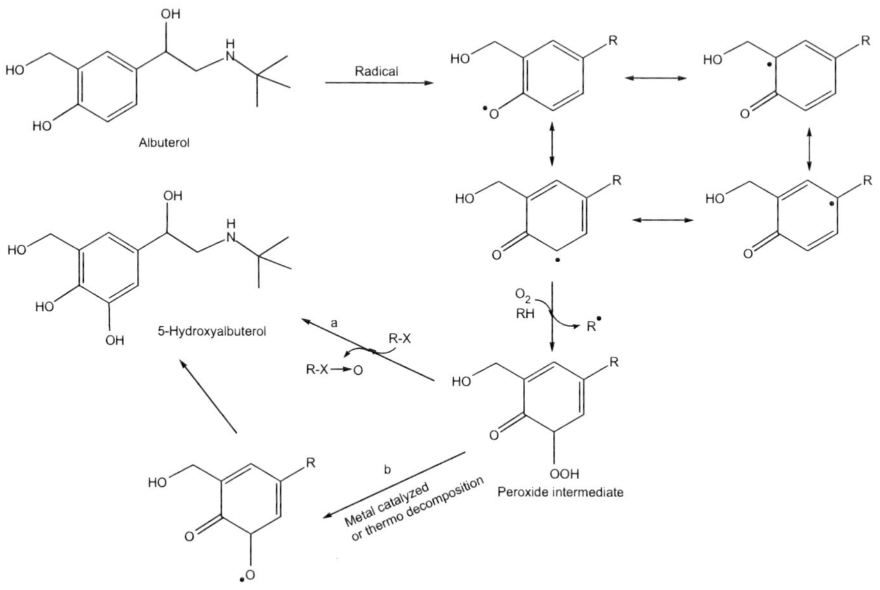

Scheme 3.51

autooxidation leading to the formation of polyhydroxylated products such as catechols. For example, the widely used bronchial smooth muscle relaxant, albuterol (salbutamol), contains a 2,4-dialkylphenol moiety which is susceptible to autooxidation at the 5-position under ambient storage condition to produce 5-hydroxyalbuterol.[200] Although no specific degradation mechanism was proposed by the original authors, this autooxidation is very likely to start from the formation of the oxygen-centered phenolic radical, which resonates with three carbon-centered radicals (Scheme 3.51).

Reaction of the 5-position radical, the least sterically hindered carbon-centered radical, with oxygen should lead to the 5-peroxyl intermediate. From the latter intermediate, two pathways are possible for the formation of 5-hydroxyalbuterol. In the presence of a reducing impurity, the peroxide intermediate can be directly reduced to the 5-hydroxy degradant (pathway a). Alternatively, the peroxide can give rise to the 5-oxy radical either through catalysis by transition metal ions or simply *via* thermolysis (pathway b).

The polyhydroxylated species are very electron-rich compounds owing to the strong electron-donating effect exerted by the multiple hydroxyl and similar groups (*e.g.* methoxy) and hence are strong reducing agents. In particular, drugs containing 1,2- and 1,4-dihydroxyphenyl moieties can undergo facile autooxidation to produce the 1,2- and 1,4-quinones, respectively, *via* the corresponding semi-quinone intermediates. Quinones are good Michael acceptors and as such they can react with nuclephiles to form 1,4-Michael adducts; these usually quickly further autooxidize to yield 4-substituted quinones. All the above degradation pathways are summarized below in Scheme 3.52 using epinephrine as an example.[201,202]

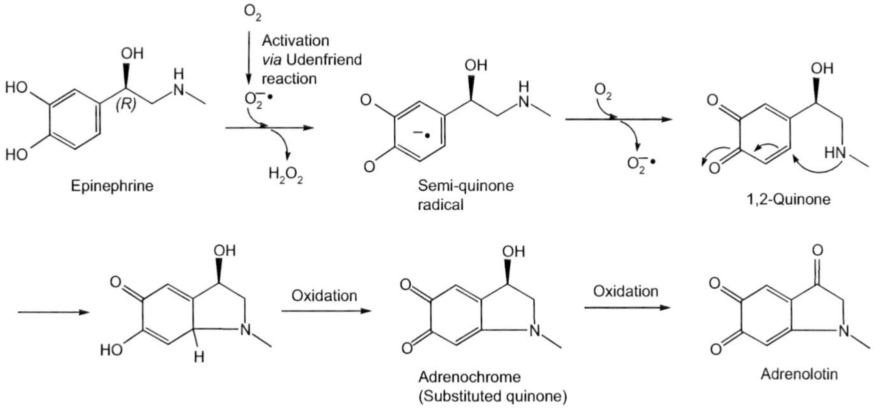

Scheme 3.52

In the above case, the 1,2-quinone formed was attacked by the neighboring amino group to produce the Michael adduct which was further oxidized to yield the final substituted quinone. When a neighboring nucleophile like this is absent, the quinone formed can react with the starting material (catechol) to produce a semi-quinone intermediate.[203] In such cases, the autooxidation becomes a self-catalytic process, that is, quinone, the intermediary oxidative degradant, promotes the oxidation of catechol by participating in compro-portionation. Consequently, such a catechol can undergo further facile autooxidation.

In another case of activated aromatic ring oxidation, Auclair and Paoletti found that the antitumor drug 9-hydroxyellipticine undergoes facile auto-oxidation in alkaline aqueous solutions producing, respectively, the corre-sponding quinone imine (9-oxoellipticine), a dimer of 9-hydroxyellipticine, and hydrogen peroxide (Scheme 3.53).[204] EPR experiments suggested that the autooxidation process involves the initial formation of a free radical of the drug and $O_2^{-\bullet}$. Both species could undergo their own disproportionation to yield hydrogen peroxide and the quinone imine 9-oxoellipticine, respectively. The latter compound is capable of reacting with the drug 9-hydroxyellipticine itself to form the dimer.

This autooxidation mechanism is somewhat different from that shown in Scheme 3.52 in that in the current case molecular oxygen acts "directly" as the oxidant (*via* catalysis through trace redox transition metal ions), while in Scheme 3.51 $O_2^{-\bullet}$ was proposed as the oxidant. In the current case, the tran-sition metal ions may directly chelate with the drug 9-hydroxyellipticine. Hence, this may be a Udenfriend type of autooxidation in which the oxidation substrate (the drug) serves as both the chelator and the reducing agent. This hypothesis by the current author is consistent with the observation made by Auclair and Paoletti[204] that use of EDTA dramatically decreased the auto-oxidation of the drug, because EDTA should disrupt the complex formed between the drug, transition metal ions, and oxygen. The reactive species, that

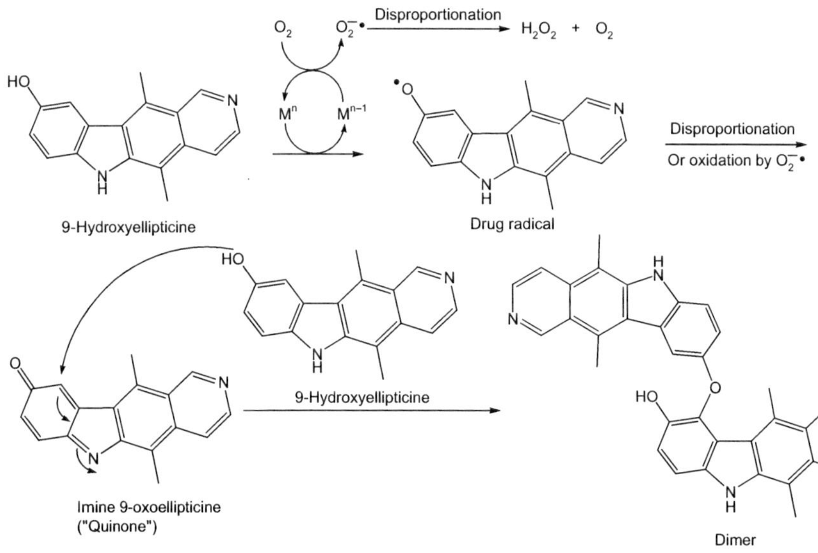

Scheme 3.53

is, $O_2^{-\bullet}$, the drug free radical, and the quinone, formed in the autooxidation process, as illustrated in Scheme 3.53 are potentially cytotoxic and their formation has been hypothesized to be responsible for the activity of the antitumor drug.

Drugs like morphine and those related to it, for example, naloxone, nalbuphine, and oxymorphone,[205] contain an *ortho*-alkyoxyphenolic moiety that can undergo autooxidative carbon–carbon coupling to yield primarily the respective 2,2'-dimers. In the case of morphine, the 2,2'-morphine dimer is also called pseudomorphine. Autooxidative dimerization most probably starts from the formation of the phenolic radical which can be resonated at the 2- and 4-positions of the aromatic ring. The 2-phenolic radical can attack a morphine molecule to give the dimer radical which should readily autooxidize, leading to the formation of the dimer. The mechanism postulated[206] is illustrated in Scheme 3.54 using morphine as the example.

3.5.10   Oxidation of Heterocyclic Aromatic Rings

Heterocyclic aromatic rings, such as imidazole, indole, pyridine, and their analogs including polycyclic rings that contain fused heterocyclic rings, are common structural features in various drug molecules. Examples of small molecule drugs that contain an imidazole ring include several azole antifungal drugs, clotrimazole, econazole, ketoconazole, isoconazole, and miconazole. For protein and peptide drugs, the imidazole-containing histidine residue is a key component that is frequently involved in metal ion binding of these drugs. As a result, the histidine residue of protein and peptide drugs becomes susceptible to a transition metal ion-mediated autooxidation, in particular

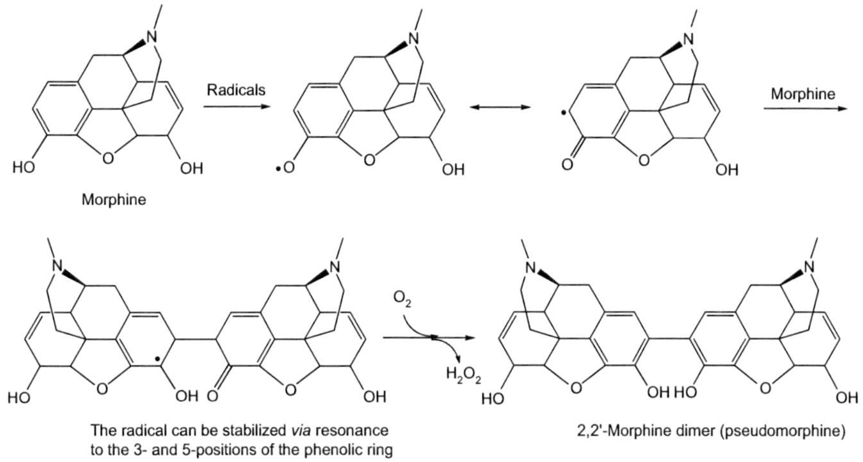

Scheme 3.54

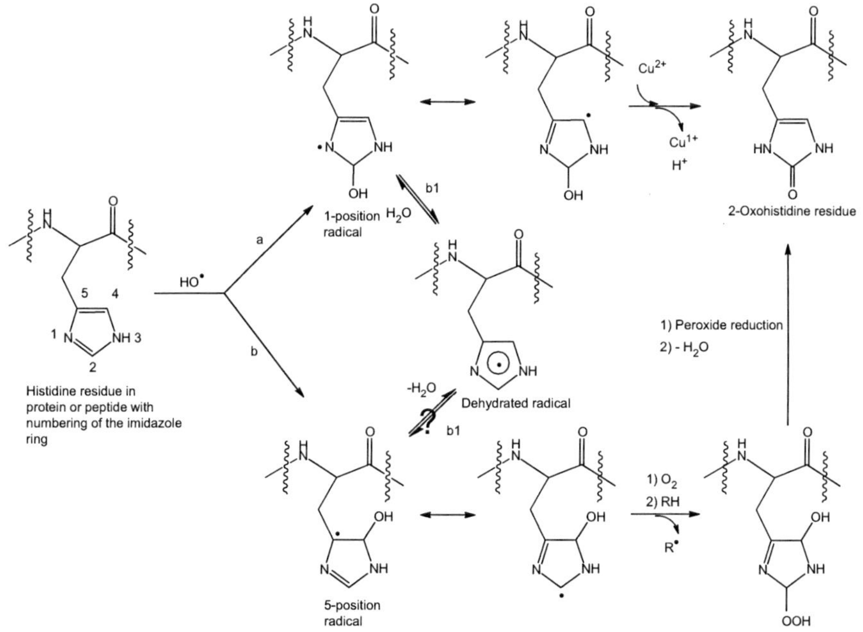

Scheme 3.55

cupric ion. According to the results obtained by several groups, the oxidative degradant resulting from the transition metal ion-mediated process, *via* the Fenton chemistry, is 2-oxohistidine.[207–209] In this metal ion-mediated oxidation, HO• was postulated as the reactive oxygen species that attacks the imidazole ring at the 2-position (pathway a, Scheme 3.55).[210] Nevertheless, a more recent EPR study of the histidine oxidation indicated that the hydroxyl radical predominantly attacks the 4-position to form the 5-position radical rather than the 2-position radical (pathway b).[211] To reconcile with the ultimate formation of 2-oxohistidine residue, Schöneich[210] proposed a dehydrated radical intermediate produced from the 5-position radical (pathway b1). Alternatively, the current author postulates another possible pathway leading to the formation of the 2-oxohistidine (pathway b). All the proposed mechanisms are summarized in Scheme 3.55.

How pathway b could be preferred over pathway a may be rationalized based on the fact that the 5-position radical is a tertiary radical which should be more stable than the alternative, secondary radicals. On the other hand, pathway b1 seems quite unlikely according to the isotope experiment with $H_2^{18}O$ in which no ^{18}O was found to be incorporated into 2-oxohistidine.[210] This result mitigates strongly against pathway b1 where rehydration of the dehydrated radical intermediate by water should lead to the incorporation of the water oxygen.

For the azole antifungal drugs mentioned above, the imidazole ring is linked to the rest of the drug molecule through the 1-nitrogen position. As such, oxidation of the imidazole ring would probably be different from that of the histidine residue. A search of the literature revealed no results for the autooxidation of the imidazole ring in these drugs, except a stress study of miconazole by the free radical initiator, azobisisobutyronitrile (AIBN),[212] which is frequently used to simulate autooxidative degradation of drug substances. In this study, the 1-nitrogen substituted imidazole ring was oxidized to 2,4,5-trioxoimidazole. The original authors hypothesized that triplet molecular oxygen reacts with the imidazole ring in the same way as singlet oxygen based on the fact that the latter is known to react with imidazole to produce unstable dioxygenated intermediates.[213–215] Nevertheless, the formation of 2,4,5-trioxoimidazole may also be explained by a mechanism that only involves triplet molecular oxygen, as illustrated in Scheme 3.56, that does not violate the spin conservation rule.

In the above mechanism, the abundant AIBN-generated peroxyl radicals can attack either the 2-position (pathway a) or the 5-position (pathway b) to yield radicals that are more stable than the alternatives owing to the resonances shown in Scheme 3.56. Reaction of the radicals with triplet molecular oxygen would give the corresponding peroxyl radicals, which in turn can produce the peroxides by abstracting hydrogens from a general hydrogen donor (RH). Elimination of a water and alcohol from the two peroxides in pathways a and b should form the same 2,5-dioxoimidazolyl intermediate. Further oxidation of the latter by another AIBN-generated peroxyl radical can lead to the formation of 2,4,5-trioxoimidazole degradant.

The indole ring is a common structural feature present in many natural products,[216] which are important sources of new drugs.[217] It is also a side chain of the most hydrophobic protein amino acid, tryptophan. Its UV absorption at 280 nm is mostly responsible for the characteristic absorption of proteins and peptides that contain tryptophan residues. The most significant oxidation of

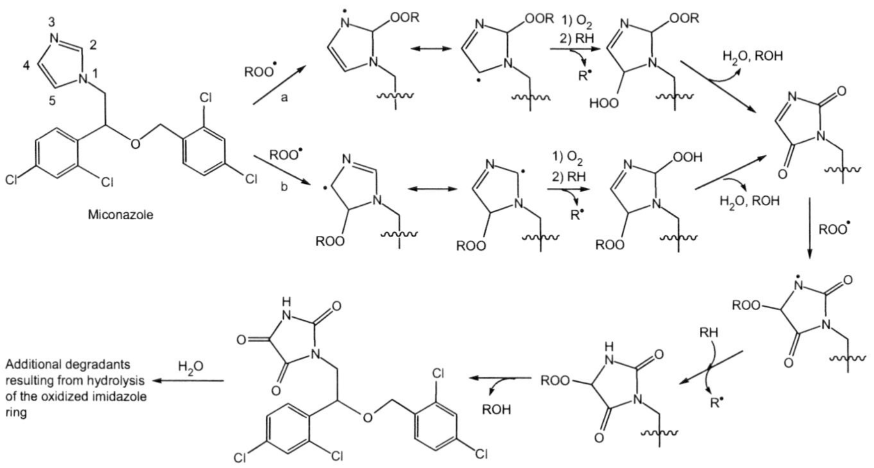

Scheme 3.56

indole ring is the oxidation of its fused pyrrole ring; an example was discussed in the case of indomethacin oxidation in Section 3.5.2. Owing to its importance in proteins and peptides, there will be further discussion of its autooxidative degradation pathways in Chapter 7, Chemical Degradation of Biological Drugs.

Towards the low end of oxidizability is the pyridine moiety which is an electron-deficient species, a weak base, as well as a weak nucleophile. Consequently, the pyridine moiety is usually stable under autooxiative conditions where electrophilic oxygen transfer is the predominant oxidation mechanism.[66] On the other hand, *N*-oxidation of the pyridine moiety is commonly observed in drug metabolism.[218]

3.5.11 Miscellaneous Oxidative Degradations

Boron is an element that people have tried to incorporate into drug molecules but with few successes. Bortezomib, a potent first-in-class dipeptidyl boronic acid 20S proteasome inhibitor used for the treatment of relapsed multiple myeloma, is perhaps the only approved boron-containing drug. During a preformulation study, Wu *et al.* found that the drug underwent facile auto-oxidation in solutions to produce two major hydroxyl degradants along with a couple of secondary and tertiary degradants.[219] The two major degradants are epimers with respect to each other with the configuration difference lying in the hydroxylated sites. To explain the formation of the two epimeric diastereomers, the original authors hypothesized the existence of an acylated Schiff base that would result from the dehydration of hydroxyl degradant 1. The epimerization could then occur by rehydration of the Schiff base to yield hydroxyl degradant 2 (Scheme 3.57).

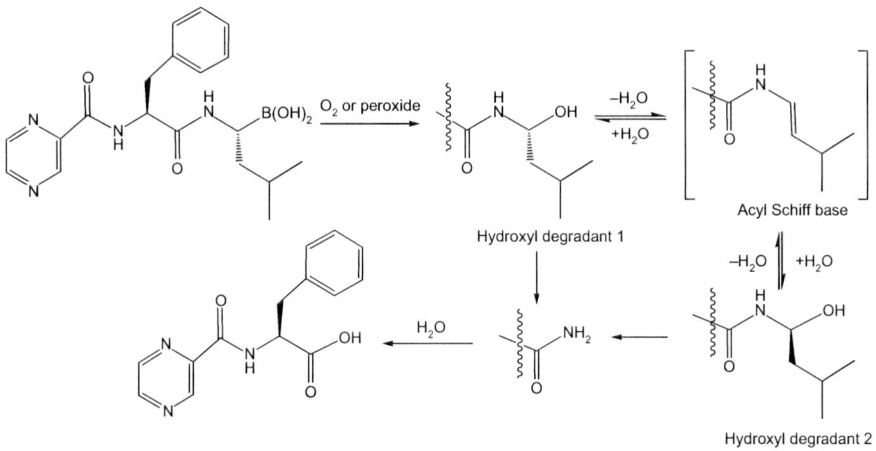

Scheme 3.57

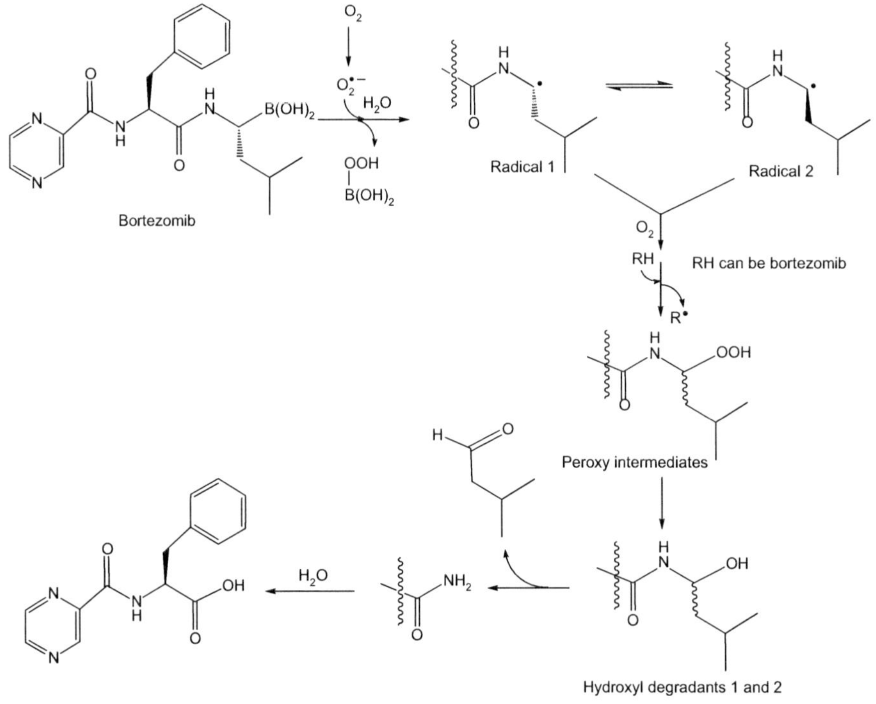

Scheme 3.58

Nevertheless, during the drug metabolism study, Labutti *et al.* observed two intermediary peroxyl degradants which can be converted into the two hydroxyl degradants.[220] Moreover, these authors also made two additional important observations: first, the oxygen atoms of the hydroxyl degradants originate solely from molecular oxygen rather than water through the use of $^{18}O_2$ and $H_2{}^{18}O$, respectively, which rules out the Schiff base mechanism as the cause for racemization. Second, stressing a solution of bortezomib with $FeSO_4$ generated a degradation profile that is very similar to those obtained from the *in vivo* drug metabolism study. These results strongly indicate that the chemical degradation and metabolic transformation of bortezomib should share very similar degradation pathways. Hence, a chemical degradation mechanism that is more consistent with all the experimental observations is illustrated in Scheme 3.58.

Since a fair amount of hydroxyl degradant 2 was observed in both cases by the two research groups, the predominant oxidative degradation of bortezomib may be rationalized by a free radical mechanism that can explain the substantial racemization that occurred. It has been reported that alkylboronates can undergo non-radical (or polar) as well as radical-mediated oxidative pathways.[221] In the non-radical (polar) pathway of the oxidative degradation, the alkyl group retains its original configuration. In the radical mechanism presented in Scheme 3.58, the H• donor, RH, can be bortezomib, in which case radical 1 may be generated again. It is also worth noting that during the preformulation study performed by Wu *et al.*,[219] use of either ascorbic acid or EDTA in the formulation was found to promote the oxidation of bortezomib. This may be another case in which the Udenfriend chemistry (see Section 3.2.1) may be responsible for initiating the autooxidative degradation.

References

1. W. O. Lumberg (ed), *Autoxidation and Antioxidants* John Wiley & Sons, 1961, Vol. 1, p. 2.
2. R. Willstatter and A. Stoll, *Justus Liebigs Ann. Chem.*, 1911, **387**, 317.
3. R. Willstatter and A. Stoll, in *Untersuchungen uber Chlorophyll*, Springer, Berlin, 1913.
4. D. M. Miller, G. R. Buettner and S. D. Aust, *Free Rad. Biol. Med.*, 1990, **8**, 95.
5. M. K. Eberhardt, *Reactive Oxygen Metabolites*, CRC Press, Boca Raton, FL, 2001.
6. B.-Z. Zhu and G.-Q. Shan, *Chem. Res. Toxicol.*, 2009, **22**, 969.
7. J. S. Edmonds, M. Morita, P. Turner, B. W. Skelton and A. H. White, *Steroids*, 2006, **71**, 34.
8. P. A. Harmon, S. Biffar, S. M. Pitzenberger and R. A. Reed, *Pharm. Res.*, 2005, **22**, 1716.
9. H. J. H. Fenton, *J. Chem. Soc. Trans.*, 1894, **65**, 899.
10. F. Haber and J. J. Weiss, *Proc. R.. Soc. London, Ser. A*, 1934, **147**, 332.
11. S. Udenfriend, C. T. Clark, J. Axelrod and B. B. Brodie, *J. Biol. Chem.*, 1954, **208**, 731.
12. C. A. Reed, in *The Biological Chemistry of Iron,* ed. H.B. Dunford, D. Dolphin, K.M. Raymond and L. Sieker, D. Reidel, Dordrecht, 1982, pp. 25–42.

13. G. Schwarzenbach and J. Heller, *Helv. Chim. Acta*, 1951, **34**, 576.
14. W. H. Koppenol and J. Butler, *Adv. Free Radical Biol. Med.*, 1985, **1**, 91.
15. C. Walling, *Acc. Chem. Res.*, 1975, **8**, 125.
16. P. A. MacFaul, D. D. M. Wayner and K. U. Ingold, *Acc. Chem. Res.*, 1998, **31**, 159.
17. S. Goldstein, D. Meyerstein and G. Czapski, *Free Rad. Biol. Med.*, 1993, **15**, 435.
18. D. T. Sawyer, A. Sobkowiak and T. Matsushita, *Acc. Chem. Res.*, 1996, **29**, 409.
19. O˙ Pestovsky, S. Stoian, E. L. Bominaar, X. Shan, E. Münck, L. J. Que and A. Bakac, *Angew. Chem., Int. Ed.*, 2005, **44**, 6871.
20. J. England, M. Martinho, E. R. Farquhar, J. R. Frisch, E. L. Bominaar, E. Münck and L. J. Que, *Angew. Chem., Int. Ed.*, 2009, **48**, 3622.
21. A. Bakac, *Inorg. Chem.*, 2010, **49**, 3584.
22. J. T. Groves, *Proc. Natl. Acad. Sci. U.S.A.*, 2003, **100**, 3569.
23. J. Rittle and M. T. Green, *Science*, 2010, **330**, 933.
24. E. R. Stadtman, *Free Rad. Biol. Med.*, 1990, **9**, 315.
25. E. R. Stadtman and C. N. Oliver, *J. Biol. Chem.*, 1991, **266**, 2005.
26. D. R. Dufield, G. S. Wilson, R. S. Glass and C. Schoneich, *J. Pharm. Sci.*, 2004, **93**, 1122.
27. M. K. Eberhardt, *Trends Org. Chem.*, 1995, **5**, 115.
28. H. Kasai and S. Nishimura, *Nucleic Acids Res.*, 1984, **12**, 2137.
29. M. Li, S. Carlson, J. A. Kinzer and H. J. Perpall, *Biochem. Biophys. Res. Commun.*, 2003, **312**, 316.
30. M. D. Engelmann, R. T. Bobier, T. Hiatt and I. F. Cheng, *BioMetals*, 2003, **16**, 519.
31. G. Schwarzenbach and J. Heller, *Helv. Chim. Acta*, 1951, **34**, 1889.
32. E. Bottari and G. Anderegg, *Helv. Chim. Acta*, 1967, **50**, 2349.
33. G. H. Buettner, *Arch. Biochem. Biophys.*, 1993, **300**, 535.
34. K. M. Ko, P. K. Yick, M. K. T. Poon and S. P. Ip, *Mol. Cell. Biochem.*, 1994, **141**, 65.
35. A. Aguiar and A. Ferraz, *Chemosphere*, 2007, **66**, 947.
36. J. E. Biaglow and A. V. Kachur, *Radiat. Res.*, 1997, **148**, 181.
37. B. W. Alderman, A. E. Ratliff and J. I. Wirgau, *Inorg. Chim. Acta*, 2009, **362**, 1787.
38. J. Hong, E. Lee, J. C. Carter, J. A. Masse and D. A. Oksanen, *Pharm. Dev. Technol.*, 2004, **9**, 171.
39. J. W. McGinity, T. R. Patel and A. H. Naqvi, *Drug Dev. Commun.*, 1976, **2**, 505.
40. T. Huang, M. E. Garceau and P. Gao, *J. Pharm. Biomed. Anal.*, 2003, **31**, 1203.
41. W. R. Wasylaschuk, P. A. Harmon, G. Wagner, A. B. Harman, A. C. Templeton, H. Xu and R. A. Reed, *J. Pharm. Sci.*, 2007, **96**, 106.
42. V. L. Antonovskii and S. L. Khursan, *Russ. Chem. Rev.*, 2003, **72**, 939.
43. P. A. Harmon, K. Kosuda, E. Nelson, M. Mowery and R. A. Reed, *J. Pharm. Sci.*, 2006, **95**, 2014.

44. A. J. Bard and L. R. Faulkner in *Electrochemical Methods–Fundamentals and Applications*, John Wiley and Sons, New York, 1980.
45. W. H. Koppenol, *FEBS Lett.*, 1990, **264**, 165.
46. K. U. Ingold, *Acc. Chem. Res.*, 1969, **2**, 1.
47. J. A. Howard and K. U. Ingold, *Can. J. Chem.*, 1967, **45**, 793.
48. G. A. Russell, *J. Am. Chem. Soc.*, 1957, **79**, 3871.
49. G. W. Burton and K. U. Ingold, *J. Am. Chem. Soc.*, 1981, **103**, 6472.
50. K. C. Waterman, R. C. Adami, K. M. Alsante, J. Hong, M. S. Landis, F. Lombardo and C. J. Roberts, *Pharm. Dev. Technol.*, 2002, **7**, 1.
51. *Handbook of Chemistry and Physics*, 75th edn, CRC Press, Boca Raton, FL. 1995.
52. M. Jonsson, *J. Phys. Chem.*, 1996, **100**, 6814.
53. N. Sebbar, J. W. Bozzelli and H. Bockhorn, *J. Phys. Chem. A*, 2004, **108**, 8353.
54. D. F. McMillen and D. M. Golden, *Ann. Rev. Phys. Chem.*, 1982, **33**, 493.
55. F. R. Cruickshank and S. W. Benson, *J. Phys. Chem.*, 1969, **73**, 733.
56. F. R. Cruickshank and S. W. Benson, *Int. J. Chem. Kinet.*, 1969, **1**, 381.
57. W. S. Nip and G. Paraskevopoulos, *J. Chem. Phys.*, 1979, **71**, 2170.
58. G. A. Russell, *J. Am. Chem. Soc.*, 1955, **78**, 1035.
59. C. J. Norton, F. L. Dormish, M. J. Reuter, N. F. Seppi and P. M. Beazley, *Ind. Eng. Chem. Prod. Res. Dev.*, 1972, **11**, 27.
60. M. A. Freyaldenhoven, P. A. Lehman, T. J. Franz, R. V. Lloyd and V. M. Samokyszyn, *Chem. Res. Toxicol.*, 1998, **11**, 102.
61. A. M. Arafat, S. K. Mathew, S. O. Akintobi and A. A. Zavitsas, *Helv. Chim. Acta*, 2006, **89**, 2226.
62. C. W. Capp and E. G. E. Hawkins, *J. Chem. Soc.*, 1955, 4106.
63. C. J. Toney, F. E. Friedli and P. J. Frank, *J. Am. Oil Chem. Soc.*, 1994, **71**, 793.
64. W. R. Thiel, *Coord. Chem. Rev.*, 2003, **245**, 95.
65. Z. Zhu and J. H. Espenson, *J. Org. Chem.*, 1995, **60**, 1326.
66. W. J. Szczepek, B. Kosmacinska, A. Bielejewska, W. Luniewski, M. Skarzynski and D. Rozmarynowskaa, *J. Pharm. Biomed. Anal.*, 2007, **43**, 1682.
67. A. L. Freed, H. E. Strohmeyer, M. Mahjour, V. Sadineni, D. L. Reid and C. A. Kingsmill, *Int. J. Pharm.*, 2008, **357**, 180.
68. R. S. Drago, A.L.M.L. Mateus and D. Patton, *J. Org. Chem.*, 1996, **61**, 5693.
69. G. B. Payne, P. H. Deming and P. H. Williams, *J. Org. Chem.*, 1961, **26**, 659.
70. Y. Sawaki and Y. Ogata, *Bull. Chem. Soc. Japan*, 1981, **54**, 793.
71. S. W. Hovorka, M. J. Hageman and C. Schöneich, *Pharm. Res.*, 2002, **19**, 538.
72. B. Balagam and D. E. Richardson, *Inorg. Chem.*, 2008, **47**, 1173.
73. R. D. Bach, M. N. Glukhovtsev and C. Gonzalez, *J. Am. Chem. Soc.*, 1998, **120**, 9902.
74. H. Yao and D. E. Richardson, *J. Am. Chem. Soc.*, 2000, **122**, 3220.

75. E. Weitz and A. Scheffer, *Chem. Ber.*, 1921, **54**, 2327.
76. C. A. Bunton and G. J. Minkoff, *J. Chem. Soc.*, 1949, 665.
77. G. Sosnovsky and E. H. Zaret in *Organic Peroxides*, ed. D. Swern, Wiley, New York, 1970, Vol. 1, p. 517.
78. I. P. Skibida and A. M. Sakharov, *Catal. Today*, 1996, 187.
79. G. A. Russell, *J. Am. Chem. Soc.*, 1954, **76**, 1595.
80. G. A. Russell and A. G. Bemis, *J. Am. Chem. Soc.*, 1966, **88**, 5491.
81. D. H. R. Barton and D. W. Jones, *J. Chem. Soc.*, 1965, 3563.
82. M. Li, B. Chen, S. Monteiro and A. M. Rustum, *Tetrahedron Lett.*, 2009, **50**, 4575.
83. J. Hansen and H. Bundgaard, *Int. J. Pharmaceut.*, 1980, **6**, 307.
84. E. V. Bejan, E. Font-Sanchis and J. C. Scaiano, *Org. Lett.*, 2001, **3**, 4059.
85. C. Pan, F. Liu, Q. Ji, W. Wang, D. Drinkwater and R. Vivilecchia, *J. Pharm. Biomed. Anal.*, 2006, **40**, 581.
86. A. Mohan, M. Hariharan, E. Vikraman, G. Subbaiah, B. R. Venkataraman and D. Saravanan, *J. Pharm. Biomed. Anal.*, 2008, **47**, 183.
87. *United States Pharmacopoeia 30*, United States Pharmacopoeial Convention, Rockville, MD, p. 1802.
88. B. Proksa, *J. Pharm. Biomed. Anal.*, 1999, **20**, 179.
89. H. Farsam, S. Eiger, J. Lameh, A. Rezvani, B. W. Gibson and W. Sadee, *Pharm. Res.*, 1990, **7**, 1205.
90. S. S. Kelly, P. M. Glynn, S. J. Madden and D. H. Grayson, *J. Pharm. Sci.*, 2003, **92**, 485.
91. A. M. Kamel, K. S. Zandi and W. W. Massefski, *J. Pharm. Biomed. Anal.*, 2003, **31**, 1211.
92. J. D. Stong, J. V. Pivnichny, H. Mrozik and F. S. Waksmunski, *J. Pharm. Sci.*, 1992, **81**, 1000.
93. Q. Wang, J. D. Stong, P. Demontigny, J. M. Ballard, J. S. Murphy, J.-S. K. Shim and A. J. Faulkner, *J. Pharm. Sci.*, 1996, **85**, 446.
94. S. Javernik, S. Kreft, B. Strukelj and F. Vrecer, *Pharmazie*, 2001, **56**, 738.
95. G. B. Smith, L. DiMichele, L. F. Colwell, Jr., G. C. Dezeny, A. W. Douglas, R. A. Reamer and T. R. Verhoeven, *Tetrahedron Lett.*, 1993, **49**, 4447.
96. R. S. Tomar, T. J. Joseph, A. S. R. Murthy, D. V. Yadav, G. Subbaiah and K.V.S.R. Krishna Reddy, *J. Pharm. Biomed. Anal.*, 2004, **36**, 231.
97. M. L. Huang, A. V. Peer, W. Robert, R. D. Coster, D. V. M. J. Heykants and A. A. I. Jonkman, *Pharm. Drug Dispos.*, 1993, **54**, 257.
98. R. P. Enever, A. Li Wan and Po and E. Shotton, *J. Pharm. Sci.*, 1979, **68**, 169.
99. R. P. Enever, A. Li Wan, Po, B. J. Millard and E. Shotton, *J. Pharm. Sci.*, 1975, **64**, 1497.
100. G. Callen, M. S. Chorghade, E. C. Lee, P. G. Nilsen, H. Petersen and A. Rustum, *Heterocycles*, 1994, **39**, 293.
101. X. Zhang and C. S. Foote, *J. Am. Chem. Soc.*, 1993, **115**, 8867.
102. M. Li, B. Conrad, R. G. Maus, S. M. Pitzenberger, R. Subramanian, X. Fang, J. A. Kinzer and H. J. Perpall, *Tetrahedron Lett.*, 2005, **46**, 3533.

103. G. Boccardi, C. Deleuze, M. Gachon, G. Palmisano and J. P. Vergnaud, *J. Pharm. Sci.*, 1992, **81**, 183.

104. http://drugbank.wishartlab.com (last accessed March 2011).

105. X. Wang, M. Li and A. M. Rustum, *Rapid Commun. Mass Spectrom.*, 2010, **24**, 2805.

106. X. Li, F. E. Blondino, M. Hindle, W. H. Soine and P. R. Byron, *Int. J. Pharm.*, 2005, **303**, 113.

107. S. B. Karki and J. P. Dinnocenzo, *Xenobiotica*, 1995, **25**, 711.

108. S. B. Karki, J. P. Dinnocenzo, J. P. Jones and K. R. Korzekwa, *J. Am. Chem. Soc.*, 1995, **117**, 3657.

109. H. Matsumoto, M. Fukumoto and A. Ogamo, *Jpn. J. Forensic Toxicol.*, 1998, **16**, 164.

110. Z. Z. Zhao, X.-Z. Qin, A. Wu and Y. Yuan, *J. Pharm. Sci.*, 2004, **93**, 1957.

111. J. Dong, S. B. Karki, M. Parikh, J. C. Riggs and L. Huang, *Drug Dev. Ind. Pharm.*, posted online on January 23, 2012.

112. R. Ramanathan, A.-D. Su, N. Alvarz, N. Blumenkrantz, S. K. Chowdhury and J. E. Patrick, *Anal. Chem.*, 2000, **72**, 1352.

113. S. K. Chowdhury and K. B. Alton, *Anal. Chem.*, 2005, **77**, 3676.

114. A. C. Cope, T. T. Foster and P. H. Towle, *J. Am. Chem. Soc.*, 1949, **71**, 3929.

115. S. Ma, S. K. Chowdhury and K. B. Alton, *Anal. Chem.*, 2005, **77**, 3676.

116. J. Meisenhemimer, *Chem. Ber.*, 1919, **52**, 1667.

117. H. E. De La Mare, *J. Org. Chem.*, 1960, **25**, 2114.

118. A. L. J. Beckwith, P. H. Eichinger, B. A. Mooney and R. H. Prager, *Aust. J. Chem.*, 1983, **36**, 719.

119. Z. Földi, T. Földi and A. Földi, *Chem. Ind.*, 1955, 1297.

120. J.-E. Belgaied and H. Trabelsi, *J. Pharm. Biomed. Anal.*, 2003, **33**, 991.

121. W. D. Emmons, *J. Am. Chem. Soc.*, 1957, **79**, 5528.

122. K. M. Ibne-Rasa and J. O. Edwards, *J. Am. Chem. Soc.*, 1962, **84**, 763.

123. B. C. Challis and A. R. Butler, in *The Chemistry of the Amino Group,* ed. S. Patai, Wiley and Sons, London, 1968, pp. 320–338.

124. D. H. Rosenblatt and E. P. Burrows, in *Supplement F: The Chemistry of Amino, Nitroso, and Nitro Compounds and Their Derivatives, Part 2,* ed. S. Patai, Wiley and Sons, Chichester, 1982, pp. 1085–1149.

125. J. D. Fields and P. J. Kropp, *J. Org. Chem.*, 2000, **65**, 5937.

126. C. Zonta, E. Cazzola, M. Mba and G. Licinia, *Adv. Synth. Catal.*, 2008, **350**, 2503.

127. H.-C. Shi and Y. Li, *J. Mol. Catal. A: Chem.*, 2007, **271**, 32.

128. J. Zheng and A. M. Rustum, *J. Pharm. Biomed. Anal.*, 2010, **51**, 146.

129. J. F. W. Keana, *Chem. Rev.*, 1978, **78**, 37.

130. W. Jahnke, S. Rüdisser and M. Zurini, *J. Am. Chem. Soc.*, 2001, **123**, 3149.

131. M. F. Semmelhack, C. S. Chou and D. A. Cortes, *J. Am. Chem. Soc.*, 1983, **105**, 4492.

132. M. R. Leanna, T. J. Sowin and H. E. Morton, *Tetrahedron Lett.*, 1992, **33**, 5029.

133. S. K. Malhotra, J. J. Hostynek and A. F. Lundin, *J. Am. Chem. Soc.*, 1968, **90**, 6565.

134. G. Modena and P. E. Todesco, *J. Chem. Soc.*, 1962, 4920.

135. J. W. Chu and B. L. Trout, *J. Am. Chem. Soc.*, 2004, **126**, 900.

136. G. R. Krow in *Organic Reactions*, Vol. 43, ed. L. A. Paquette *et al.*, John Wiley & Sons, 1993, pp. 251–798.

137. F. Di Furia and G. Modena, *Pure Appl. Chem.*, 1982, **54**, 1853.

138. B. L. Miller, T. D. Williams and C. Schöneich, *J. Am. Chem., Soc.*, 1996, **118**, 11014.

139. B. L. Miller, K. Kuczera and C. Schöneich, *J. Am. Chem. Soc.*, 1998, **120**, 3345.

140. K.-D. Asmus, in *Sulfur-centered Reactive Intermediates in Chemistry and Biology*, ed. C. Chatgilialoglu and K.-D. Asmus, NATO ASI Series 197, Plenum Press, New York, 1990, pp. 155–172.

141. C. Schöneich, A. Aced and K.-D. Asmus, *J. Am. Chem. Soc.*, 1993, **115**, 11376.

142. K. Bobrowski, G. L. Hug, D. Pogocki, B. Marciniak and C. Schöneich, *J. Phys. Chem. B*, 2007, **111**, 9608.

143. C. Schöneich, *Biochim. Biophys. Acta*, 2005, **1703**, 111.

144. I. Fourré and J. Bergès, *J. Phys. Chem. A*, 2004, **108**, 898.

145. M. L. Huang and A. Rauk, *J. Phys. Chem. A*, 2004, **108**, 6222.

146. P. Brunelle and A. Rauk, *J. Phys. Chem. A*, 2004, **108**, 11032.

147. C. Schöneich, *Arch. Biochem. Biophys.*, 2002, **397**, 370.

148. A. Rauk, D. A. Armstrong and D. P. Fairlie, *J. Am. Chem. Soc.*, 2000, **122**, 9761.

149. R. S. Glass, *Top. Curr. Chem.*, 1999, **205**, 1.

150. E. D. Nelson, P. A. Harmon, R. C. Szymanik, M. G. Teresk, L. Li, R. A. Seburg and R. A. Reed, *J. Pharm. Sci.*, 2006, **95**, 1527.

151. M. M. Al Omari, R. M. Zoubi, E. I. Hasan, T. Z. Khader and A. A. Badwan, *J. Pharm. Biomed. Anal.*, 2007, **45**, 465.

152. H. A. Rosenberg, J. T. Dougherty, D. Mayron and J. G. Baldinus, *Am. J. Hosp. Pharm.*, 1980, **37**, 390.

153. S. E. Walker, T. W. Paton, T. M. Fabian, C. C. Liu and P. E. Coates, *Am. J. Hosp. Pharm.*, 1981, **38**, 881.

154. P. M. G. Bavin, A. Post and J. E. Zarembo, in *Analytical Profiles of Drug Substances*, ed. K. Florey, Academic Press, Orlando, 1984, Vol. 13, pp. 127–183.

155. G. W. Mihaly, O. H. Drummer, A. Marshall, R. A. Smallwood and W. J. Louis, *J. Pharm. Sci.*, 1980, **69**, 1155.

156. N. Beaulieu, P. A. Lacroix, R. W. Sears and E. G. Lovering, *J. Pharm. Sci.*, 1988, **77**, 889.

157. M. J. Puz, B. A. Johnson and B. J. Murphy, *Pharm. Dev. Technol.*, 2005, **1**, 115.

158. V. Caplar, S. Rendic, F. Kajfez, H. Hofman, J. Kuftinec and N. Blazevic, *Acta Pharm. Jugosl.*, 1982, **32**, 125.

159. K. A. Connors, G. L. Amidon and V. J. Stella, *Chemical Stability of Pharmaceuticals: A handbook for pharmacists*. John Wiley & Sons, New York, 2nd edn, 1986.

160. M. Benrahmoune, P. Thérond and Z. Abedinzadeh, *Free Radicals Biol. Med.*, 2000, **29**, 775.

161. L. Gu, H.-S. Chiang and A. Becker, *Int. J. Pharm.*, 1988, **41**, 95.

162. B. Mao, A. Abrahim, Z. Ge, D. K. Ellison, R. Hartman, S. V. Prabhu, R. A. Reamer and J. Wyvratt, *J. Pharm. Biomed. Anal.*, 2002, **28**, 1101.

163. L. R. Reddy and E. J. Corey, *Tetrahedron Lett.*, 2005, **46**, 927.

164. H. Bundgaard and J. Hansen, *Arch. Pharm. Chemi., Sci. Ed.*, 1980, **8**, 187.

165. T. Chulski and A. A. Forist, *J. Am. Pharm. Assoc., Sci. Ed.*, 1958, **47**, 553.

166. R. E. Conrow, G. W. Dillow, L. Bian, L Xue, O. Papadopoulou, J. K. Baker and B. S. Scott, *J. Org. Chem.*, 2002, **67**, 6835.

167. D. V. C. Awang, A. Vincent and F. Matsui, *J. Pharm. Sci.*, 1973, **62**, 1673.

168. H. E. Williams, V. C. Loades, M. Claybourn and D. M. Murphy, *Anal. Chem.*, 2006, **78**, 604.

169. *Handbook of Chemistry and Physics*, ed. R. C. Weast, 61st edn, CRC Press, Boca Raton, FL, pp. 1980–1981.

170. S. J. Blanksby, T. M. Ramond, G. E. Davico, M. R. Nimlos, S. Kato, V. M. Bierbaum, W. C. Lineberger, G. B. Ellison and M. Okumura, *J. Am. Chem. Soc.*, 2001, **123**, 9585.

171. D. M. Johnson and L. C. Gu, in *Encyclopedia of Pharmaceutical Technology*, ed. J. Swarbrick and J.C. Boylan, Marcel Dekker, New York, 1988, Volume 1, pp. 415–449.

172. R. A. Sheldon and J. K. Kochi, *Metal-Catalyzed Oxidations of Organic Compounds*, Academic Press, New York, 1981, p. 359.

173. M. Matsumoto, H. Kobayashi and Y. Hotta, *J. Org. Chem.*, 1984, **49**, 4740.

174. G. Brenner, F. E. Roberts, A. Hoinowski, J. Budavari, B. Powell, D. Hinkley and E. Schoenewaldt, *Angew. Chem., Int. Ed.*, 1969, **8**, 975.

175. C.-T. Lin, P. Perrier, G. G. Clay, P. A. Sutton and S. R. Byrn, *J. Org. Chem.*, 1982, **47**, 2978.

176. A. H. Kibbe, *Handbook of Pharmaceutical Excipients*, American Pharmaceutical Association, 2000, pp. 41–43.

177. S. Korcek, J. H. B. Chenier, J. A. Howard and K. U. Ingold, *Can. J. Chem.*, 1972, **50**, 2285.

178. A. B. Levina, S. R. Trusov and Z. Obshchei, *Khimii*, 1990, **60**, 1932.

179. V. R. Choudhary, P. A. Chaudhari and V. S. Narkhede, *Catal. Commun.*, 2003, **4**, 171.

180. A. M. Abend, L. Chung, R. Todd Bibart, M. Brooks and D. G. McCollum, *J. Pharm. Biomed. Anal.*, 2004, **34**, 957.

181. W. G. Llyod, *J. Chem. Eng. Data*, 1961, **6**, 541.

182. R. Hamburger, E. Azaz and M. Donbrow, *Pharm. Acta Helv.*, 1975, **50**, 10.

183. M. Donbrow, E. Azaz and A. Pillersdorf, *J. Pharm. Sci.*, 1978, **67**, 1676.
184. K. C. Waterman, W. B. Arikpo, M. B. Fergione, T. W. Graul, B. A. Johnson, B. C. MacDonald, M. C. Roy and R. J. Timpano, *J. Pharm. Sci.*, 2008, **97**, 1499.
185. M. Bergh, L. P. Shao, K. Magnusson, E. Gäfvert, J. L. G. Nilsson and A.-T. Karlberg, *J. Pharm. Sci.*, 1999, **88**, 483.
186. B. Plesnicar, in *The Chemistry of Peroxides,* ed. S. Patai, John Wiley & Sons, Chichester, UK, 1983, p. 559.
187. M. Renz and B. Meunier, *Eur. J. Org. Chem.*, 1999, 737.
188. A. Nishinaga, K. Rindo and T. Matsuura, *Synthesis*, 1986, 1038.
189. H. Yuasa, M. Matsuno and H. Imai, *Eur. Pat. Appl.*, EP 103099 A2, 1984.
190. B. G. Snider, T. A. Runge, P. E. Fagerness, R. H. Robins and B. D. Kaluzny, *Int. J. Pharm.*, 1990, **66**, 63.
191. C. A. Grob, *Angew. Chem., Int. Ed.*, 1969, **8**, 535.
192. D. Wenkert, K. M. Eliasson and D. Rudisill, *J. Chem. Soc., Chem. Commun.*, 1983, 392.
193. E. Caspi, Y. W. Chang and R. I. Dorfman, *J. Med. Pharm. Chem.*, 1962, **5**, 714.
194. J. T. Pinhey and K. Schaffner, *Aust. J. Chem.*, 1968, **21**, 1873.
195. G. A. Krafft and J. A. Katzenellenbogen, *J. Am. Chem. Soc.*, 1981, **103**, 5459.
196. M. S. Ahmad and A. R. Siddiqi, *Indian J. Chem., Sect. B: Org. Chem. Incl. Med. Chem.*, 1978, **16**, 963.
197. S. D. Levine, *J. Org. Chem.*, 1966, **31**, 3189.
198. B. B. Brodie, J. R. Gillette and B. N. La Du, *Annu. Rev. Biochem.*, 1958, **27**, 427.
199. Y. Wu, X. Chen, L. Gier, O. Almarsson, D. Ostovic and A. E. Loper, *J. Pharm. Biomed. Anal.*, 1999, **20**, 471.
200. N. K. Yee, L. J. Nummy and G. P. Roth, *Bioorg. Med. Chem. Lett.*, 1996, **6**, 2279.
201. H. P. Misra and I. Fridovich, *J. Biol. Chem.*, 1972, **247**, 3170.
202. K. A. Connors, G. L. Amidon and V. J. Stella, *Chemical Stability of Pharmaceuticals: A handbook for pharmacists*, John Wiley & Sons, New York, 2nd edn, 1986, pp. 438–447.
203. G. Li, H. Zhang, F. Sader, N. Vadhavkar and D. Njus, *Biochemistry*, 2007, **46**, 6978.
204. C. Auclair and C. Paoletti, *J. Med. Chem.*, 1981, **24**, 289.
205. M. A. Quarry, D. S. Sebastian and F. Diana, *J. Pharm. Biomed. Anal.*, 2002, **30**, 99.
206. S.-Y. Yeh and J. L. Lach, *J. Pharm. Sci.*, 1961, **50**, 35.
207. K. Uchida and S. Kawakishi, *Arch. Biochem. Biophys.*, 1990, **283**, 20.
208. K. Uchida and S. Kawakishi, *FEBS Lett.*, 1993, **332**, 208.
209. F. Zhao, E. Ghezzo-Schöneich, G. I. Aced, J. Hong, T. Milby and C. Schöneich, *J. Biol. Chem.*, 1997, **272**, 9019.
210. C. Schöneich, *J. Pharm. Biomed. Anal.*, 2000, **21**, 1093.

211. G. Lassmann, L. A. Eriksson, F. Himo, F. Lendzian and W. Lubitz, *J. Phys. Chem. A*, 1999, **103**, 1283.
212. A. R. Oyler, R. E. Naldi, K. L. Facchine, D. J. Burinsky, M. H. Cozine, R. Dunphy, J. D. Alves-Santana and M. L. Cotter, *Tetrahedron*, 1991, **47**, 6549.
213. H. Wasserman, K. Stiller and M. Floyd, *Tetrahedron Lett.*, 1968, **29**, 3277.
214. H. H. Wasserman, M. S. Wolff, K. Stiller, I. Saito and J. E. Pickett, *Tetrahedron*, 1981, **37**, 191.
215. H.-S. Ryang and C. S. Foote, *J. Am. Chem. Soc.*, 1979, **101**, 6683.
216. J. W. Blunt, B. R. Copp, W.-P. Hu, M. H. G. Munro, P. T. Northcote and M. R. Prinsep, *Nat. Prod. Rep.*, 2007, **24**, 31.
217. D. J. Newman and G. M. Cragg, *J. Nat. Prod.*, 2007, **70**, 461.
218. P. Kulanthaivel, R. J. Barbuch, R. S. Davidson, P. Yi, G. A. Rener, E. L. Mattiuz, C. E. Hadden, L. A. Goodwin and W. J. Ehlhardt, *Drug Metab. Dispos.*, 2004, **32**, 966.
219. S. Wu, W. Waugh and V. J. Stella, *J. Pharm. Sci.*, 2000, **89**, 758.
220. J. Labutti, I. Parsons, R. Huang, G. Miwa, L.-S. Gan and J. S. Daniels, *Chem. Res. Toxicol.*, 2006, **19**, 539.
221. C. Cadot, P. I. Dalko and J. Cossy, *J. Org. Chem.*, 2002, **67**, 719.

CHAPTER 4

Various Types and Mechanisms of Degradation Reactions

4.1 Elimination

An elimination reaction is one in which two substituents are removed or eliminated from the parent molecule. The elimination can proceed either *via* a one- or two-step mechanism, known as E1 and E2 elimination, respectively (Scheme 4.1). The numbers 1 and 2 here refer to the orders, rather than the steps, of the elimination reactions. In the E1 elimination, an intermediate is formed by the elimination of the first substituent, which is then followed by the elimination of the second substituent. The first step of the E1 elimination is usually the rate-liming step and hence the reaction is of first order relative to the elimination substrate. In the E2 elimination, a concerted mechanism is operative; frequently, a base is required to effect the concerted elimination of the two substituents. Therefore, the reaction is second order owing to the involvement of the two reactants.

In both types of the elimination reaction, the double bond equivalency (DBE) of the resulting degradation product increases by one, that is, the elimination causes the formation of a double bond (from a single bond), a triple bond (from a double bond), or a ring. The most pharmaceutically relevant drug degradation *via* elimination is probably dehydration. Other elimination reactions that have been observed in pharmaceutical products include dehydrohalogenation (elimination of HX, where X = halogen such as Cl) and the Hofmann elimination. Decarboxylation is a degradation reaction that is closely related to, but is not elimination. They will be discussed in the following sub-sections, respectively.

4.1.1 Dehydration

Dehydration is probably the most commonly seen degradation type due to an elimination reaction. Several corticosteroids containing the

RSC Drug Discovery Series No. 29
Organic Chemistry of Drug Degradation
By Min Li
© Min Li 2012
Published by the Royal Society of Chemistry, www.rsc.org

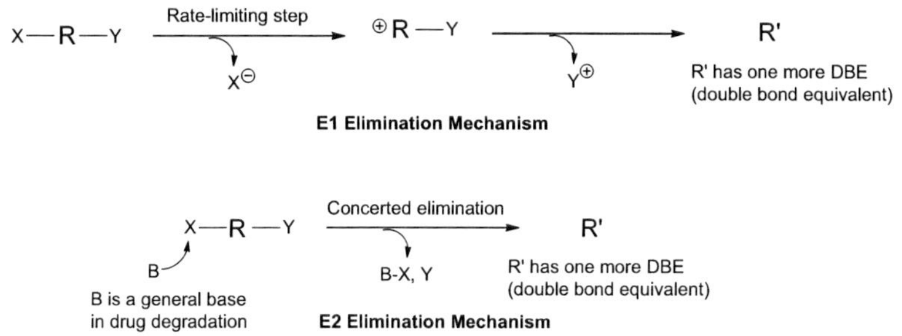

E1 Elimination Mechanism

E2 Elimination Mechanism

Scheme 4.1

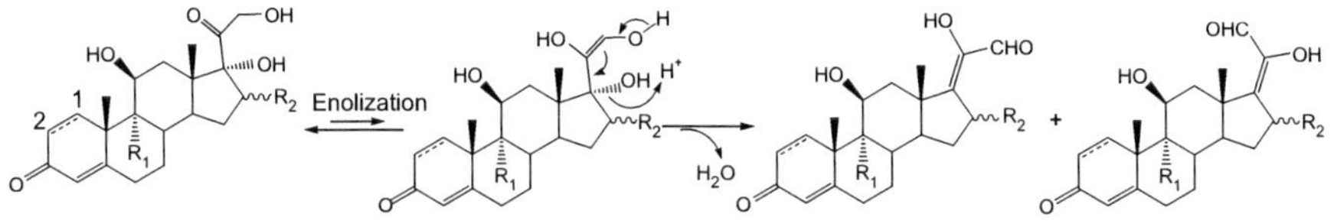

Dexamethasone, 1-2 double bond,
R_1 = F, R_2 = CH_3 at α-position;
Betamethasone, 1-2 double bond,
R_1 = F, R_2 = CH_3 at β-position;
Prednisolone, 1-2 double bond,
R_1 = H, R_2 = H
Cortisol, 1-2 single bond,
R_1 = H, R_2 = H

Enol aldehyde Z-Isomer   **Enol aldehyde E-Isomer**

Scheme 4.2

1,3-dihydroxyacetone side chain on the D-ring, such as dexamethasone, beta-methasone, prednisolone, and cortisol undergo a dehydration reaction known as the Mattox rearrangement (which is actually not a rearrangement), in particular under acidic conditions.[1,2] During the Mattox process as shown in Scheme 4.2, the side chains of the corticosteroids are presumed to enolize prior to the dehydration, which results in the formation of two regio-isomeric enol aldehydes as the degradants.

In the case of betamethasone, a competing dehydration process, in which the 17-hydroxyl group is eliminated along with the neighboring 16-hydrogen, producing $^{16}\Delta$-betamethasone, was also observed.[3] It appears that the nature of the counter ion in the acid and the solvent used in the forced degradation has an impact on the distribution of the dehydration products: when stressed with HCl in dioxane, an appreciable amount of $^{16}\Delta$-betamethasone was formed, while stressing with sulfuric acid in a mixture of acetonitrile and water produced no detectable level of the latter degradant[4] (Scheme 4.3).

The 17,21-di-esters of betamethasone, betamethasone 9,11-epoxide, and dexamethasone also undergo elimination, analogous to the Mattox rearrangement, in which the equivalent of a molecule of carboxylic anhydride is removed.[4,5] This variation of the Mattox degradation process was shown to be catalyzed by a base or nucleophile (Scheme 4.4). Both the original Mattox

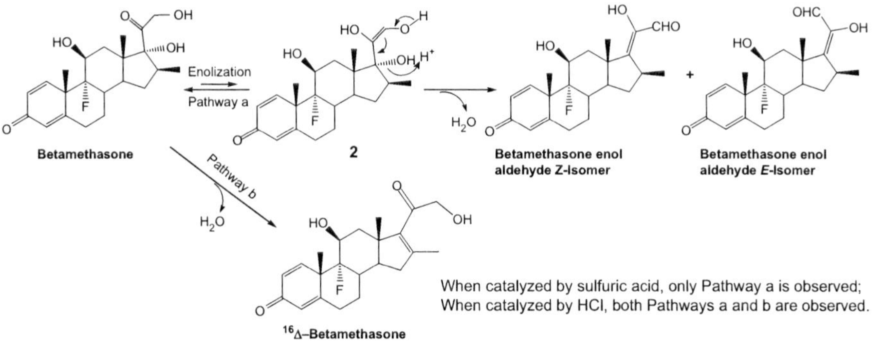

Scheme 4.3

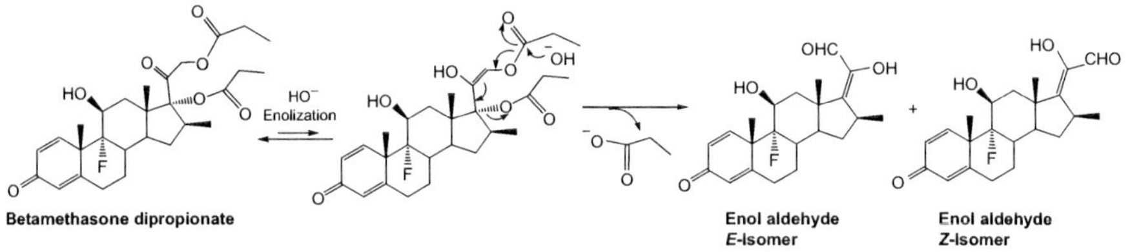

Scheme 4.4

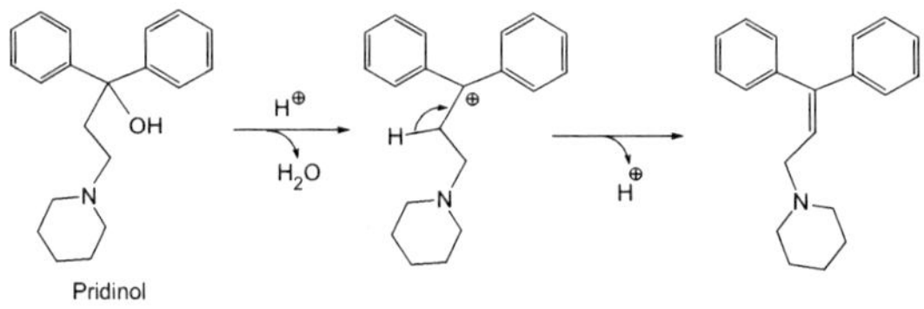

Scheme 4.5

process[1,2] and its variation with the di-esters[5] are likely to be concerted processes, that is, both should proceed *via* the E2 elimination mechanism.

In a number of products formulated with betamethasone dipropionate, betamethasone enol aldehyde has been observed to be a degradant, especially in liquid formulations.[6-8]

For a drug like pridinol which contains a diphenylalkylcarbinol moiety, it would be expected that dehydration is at least a potential degradation pathway. Indeed, when a sample of pridinol was stressed with 0.1 N HCl at room temperature for 6 days, Bianchini *et al.* found that ~0.3% of the dehydrated product was formed.[9] These workers also determined that the dehydration reaction is first order with an activation energy of 25.5 kcal mol^{-1}. This elimination reaction probably takes place *via* an E1 mechanism, owing to the stabilization of the initially formed carbocation by the two phenyl groups (Scheme 4.5).

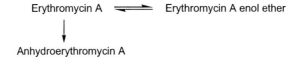

Mechanism proposed by Cachet *et al*.

Erythromycin A ⟶ Erythromycin A enol ether ⟶ Anhydroerythromycin A

Original mechanism proposed by Atkins *et al*.

Scheme 4.6

For drug molecules that contain multiple hydroxyl groups, dehydration by intramolecular condensation between two hydroxyl groups can occur. Erythromycin, an important member in the macrolide antibiotics family, possesses a number of hydroxyl groups. Erythromycin A, which is the major component of erythromycin, has long been known to degrade in aqueous solution, particularly in acidic solutions. According to the studies performed by Atkins *et al*.[10] and Cachet *et al*.,[11] once dissolved in acidic to neutral solutions, erythromycin A quickly converted to two dehydrated species, erythromycin A enol ether and anhydroerythromycin A. Cachet *et al*. proposed a mechanism[11] that differs from the original mechanism proposed by Atkins *et al*.[10] (Scheme 4.6).

In the new mechanism postulated by Cachet *et al*., erythromycin A directly degrades to the dehydrated degradant, anhydroerythromycin A, while the original mechanism proposed by Atkins *et al*. suggested a consecutive degradation pathway. The new mechanism was supported by the observation that erythromycin A enol ether quickly gave rise to erythromycin A in aqueous solution. In addition, the kinetic model derived from the new mechanism fits the experimental data better.[11] In the above studies conducted by the two research groups, the detailed configuration and stereochemistry of the degradants and related compounds was not examined. Since the mid-1990s, efforts have been directed toward resolving this deficiency; Alam *et al*. confirmed the structure and stereochemistry of erythromycin A enol ether using 2D nuclear magnetic resonance (NMR),[12] while Hassanzadeh *et al*. established the stereochemistry of anhydroerythromycin A using 2D NMR, X-ray crystallography, and computer modeling.[13] Based on all the key results reported in the literature,[10–13] the overall degradation behavior of erythromycin A can be summarized in Scheme 4.7.

In either of the two pathways illustrated in Scheme 4.7, it is apparent that a hemiacetal (6,9- or 9,12-hemiacetal), formed when either 6-hydroxyl or 12-hydroxyl attack the 9-keto position, should be the transient intermediate respectively. It should be noted that erythromycin A enol ether was sometimes incorrectly referred to as its intermediate, erythromycin A 6,9-hemiacetal. According to a study on solution stability of erythromycin A by Barber *et al*.,[14] the ratio of the native 9-keto form of erythromycin A *versus* the therapeutically inactive enol ether form is 5:2 in neutral solution at ambient temperature.

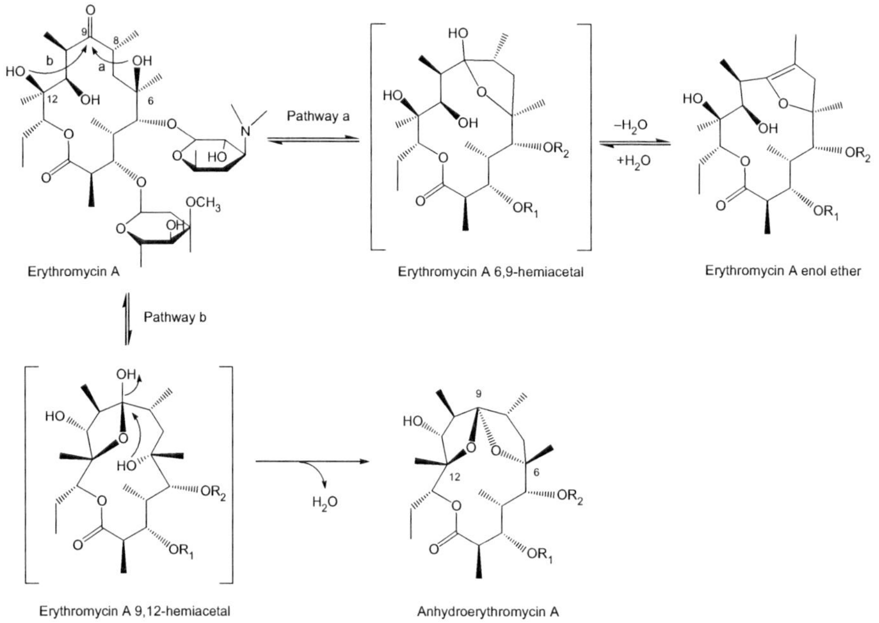

Scheme 4.7

Azithromycin, one of the best selling antibiotic drugs, was made by structure modification of erythromycin through the replacement of the 9-keto functionality by a methyl-substituted nitrogen moiety. Since the 9-keto functionality, which is essential for the formation of the two major degradants shown in Scheme 4.7, no longer exits, the stability of azithromycin in acidic solution is markedly improved: its main degradation pathway is the much slower hydrolysis of the ether linkage to the cladinose sugar substituent at the 3-position of the macrolide ring.[15]

Another frequently observed dehydration reaction is the formation of an intra-molecular amide or ester linkage between an amino or hydroxyl group and a carboxyl group, giving rise to a new ring in the degradant formed. A number of drugs based on dipeptides or dipeptide analogs tend to undergo this kind of dehydration yielding a type of degradants collectively known as diketopiperazines (DKPs). Since all the degradants are cyclic compounds, this degradation will be discussed as part of Section 4.6.

4.1.2   Dehydrohalogenation

Mometasone furoate, a corticosteroid pro-drug widely used for a variety of anti-inflammation indications, possesses a 9α-chloro,11β-hydroxyl moiety that is susceptible to dehydrochlorination. Two research groups studied the stability of mometasone furoate in simulated lung fluid, an aqueous solution made up of various salts and buffers with pH adjusted to 7.4.[16,17] Both groups found that mometasone furoate underwent dehydrochlorination rather quickly at 37 °C to yield mometasone furoate 9,11-epoxide; the half-life of this degradation was determined to be 1.3 hours by Sahasranaman *et al.*[17] The rate of dehydro-chlorination increased as the pH of the solution was raised. The relatively facile elimination and its pH dependence may be attributed to the following factors: first, the two reacting groups, 11β-hydroxyl and 9α-chloro groups are already in the *trans* orientation that would be favored by an E2 elimination mechanism. Second, chloride is a reasonably good leaving group. Third, under higher pH, the nucleophilicity of the 9-hydroxyl group is enhanced.

The latter authors also found that the 9,11-epoxide degradant underwent a further condensation reaction between the 20-methylene and the carbonyl group of the furoate moiety to give a dehydrated degradant.[17] The kinetic model fitting consistent with this overall sequential degradation pathway is illustrated in Scheme 4.8.

In a stability study performed by Teng *et al.*,[16] a small amount of dehydrated mometasone furoate was observed; hence, the alternate pathway (dotted arrows) could not be ruled out as a minor contributor to the final degradant. The major sequential pathway seems to suggest that the conformation change in the steroid core structure induced by the formation of the 9,11-epoxide bond renders the subsequent condensation much easier.

A few other corticosteroids including mometasone (the precursor of its pro-drug, mometasone furoate), beclomethasone, and clocortolone, also possess the 9α-chloro,11β-hydroxyl moiety and hence could be susceptible to the same dehydrochlorination degradation,[5] especially in liquid formulations.[18] On the other hand, corticosteroids containing the 9α-fluoro,11β-hydroxyl moiety, such as betamethasone, dexamethasone, triamcinolone, and desoximetasone (Figure 4.1), would be less likely to undergo the corresponding dehydrofluorination reaction because fluoride is a poorer leaving group compared to chloride.

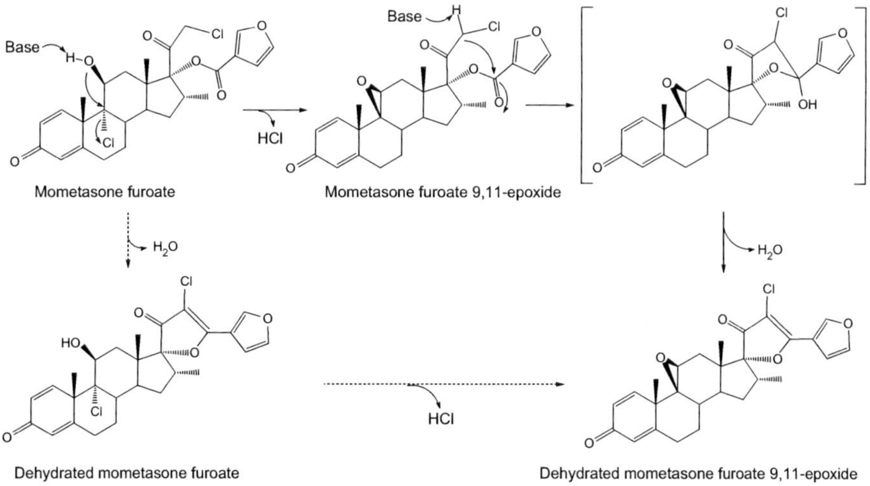

Mometasone furoate

Mometasone furoate 9,11-epoxide

Dehydrated mometasone furoate

Dehydrated mometasone furoate 9,11-epoxide

Scheme 4.8

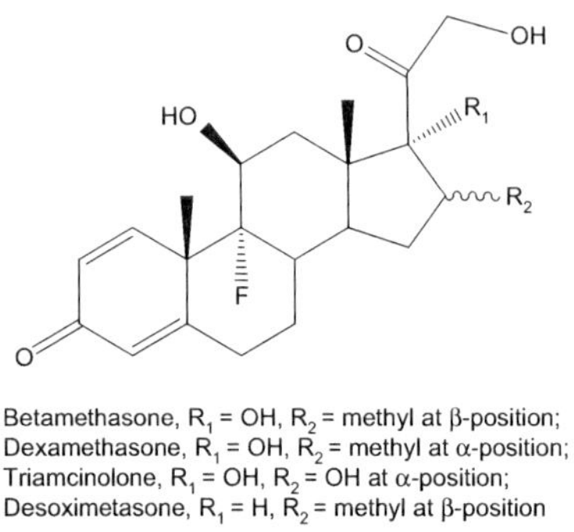

Betamethasone, R$_1$ = OH, R$_2$ = methyl at β-position;
Dexamethasone, R$_1$ = OH, R$_2$ = methyl at α-position;
Triamcinolone, R$_1$ = OH, R$_2$ = OH at α-position;
Desoximetasone, R$_1$ = H, R$_2$ = methyl at β-position

Figure 4.1   Betamethasone, dexamethasone, triamcinolone, and desoximetasone.

4.1.3   Hofmann Elimination

The Hofmann elimination, also known as Hofmann degradation or exhaustive methylation, is a reaction in which an amine is converted to the corresponding quaternary salt intermediate by methyl iodide and then pyrolyzed to yield an olefin through elimination of the amine moiety. In drug degradation chemistry, the relevance of the Hoffmann elimination is limited to those drug substances that already have a quaternary salt functionality. For example, widely used clinically skeletal muscle relaxants, atracurium and its 1R-*cis*, 1R′-*cis* isomer, cisatracurium contain two symmetric quaternary ammonium salt units which are linked together by a diester spacer. This spacer was purposely designed to be biodegradable *in vivo*, so that the drug has a shorter duration of action compared to the metabolically stable precursor drug, laudexium, upon which atracurium was developed.[19] Lability towards *in vivo* degradation is achieved through Hofmann elimination of the quaternary salt moieties. Unfortunately, this degradation pathway also occurs in aqueous buffers as shown in Scheme 4.9.

The rate of the atracurium degradation was found to increase as the pH became higher, which is consistent with the above mechanism where deprotonation of the activated methylene moiety α to the carbonyl group triggers the Hofmann elimination.[20]

Another example of drug degradation by Hofmann elimination can be found in the case of a drug candidate in the family of carbapenem. This drug candidate contains a side chain that is releasable upon the cleavage of the β-lactam linkage, in order to minimize the potential immunogenic side effect of the intact drug.[21] The side chain possesses a twin quaternary ammonium salt moiety, which is susceptible to Hofmann elimination and related degradation in both the solid and solution states (Scheme 4.10).[22,23]

As discussed above, one of the key factors triggering the Hofmann elimination is the availability of an "activated" methylene moiety β to the quaternary salt, which is also true in this case. The second key factor is the nature of the leaving group: a tertiary amine can be a reasonably good leaving group. In pathway b of Scheme 4.10, the Hofmann elimination is not possible due to

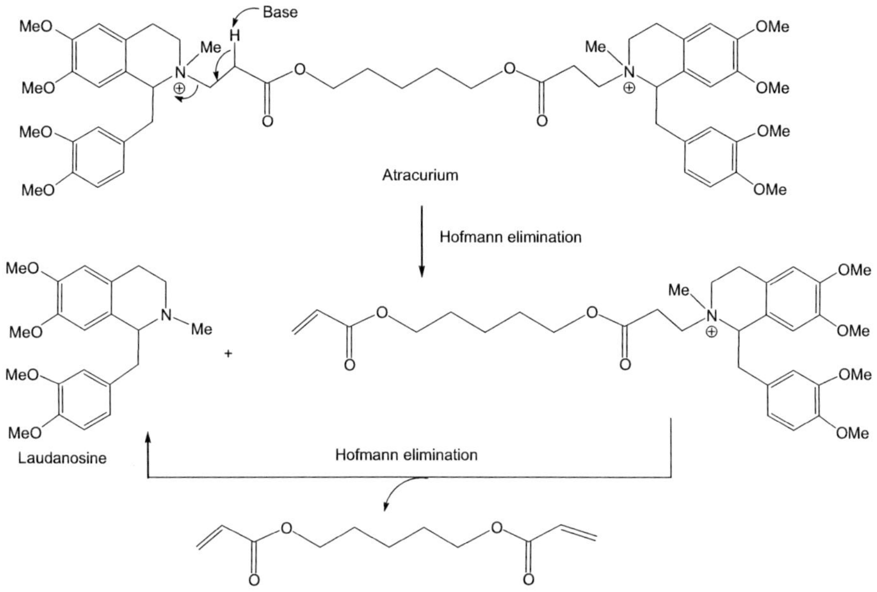

Base
Atracurium
Hofmann elimination
Laudanosine
Hofmann elimination

Scheme 4.9

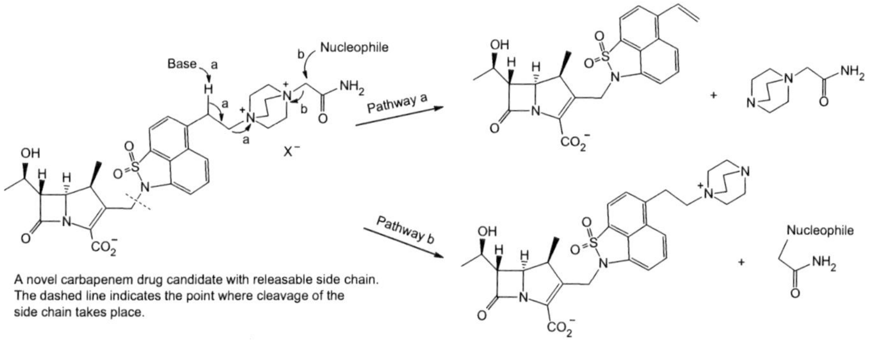

Base
Nucleophile
Pathway a
Pathway b
Nucleophile
A novel carbapenem drug candidate with releasable side chain. The dashed line indicates the point where cleavage of the side chain takes place.

Scheme 4.10

the lack of a β-methylene moiety. However, attack on the α-methylene functionality by a nucleophile could lead to the removal of the acetamide moiety. The latter pathway (b) was observed in a stress study of a chloride salt of the drug (X = Cl), while it was absent in a study of a benzenesulfonate salt.[22] These results suggest that the nucleophile in pathway b could be the chloride since it is a stronger nucleophile than the benzenesulfonate.

4.1.4 Miscellaneous Eliminations

For keto-compounds containing a β-hetero atom such as oxygen, nitrogen, and sulfur, elimination can occur *via* the mechanism of retro-nucleophilic conjugate addition, sometimes also referred to as retro-Michael addition. This type of drug degradation will be discussed in Section 4.3.

4.2 Decarboxylation

In general, carboxyl-containing compounds, except for those that contain an "activator" at the position β to the carboxyl group, do not undergo decarboxylation easily unless they are treated under fairly harsh conditions. In the transition state of the decarboxylation process, a negative charge develops at the position α to the carboxyl group, while a positive charge develops at the carboxyl hydrogen. In order for the decarboxylation to proceed, the negative charge needs to be neutralized or released by the "activator" at the β position, while the positive charge is captured so that the transition state can be stabilized. This explains why a carboxyl-containing compound that has a β-double bond is quite susceptible to decarboxylation, because the double bond fosters the formation of a stabilized transition state, from which the negative charge can reunite with the carboxyl proton according to the concerted mechanism in Scheme 4.11.

A great number of non-steroidal anti-inflammatory drugs (NSAIDs) contain a carboxyl group. 4-Aminosalicylic acid, one of the NSAIDs used for treating tuberculosis and inflammatory bowel diseases, is known to decarboxylate upon heating to produce CO_2 and 3-aminophenol.[24] The drug can tautomerize into the β-keto form, which is more susceptible to decarboxylation by the concerted mechanism discussed above. Scheme 4.12 shows tautomerization of the drug followed by decarboxylation.

Diflunisal [4-(2,4-difluoro)salicylic acid, Figure 4.2], structurally analogous to 4-aminosalicylic acid, also contains a β-hydroxyl group on the phenyl ring which can tautomerize as well to facilitate decarboxylation.[25]

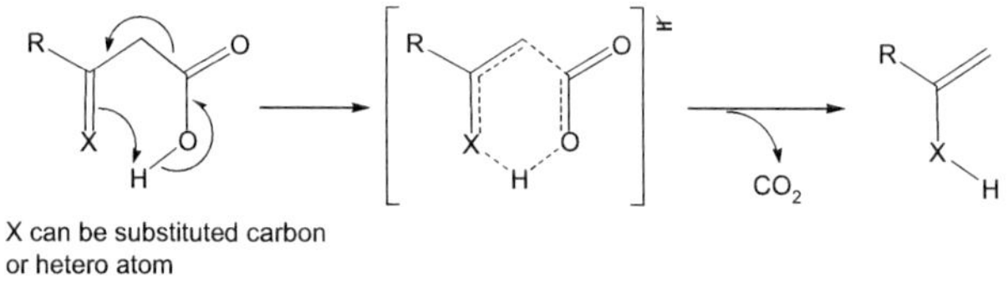

X can be substituted carbon
or hetero atom

Scheme 4.11

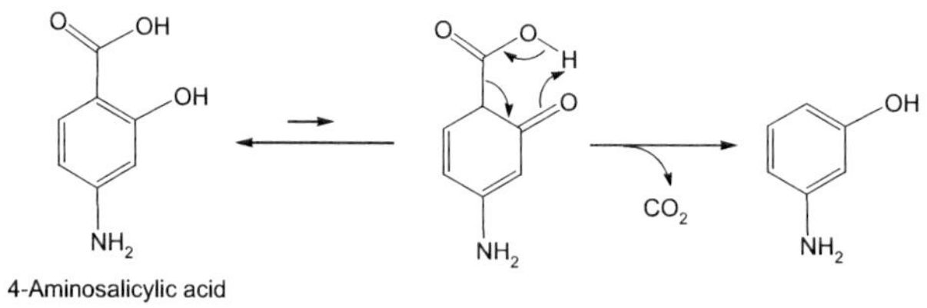

4-Aminosalicylic acid

Scheme 4.12

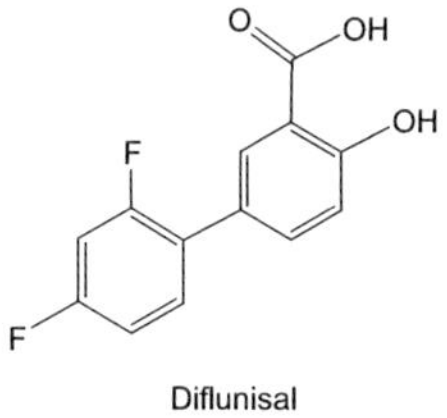

Diflunisal

Figure 4.2 Diflunisal [4-(2,4-difluoro)salicylic acid].

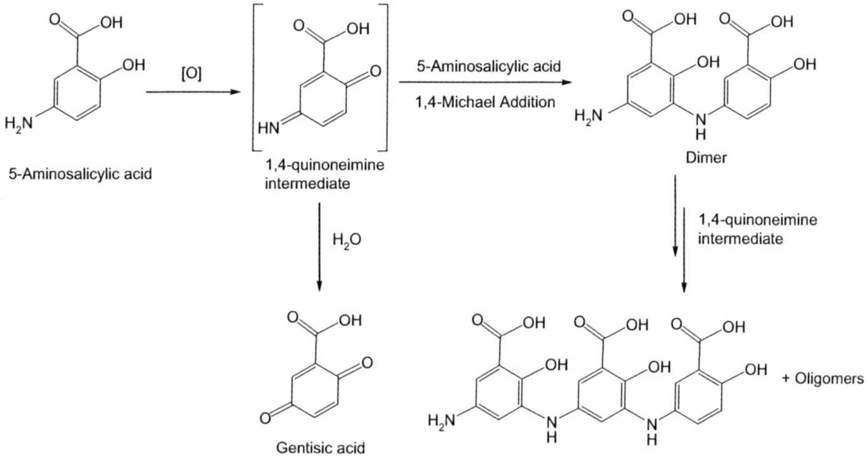

Scheme 4.13

On the other hand, 5-aminosalicylic acid, which is also an NSAID, does not readily undergo decarboxylation, despite being structurally related to 4-amino-salicylic acid. Instead, 5-aminosalicylic acid was shown to degrade primarily *via* oxidative pathways.[25] This is due to the fact that 5-aminosalicylic acid contains a 4-aminophenol moiety which can readily autooxidize to produce a substituted 1,4-quinoneimine intermediate (for a detailed discussion of auto-oxidation, see Chapter 3, Oxidative Degradation). The latter can be further decomposed by at least two pathways: (1) hydrolysis of the imine moiety leads to the formation of gentisic acid, and (2) attack by the active pharmaceutical ingredient (API) *via* a Michael-type addition results in the formation of a dimer which can further react with the 1,4-quinoneimine intermediate to produce oligomers (Scheme 4.13).

As discussed above, carboxyl-containing compounds that have a β-double bond as the "activator" tend to undergo decarboxylation relatively easily. Other structural moieties at the β-position such as a β-epoxide or a β-leaving group can play a similar role as the β-double bond in terms of stabilizing the cyclic transition state similar to the one shown in Scheme 4.11. Hence, initial degradation which imparts these structural moieties into a carboxyl-containing

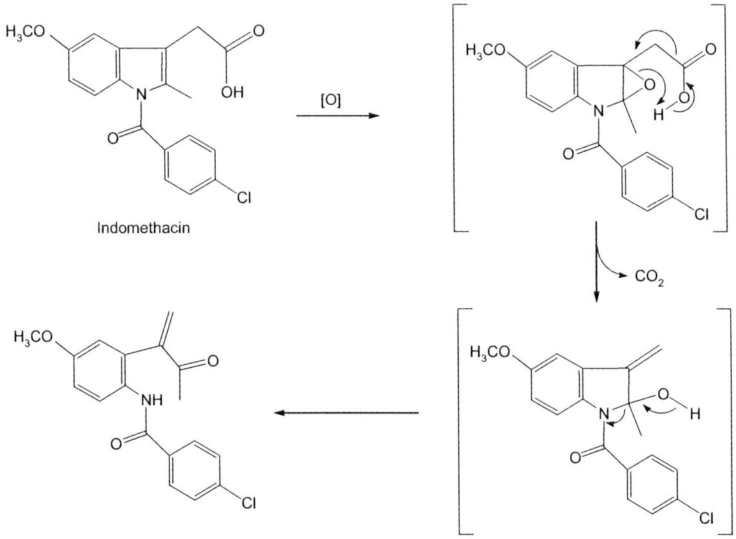

Scheme 4.14

drug can induce decarboxylation as one of the degradation pathways of the drug. For example, oxidative degradation of indomethacin, another NSAID, is believed to go through initial epoxidation on the 2,3-double bond of the indole ring.[26] From the epoxide intermediate, decarboxylation can occur as shown in Scheme 4.14.[27]

   Etodolac is another NSAID that also possesses an indole core structure. At the position β to the carboxyl group lies an ether functionality. According to a stability study performed by Lee *et al.*, etodolac is susceptible to decarboxylation under neutral and acidic stress conditions.[28] Several possible scenarios (including pathways a and b in Scheme 4.15) were proposed by the original researchers. In pathway a, the key intermediate prior to the decarboxylation was proposed to be a carbocation, which is α to the indole ring but β to the carboxyl group, resulting from the protonation of the ether functionality and subsequent leaving as a hydroxyl group.

   Once the carbocation is formed, which should be stabilized by the indole ring, decarboxylation can readily occur to give the decarboxylated product (**2A**). Alternatively, a concerted mechanism (pathway b) was also proposed where decarboxylation occurs simultaneously when the ether functionality leaves (as a hydroxyl group) to yield **2A**. Acidification of **2A** would then lead to the formation of **3** and **2B**, respectively.

   Based on the degradation product distributions under both the acidic and neutral conditions as presented by Lee *et al.*, the present author believes that a third degradation pathway (pathway c) is also very likely. In this mechanism, decarboxylation would go *via* the stabilized six-membered transition state to form a 1,3-dipolar intermediate. Under neutral conditions, pathway c1 would predominate, giving **2A**. According to the experimental results,[28] **2A** is the

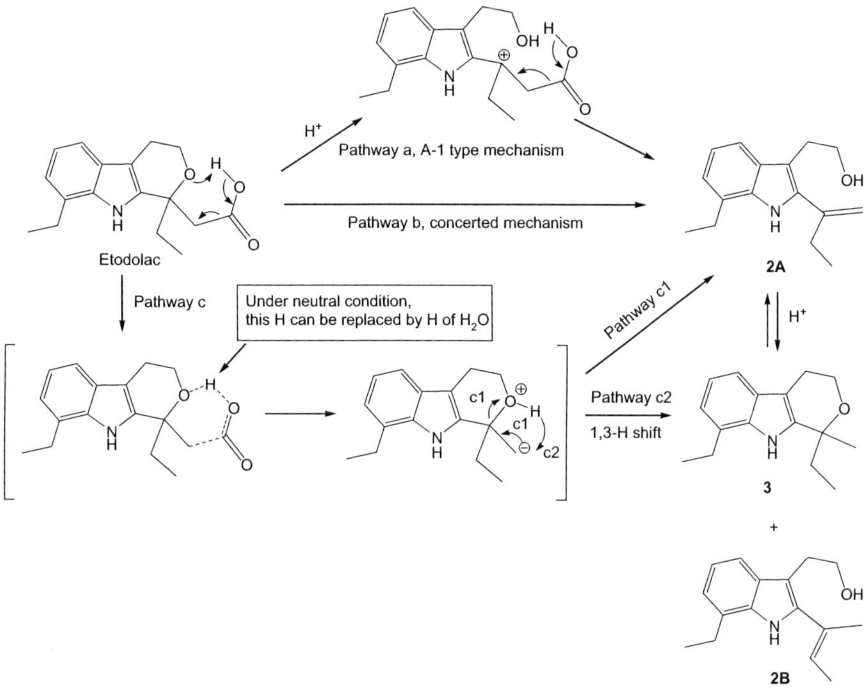

Scheme 4.15

predominant degradant when the overall degradation of etodolac is below 10%. Under the acidic condition, nevertheless, a 1,3-proton shift could be preferred. Alternatively, the carbanion could be quenched very quickly by nearby protons under acidic conditions with the eventual release of the proton on the oxygen, which process is equivalent to a net 1,3-proton shift. Both scenarios fit well with the observation that under acidic stress conditions, **3** is the predominant degradant when the overall degradation of etodolac is less than 50%.

A large number of carboxyl-containing NSAIDs also undergo photo-catalyzed decarboxylation. The mechanisms of photochemical decarboxyla-tion, which are different from those mentioned above, will be discussed in detail in Chapter 6, Photochemical Degradation.

4.3 Nucleophilic Conjugate Addition and Retro-nucleophilic Conjugate Addition

Nucleophilic conjugate addition is a type of synthetically useful reaction between nucleophiles and α,β-unsaturated carbonyl and related compounds. The nucleophile can be an enolate, amine, alcohol, or thiol, while the carbonyl can be replaced by a nitro, cyano, sulfoxide, or sulfonyl group. When the

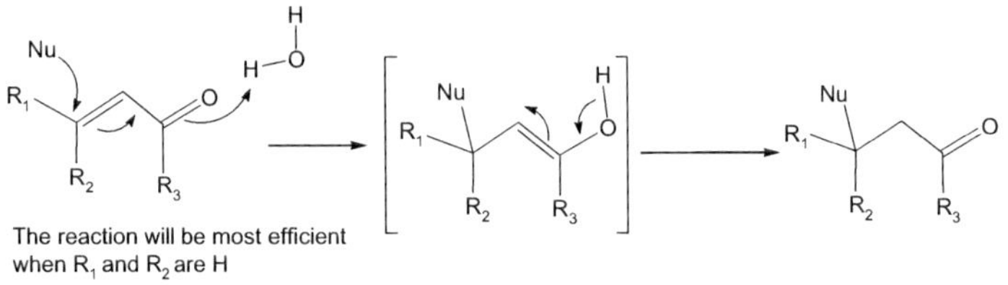

The reaction will be most efficient
when R₁ and R₂ are H

Scheme 4.16

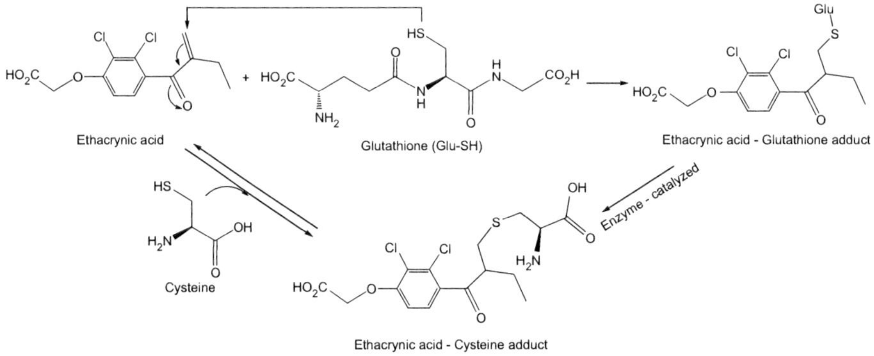

Scheme 4.17

nucleophile is an enolate, that nucleophilic conjugate addition is called Michael addition. For this reason, a nucleophilic conjugate addition is sometimes also referred to as a "Michael-type reaction or addition"; a general reaction scheme is shown in Scheme 4.16.

Since an α,β-unsaturated carbonyl compound is usually electrophilic and can potentially be mutagenic,[29] not many drugs contain such a functionality. Ethacrynic acid, a diuretic for the treatment of hypertension and congestive heart failure, has an α,β-unsaturated carbonyl moiety. It is interesting to note that this functionality is critical to the therapeutic effect of ethacrynic acid, while ticrynafen, another diuretic in the same family, does not contain this functionality. It was hypothesized that ethacrynic acid is a pro-drug and its adduct with cysteine, produced *in vivo* by a process involving initial nucleophilic conjugate addition by glutathione, is the active form.[30] Nevertheless, this hypothesis was questioned by Koechel,[31] based on the fact that not only does the *in vitro* reaction between ethacrynic acid and cysteine (*i.e. via* the expected nucleophilic conjugate addition) occur extremely fast,[32] but also so does the reverse reaction (retro-nucleophilic conjugate addition) under specific *in vitro* conditions.[33,34] These observations show that ethacrynic acid and its cysteine adduct are readily interconvertible. Hence, the possibility that the role of the ethacrynic acid–cysteine adduct is to transport and then release ethacrynic acid at the site of action cannot be ruled out. All of the above pathways are summarized in Scheme 4.17.

The same type of nucleophilic conjugate addition was also observed in a stability study of ethacrynic acid in aqueous solutions, in which both water and ammonium/ammonia can add onto the drug molecule.[35] In the former case, the resulting β-hydroxyl degradant further decomposed to produce two additional degradants *via* a retro-aldol condensation and subsequent Michael addition with another molecule of ethacrynic acid (Scheme 4.18).

Maleic acid, frequently used to form salts with basic drugs, is an α,β-unsaturated dicarboxylic acid. In the presence of a primary or secondary amine drug, nucleophilic conjugate addition can occur between the amine drug and maleate. For example, in several common cold medications containing phenylephrine hydrochloride and chlorpheniramine maleate or dexbrompheniramine maleate, a major degradant that formed between maleate and phenylephrine (a secondary amine) *via* nucleophilic conjugate addition was observed by Marin *et al.*[36] and Wong *et al.*,[37] respectively, and its structure was correctly identified by the latter group (Scheme 4.19).

When the nucleophile in nucleophilic conjugate addition is an amine, the resulting product is a β-aminoketone. The latter compound is also called a Mannich base because it may also be synthesized by the Mannich reaction, a synthetically versatile procedure involving three starting materials: a primary or secondary amine, a non-enolizable aldehyde, and an enolate.

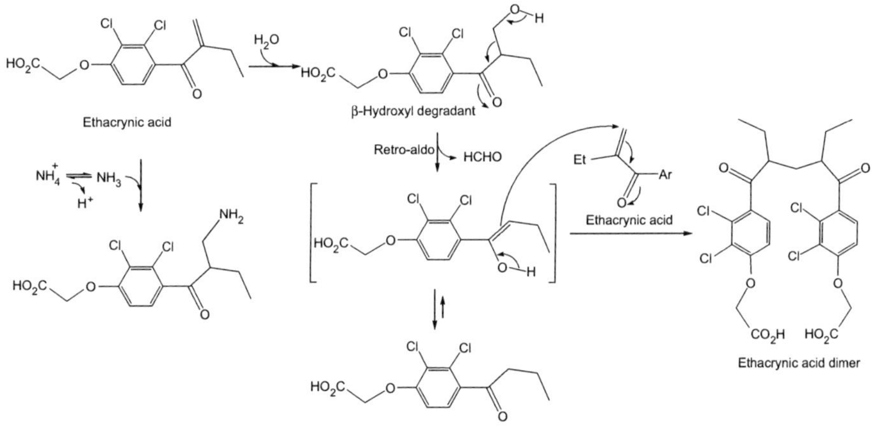

Scheme 4.18

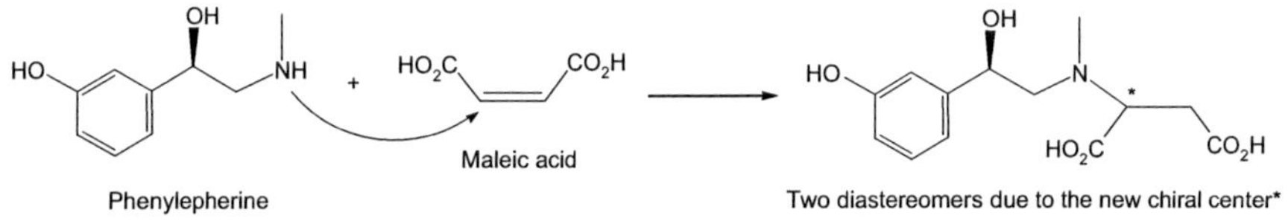

Scheme 4.19

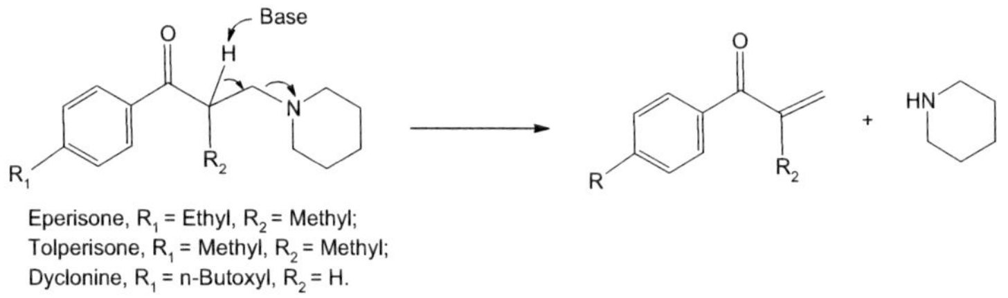

Eperisone, R$_1$ = Ethyl, R$_2$ = Methyl;
Tolperisone, R$_1$ = Methyl, R$_2$ = Methyl;
Dyclonine, R$_1$ = n-Butoxyl, R$_2$ = H.

Scheme 4.20

β-Aminoketones are susceptible to degradation *via* retro-nucleophilic conjugate addition, which is also a type of elimination reaction. For example, the muscle relaxants eperisone and tolperisone are Mannich bases. Under both stress and ambient storage conditions, the drugs were found to degrade *via* the retro-nucleophilic conjugate addition mechanism (Scheme 4.20).[38,39] The degradation occurred in neutral to basic solutions and the rate of the degradation was accelerated as the pH increased, which is consistent with the mechanism shown above. Dyclonine in Scheme 4.20 above is another β-aminoketone drug that degrades *via* the same retro-nucleophilic conjugate addition.[40]

4.4 Aldol Condensation and Retro-aldol

4.4.1 Aldol Condensation

Aldol condensation is an important organic reaction between an enol or enolate and a non-conjugated ketone or aldehyde. Since enol and enolate are generated from a ketone/aldehyde, aldol condensation can take place between the same ketone/aldehyde molecules, as well as between different ketone/aldehyde molecules (Scheme 4.21).

The reaction is catalyzed by both acid and base, and the resulting β-hydroxyketo intermediate (*i.e.* an aldol) usually undergoes dehydration to yield the final, more stable α,β-unsaturated keto product.

Although synthetically a well utilized methodology, aldol condensation does not appear to be a very common pathway in drug degradation, in particular in condensation between molecules of the same carbonyl compound. One example can be found in a veterinary antibiotic drug, tylosin. Tylosin is produced by fermentation and its main component, tylosin A, and most other related macrolides, contain an aldehyde group in the core macrocyclic ring. In a stability study of tylosin by Fish and Carr, a major degradant was observed in a liquid formulation with a pH of 9.[41] The degradation was attributed to the aldol condensation of the drug substance, although no specific structure of the degradant was shown. Nevertheless, it is conceivable that the condensation occurs between the aldehyde group and the enol/enolate generated on the

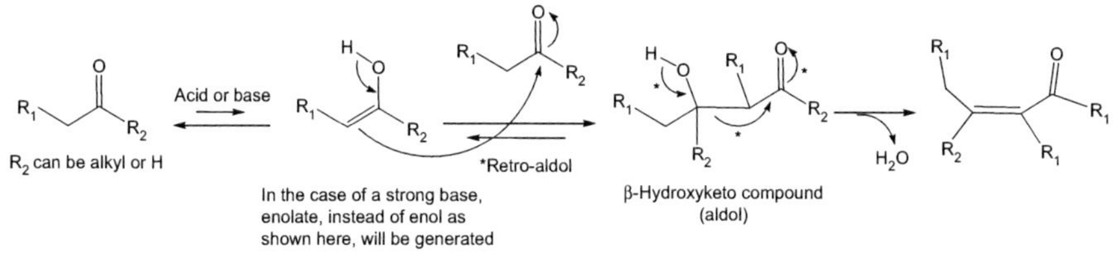

Scheme 4.21

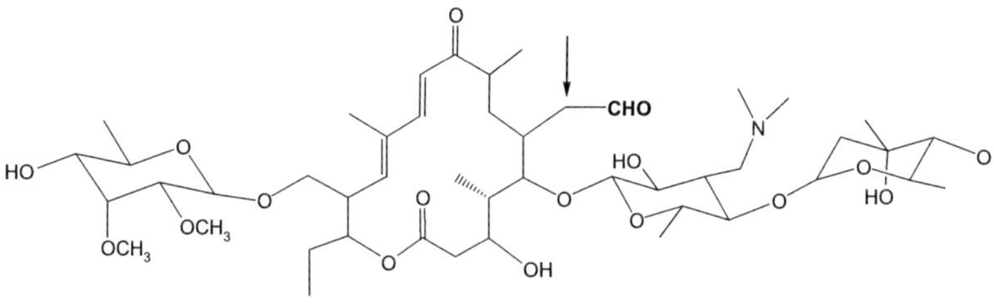

Tylosin A, arrow indicates where the enol/enolate would be generated

Figure 4.3 Tylosin A.

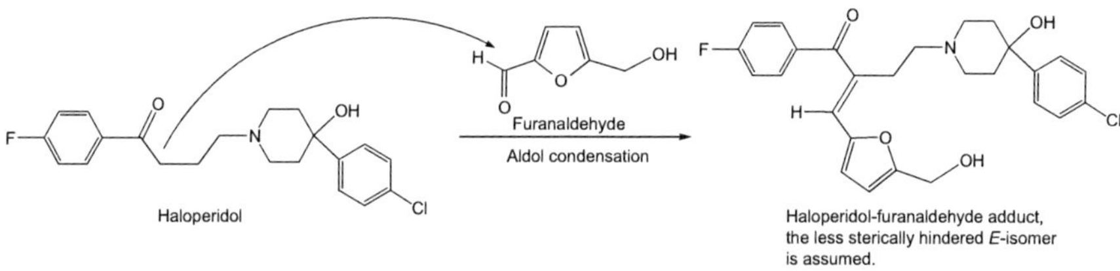

Scheme 4.22

methylene group α to the aldehyde functionality of another tylosin molecule (Figure 4.3).

Haloperidol is an antipsychotic drug in the butyrophenone family containing a phenyl-conjugated ketone functionality. During a comparability study with lactose, it was found to form an aldol condensation product with a furanaldehyde, 5-(hydroxymethyl)-2-furaldehyde (5-HMF),[42] an impurity found in a number of sugars including lactose.[43,44] The degradation reaction is shown in Scheme 4.22.

Ziprasidone is another antipsychotic drug that has a lactam rather than a ketone functionality. During a formulation study with various cyclodextrins and their derivatives, Hong *et al.* found that aldol condensation was involved in the degradation of the formulated ziprasidone both in solution and lyophilized

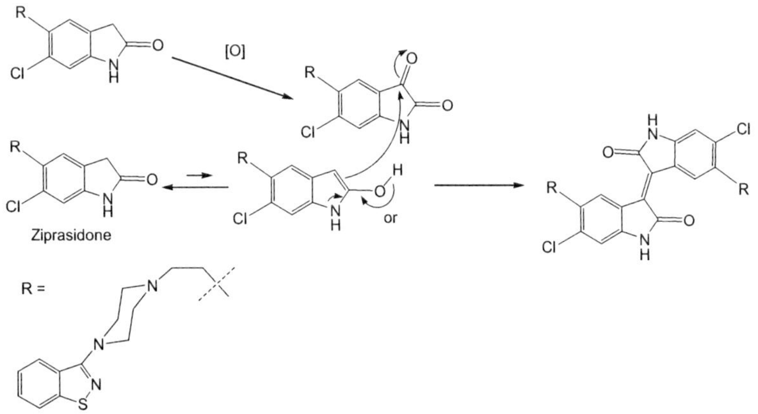

Scheme 4.23

amorphous solid.[45] What happened was that the methylene moiety next to the lactam carbonyl was first oxidized to form a new carbonyl group. This carbonyl is more reactive than a regular ketone owing to activation by the neighboring lactam carbonyl group. Hence, this oxidative degradant was capable of aldol condensation with ziprasidone in the formulation. The final degradant was found to be in the *E*-configuration as illustrated in Scheme 4.23.

4.4.2 Retro-aldol Reaction

As discussed above, the β-hydroxyketone intermediate in aldol condensation usually undergoes dehydration, if an α-hydrogen is present. On the other hand, it can also revert to the two starting molecules, that is, undergo a retro-aldol reaction. Hence, drug molecules or their intermediary degradants containing a β-hydroxyketone moiety are susceptible to degradation *via* the retro-aldol pathway, which has been observed in a few cases. For example, basic stress of corticosteroids containing a 1,3-dihydroxylacetone side chain on the D-ring, such as prednisolone,[46] hydrocortisone,[47] and betamethasone[3] under anaerobic conditions leads to the formation of the corresponding corticosteroid 17-ketones. This degradation process has been attributed to retro-aldol reaction after the drug substances tautomerize to produce the β-hydroxyketone moiety as depicted in Scheme 4.24.[48,49] In the presence of oxygen, the formation of the corticosteroid 17-ketone could also stem from the retro-aldolization of those oxidative degradants that contain a β-hydroxyketone moiety.[50]

Ginger has not only been widely used as a spice, in particular in Asian cuisines, but also as a folk medicine since ancient times. 6-Gingerol, the major pungent principle isolated from ginger oleoresin, possesses a β-hydroxyketo moiety and hence is susceptible to retro-aldol reaction to yield zingerone (Scheme 4.25).[51]

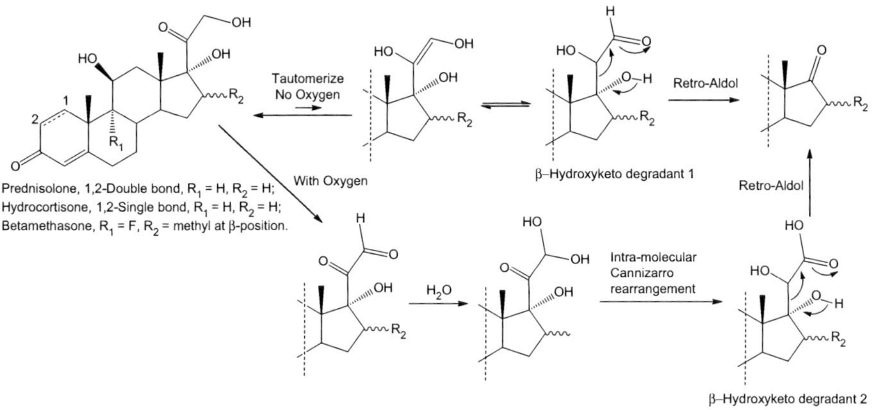

Scheme 4.24

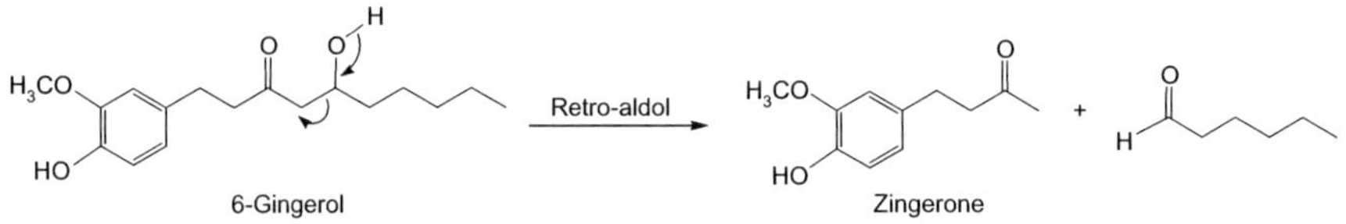

Scheme 4.25

4.5 Isomerization and Rearrangement

Isomerization is the process in which a molecule is transformed into another chemical entity with the same chemical formula. Rearrangement refers to any organic reaction in which the carbon framework of a reactant is rearranged to produce a product that is isomeric with the original molecule. According to this definition, rearrangement can be considered a subset of isomerization. Nevertheless, a broader definition of rearrangement is sometimes used, whereby the rearranged product is not isomeric but largely similar in its formula to the original molecule. A vast number of chemical transformations fall into the definitions of isomerization and rearrangement, which will be discussed in the following sections.

4.5.1 Tautomerization

Tautomerization is a key process involved in many chemical transformations and/or mechanisms, including various isomeric degradation pathways. The examples include the formation of enol/enolate in an aldol condensation and imine–enamine tautomerization. Frequently, tautomers, in particular intermediary tautomers, may not be isolatable and/or observable chromatographically. Nevertheless, in certain cases, the energy barrier between the tautomers is large enough to enable their isolation. For example, it was found

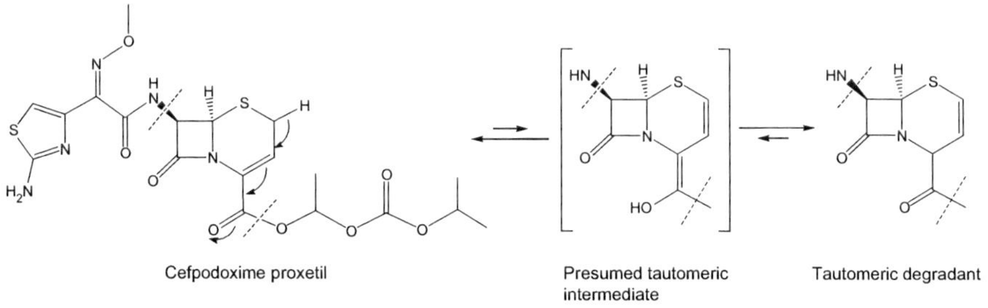

Cefpodoxime proxetil   Presumed tautomeric   Tautomeric degradant
                                       intermediate

Scheme 4.26

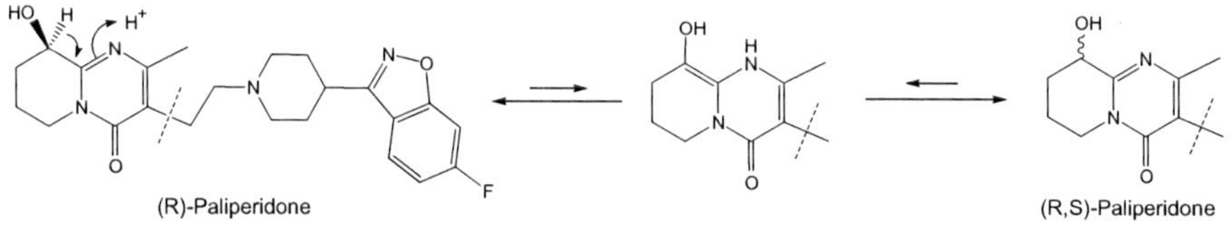

(R)-Paliperidone   (R,S)-Paliperidone

Scheme 4.27

that the main degradation product of cefpodoxime proxetil in both the solid and solution states is a tautomeric degradant resulting from a double bond shift inside the six-membered ring of the cephalosporin core structure (Scheme 4.26).[52]

When in solution, it was found that the rate of isomerization increased as the pH decreased. The isomerization is likely to proceed through the enol intermediate as illustrated above.

4.5.2 Racemization

A racemic mixture is a collection of two enantiomers and racemization is a process of chemical transformation in which one enantiomer converts to another. The only meaningful racemization, in terms of drug degradation, is usually one related to compounds that contain a single chiral center. For compounds containing multiple chiral centers, conversion of one chiral center results in the formation of a diastereomer, a process called epimerization which will be discussed in the next section.

Paliperidone, or 9-hydroxyrisperidone, is an active oxidative metabolite of risperidone. The hydroxylation creates a chiral center at the 9-position of paliperidone. Despite the two enantiomers of paliperidone having comparable pharmacological activity, they display different affinities in plasma protein binding.[53] Danel *et al.* studied the configuration stability of paliperidone and found that the drug substance undergoes racemization under both acidic and basic conditions, but a faster rate of racemization was observed under acidic conditions.[54] The mechanism of racemization was found to be mediated by an imine–enamine tautomerization process (Scheme 4.27), according to evidence

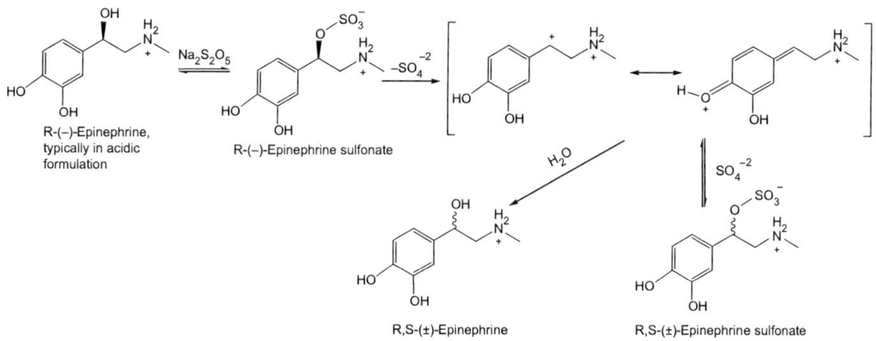

Scheme 4.28

from a H/D exchange study *via* NMR analysis. Under the physiological condition, no racemization of paliperidone was observed.

Epinephrine, also called adrenaline, is a therapeutic agent commonly used in emergency medicine to treat cardiac arrest and other cardiac dysrhythmias. The drug is typically administrated by intravenous injection, and since it is chemically a catecholamine, it is usually formulated with excipients that have antioxidation capability, such as sodium metabisulfite, in order to suppress its rather facile autooxidation. In a liquid formulation, epinephrine was found to undergo sulfonation and subsequent racemization *via* carbocation through an S_N1 mechanism (Scheme 4.28).[55]

The S_N1 mechanism is probably made possible by a combination of the following two factors: first, sulfonate is an excellent leaving group and second, the carbocation formed can be stabilized through resonance with the adjacent catechol moiety. The D- or S-epinephrine resulting from the racemization is essentially inactive.

4.5.3 Epimerization

Epimerization relates to compounds with multiple chiral centers. Epimers are those diastereomers that differ in configuration only at a single chiral center. The anticancer drug etoposide is a semi-synthetic epipodophyllotoxin derivative that contains multiple chiral centers. The chiral center α to its lactone carbonyl group is susceptible to racemization, which was found to be mediated by enolization at pH above 4.[56,57] The mechanism of the enolization-mediated epimerization is shown in Scheme 4.29.

4.5.4 *Cis-trans* Isomerization

Oxime including oxime ether is a common structural moiety utilized in drug design and optimization. Drug molecules possessing this moiety are susceptible to *cis–trans* (also called *syn–anti*) isomerization around the oxime double bond. The isomerization is reversible and the more stable *trans*-isomer (*E*-isomer)

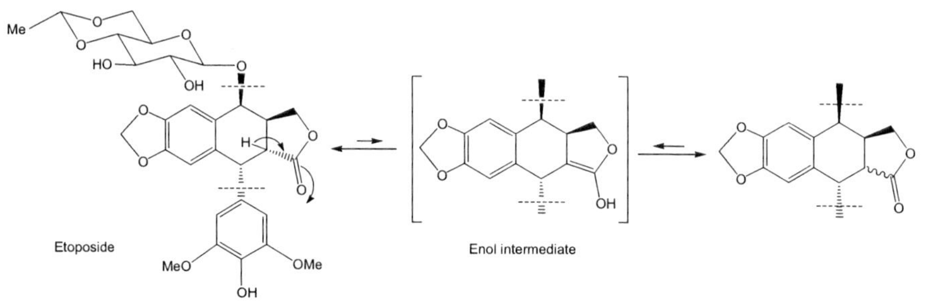

Scheme 4.29

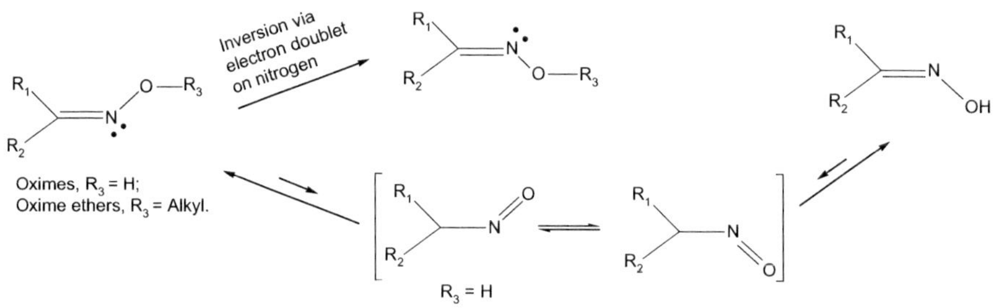

Scheme 4.30

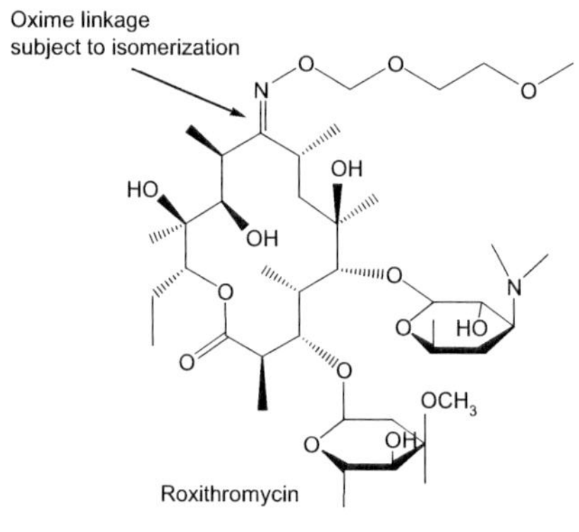

Figure 4.4   Roxithromycin.

usually predominates. The isomerization process can proceed *via* two competing pathways as illustrated in Scheme 4.30. The first is electron doublet inversion and the second catalyzed rotation involving oxime/nitroso tautomerization in both acidic and basic media.[58]

Roxithromycin (Figure 4.4) is an oxime ether derivative of erythromycin whose oxime moiety undergoes *cis–trans* isomerization at pH below 5 to give the less potent Z-isomer.[59] The ratio of roxithromycin (the *E*-isomer) *versus* the

Z-isomer appeared to be constant at a particular pH value. Between pH 1 and 3, the isomeric degradation followed pseudo first-order kinetics and the rate of isomerization and the *Z/E* ratio increased as the pH decreased.

A fairly large number of antibiotics in the cephalosporin family, including cefdinir, cefixime, cefpodoxime, ceftizoxime, and cefmenoxime, contain an oxime/oxime ether moiety and hence should be susceptible to the same degradation *via* the *cis–trans* isomerization mechanism discussed here.

Compounds containing polyconjugated double bonds undergo *cis–trans* isomerization both photochemically and non-photochemically. As photochemical isomerization will be covered in Chapter 6, Photochemical Degradation, only non-photochemical isomerization is discussed in this section.

Ro-26-9228 is a vitamin D derivative, structurally similar to calcitriol, the hormonally active form of vitamin D. Like calcitriol,[60] Ro-26-9228 also undergoes conformational as well as chemical *cis–trans* isomerization.[61] The chemical *cis–trans* isomerization was shown to proceed through a {1,7}-hydrogen shift in the minor *Z*-conformer as depicted in Scheme 4.31.

Drugs possessing carbon–carbon double bonds conjugated to a carbon–hetero atom double bond are also susceptible to geometric isomerization. For example, ceftibuten, another antibiotic in the cephalosporin family, is structurally unique in its C-7 side chain as compared to the vast majority of cephalosporin antibiotics in that the typical oxime/oxime ether moiety of the latter is replaced by a carbon–carbon double bond. The latter double bond is conjugated to both the 7-amide carbonyl group and aminothiazole ring. It was found that ceftibuten undergoes *cis–trans* isomerization around the conjugated carbon–carbon double bond in both acidic and basic solutions.[62] There was no detailed discussion regarding the mechanism of the isomerization in the original study. However, based on the structure of ceftibuten, it appears that tautomerization, which can be trigged by deprotonation of the methylene group α to the conjugated carbon–carbon double bond, could be responsible for the isomeric degradation as shown in Scheme 4.32.

The proposed mechanism seems to be consistent with the experimental results presented by the original researchers. For example, several ceftibuten analogs, in which the aminothiazole ring was replaced by rings with fewer hetero atoms (*e.g.* phenyl ring), were also prepared and their isomerization behavior was compared with that of ceftibuten. These analogs all displayed

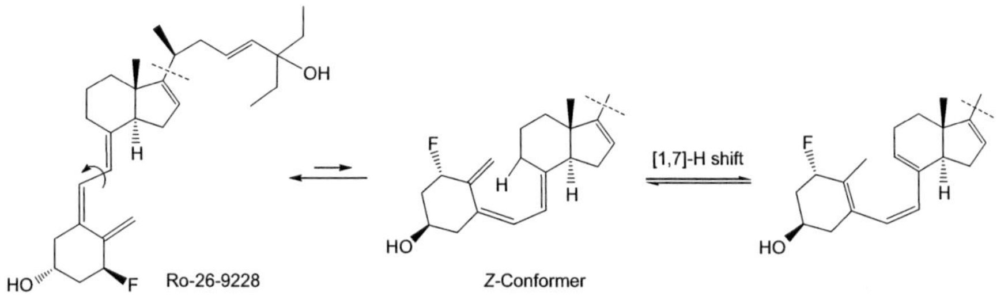

Scheme 4.31

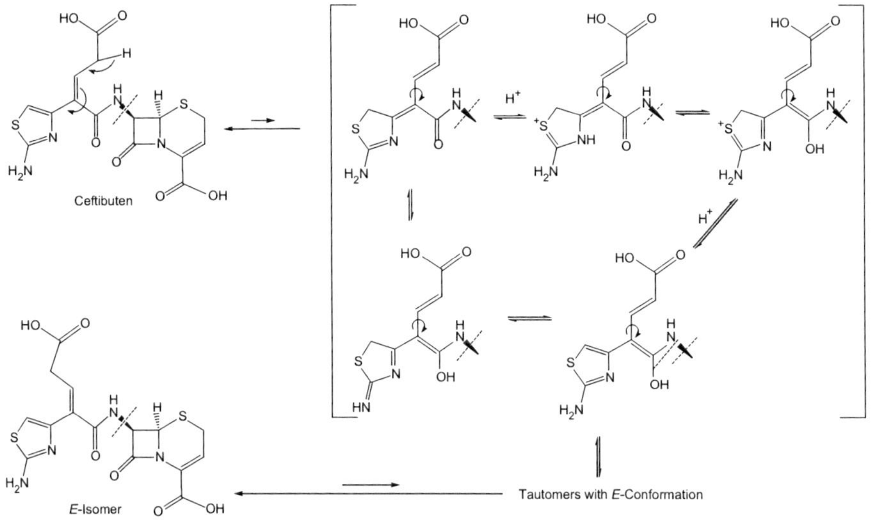

Scheme 4.32

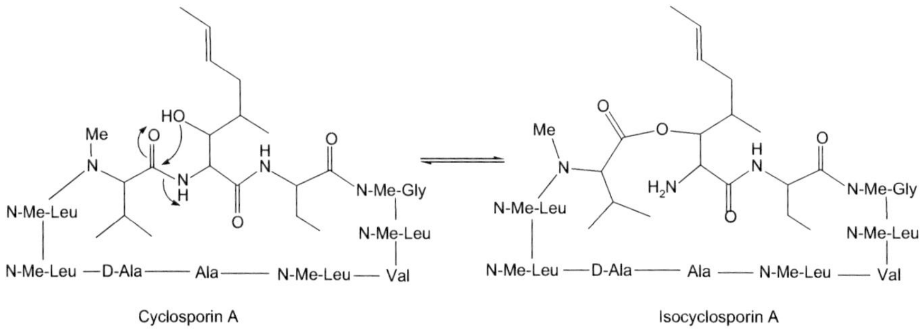

Scheme 4.33

much reduced rates of isomerization under acidic conditions, probably due to their inability to form as many tautomers as ceftibuten, which makes it more difficult for the tautomerization process to take place.

4.5.5 *N,O*-Acyl Migration

Cyclosporin A is a widely used immunosuppressive drug for patients after organ and tissue transplantation. It has a cyclic undecapeptide structure and the side chain hydroxyl group of one unusual amino acid residue can attack the carbonyl group of the neighboring valine under acidic conditions. This *N,O*-acyl migration results in the formation of isocyclosporin A (Scheme 4.33).[63,64]

The impact of this isomerization arising from possible drug degradation in the stomach was assessed.[64] Based on its half-lives at 37 °C between pH 1 to 3,

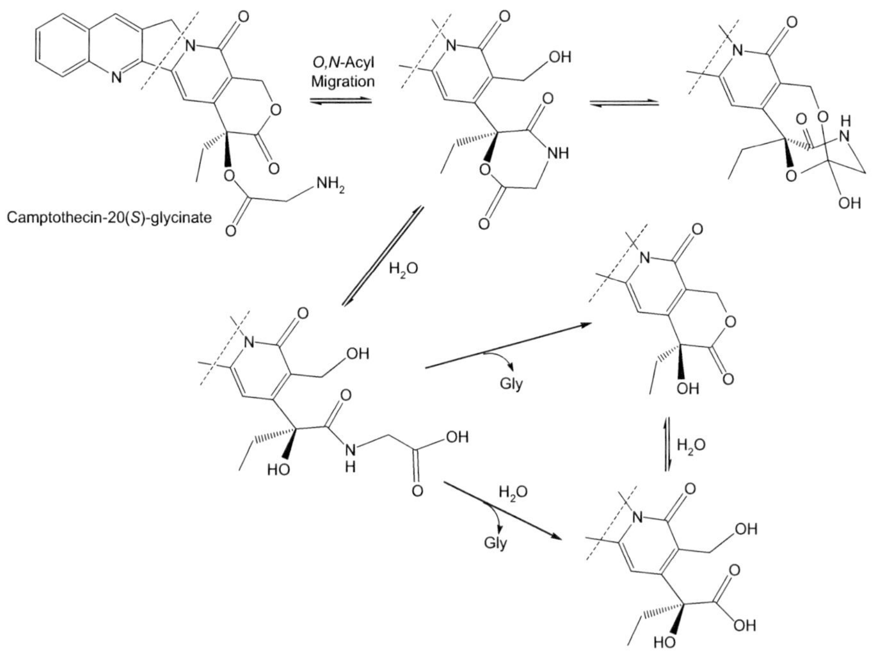

Scheme 4.34

it was estimated that only 1–2% of the ingested drug would undergo decomposition while passing through the stomach.

Another example of *N,O*-acyl migration in drug degradation was found during the development of a water soluble derivative of camptothecin, camptothecin-20(*S*)-glycinate.[65] In this case, the direction of acyl migration is from *O* to *N*, which is the opposite of the previous example. At pH 7.4, 37 °C, the glycinate was found to decompose quickly with a half-life of ~30 minutes. On the other hand, the analogous 20(*S*)-acetate, which lacks the α-amino group, showed no sign of decomposition during a period of 3 hours under identical conditions. The instability of the 20(*S*)-glycinate was attributed to a sequential degradation that is triggered by an *N,O*-acyl migration step (Scheme 4.34).

4.5.6 Rearrangement *via* Ring Expansion

It has long been known that corticosteroids containing the 17-hydroxy-20-keto moiety are susceptible to D-ring expansion, also known as D-homoannulation, under catalysis by metal ions (or Lewis acids), bases, and other factors.[66] Over the many decades that followed this study, various studies have been performed to understand the mechanism of this rearrangement reaction.[67–69] Dekker and Beijnen studied the degradation behavior of prednisolone, dexamethasone, and betamethasone at a moderately alkaline pH of 8.3 and under elevated temperature. It was found that the D-homoannulation rearrangement products

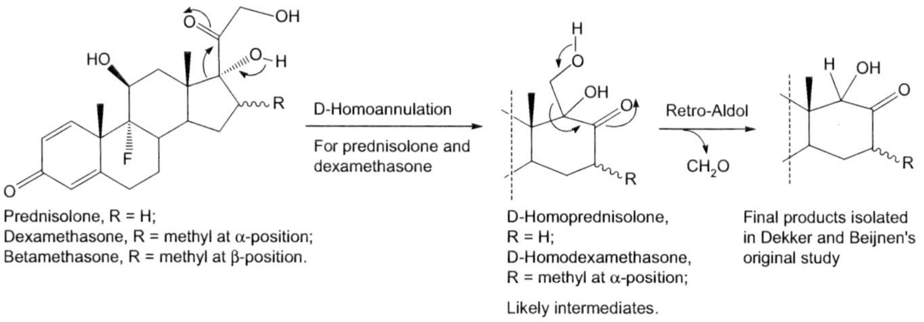

Prednisolone, R = H;
Dexamethasone, R = methyl at α-position;
Betamethasone, R = methyl at β-position.

D-Homoannulation

For prednisolone and
dexamethasone

D-Homoprednisolone,
R = H;
D-Homodexamethasone,
R = methyl at α-position;

Likely intermediates.

Retro-Aldol

CH₂O

Final products isolated
in Dekker and Beijnen's
original study

Scheme 4.35

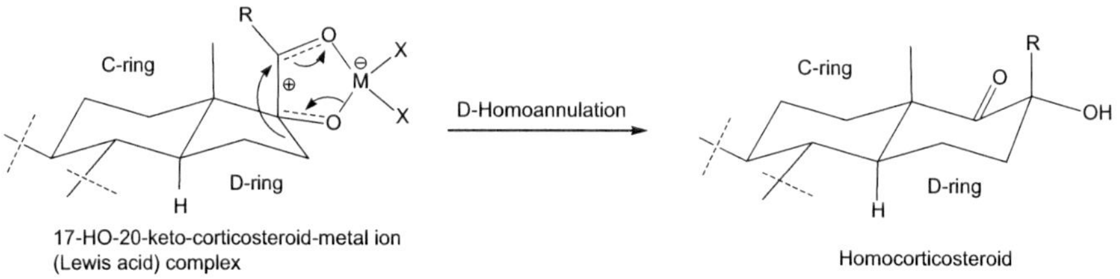

C-ring

D-ring

H

17-HO-20-keto-corticosteroid-metal ion
(Lewis acid) complex

D-Homoannulation

C-ring

D-ring

H

Homocorticosteroid

Scheme 4.36

were among the top three degradants of prednisolone and dexamethasone, respectively (Scheme 4.35).[70]

During these studies, the original researchers only identified the two final degradation products shown in Scheme 4.35. Based on current knowledge, the D-homoannulation of prednisolone and dexamethasone should first give rise to their respective ring expansion products (*i.e.* the two intermediates in Scheme 4.35). Since the original studies were performed under a relatively high temperature of 100 °C, a retro-aldol process for the initial D-homoannulation products would be very likely to occur, which should yield the final isolated products.

On the other hand, no degradation product of betamethasone resulting from the D-homoannulation was observed under the same pH and stress temperature.[71] These observations are interesting in that small differences at the 16-position of the three steroids can cause a quite dramatic difference in the distribution of their degradation products.

When catalyzed by metal ions (Lewis acids), it was proposed that the D-homoannulation proceeds through a transition complex in which the metal ion is chelated by the 17-hydroxyl and 20-keto groups, as shown in Scheme 4.36.[72]

Because of the chelation, the 17-hydroxyl and 20-keto groups are locked in a *syn*-configuration, which enables a stereo-specific attack on the 17-keto group by the migrating alkyl group.[69] As a result, only a single diastereomeric degradant is formed during the metal ion-catalyzed D-homoannulation process.

Triamcinolone is a corticosteroid that is structurally identical to prednisolone except for an additional hydroxyl group at the 16α-position.

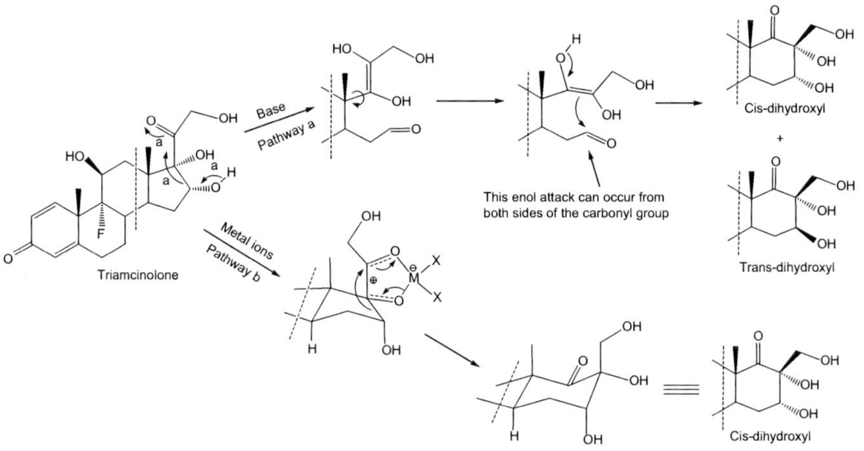

Scheme 4.37

Owing to the presence of the 16α-hydroxyl group, which is β to the 20-carbonyl group, triamcinolone appears to be more susceptible to D-ring expansion, consistent with previous observations that the 16,17-dihydroxy-20-keto moiety is more prone to rearrangement than the 17-hydroxy-20-keto moiety.[68] Furthermore, under base catalysis, the D-ring expansion of triamcinolone appears to proceed *via* a different mechanism. According to Delaney *et al.*,[73] the D-ring rearrangement begins with a retro-aldol process, producing an aldehyde at the 16-position and a 17,20-dihydroxy-enol moiety. The 17,20-enol can then attack the 16-aldehyde from both sides, yielding two epimeric degradants, that is, *cis*-dihydroxy- and *trans*-dihydroxyhomotriamcinolone, as shown in pathway a, Scheme 4.37. When the D-ring expansion of triamcinolone is catalyzed by metal ions, the rearrangement mechanism *via* the transition metal ion complex (with 17-hydroxy-20-keto moiety) as illustrated in Scheme 4.36 can explain the stereo-specific formation of the *cis*-dihydroxyhomotriamcinolone through pathway b, Scheme 4.37.

The 21-hydroxyl group of the corticosteroids containing the 17,21-dihydroxyacetone side chain is frequently phosphorylated to make water-soluble corticosteroid phosphates for injectable formulations. These pro-drugs, for example, prednisolone phosphate,[74] and betamethasone phosphate,[75] also undergo D-homoannulation and it appears that the phosphate group may play a role in facilitating the rearrangement (Scheme 4.38).[75] In the case of betamethasone phosphate, three D-homoannulation degradants, that is, BSP isomers 1 to 3, were observed in significant quantities.

Vancomycin is a glycopeptide antibiotic that is typically used for the treatment of severe infection caused by Gram-positive bacteria[76] and often is the choice of last resort. Despite of its long history of clinical application since the 1950s, its exact structure and degradation mechanism was not elucidated until the 1980s.[77] The main degradation of vancomycin is caused by the initial cyclization of its asparagine residue (a deamidation process), followed by hydrolysis

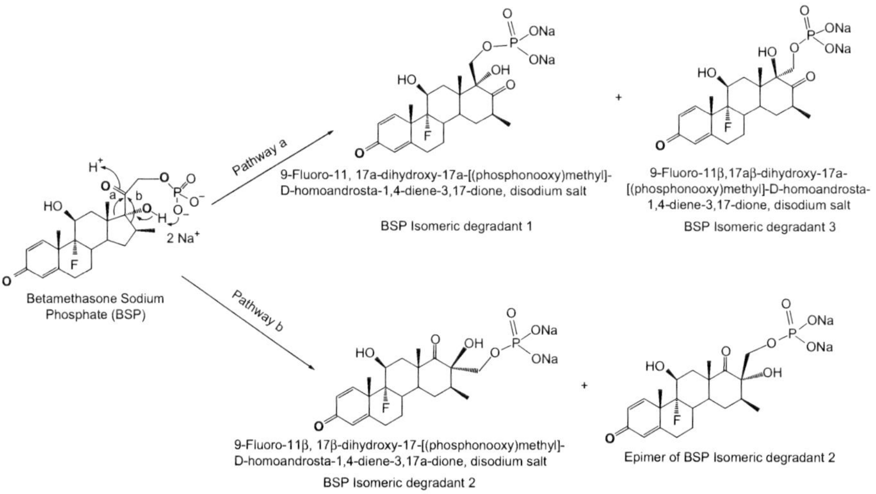

Scheme 4.38

of the succinimide intermediate formed, to yield first the minor degradant called CDI-m. Furthermore, as a consequence of the expansion of the macrocyclic ring by a CH_2 unit during the first step of the degradation, the previously restricted chlorophenol moiety in the enlarged macrocyclic ring is now capable of rotation, which produces the major degradant, CDI-M (Scheme 4.39).

This transformation is still referred to as "rearrangement", despite the change in the overall formula of vancomycin (albeit slightly: an amino is replaced by a hydroxyl). The key step in the degradation of vancomycin illustrated above is the "rearrangement" of the asparagine residue. In protein and peptide drugs containing asparagine and/or aspartic acid residues, degradation caused by this rearrangement (which is typically referred to as deamidation) is one of the main degradation pathways of these drugs. The latter will be discussed in details in Chapter 7, Chemical Degradation of Biological Drugs.

4.5.7   Intramolecular Cannizzaro Rearrangement

For corticosteroids containing a 1,3-dihydroxyacetone side chain on the D-ring, such as the ones we have just discussed in this chapter, degradation occurring due to this moiety is responsible for the vast majority of the degradation products of this important class of drugs. As we have demonstrated, various mechanisms, for example, dehydration, oxidation, and retro-aldol, are involved in different pathways of this degradation. One of these degradation mechanisms is the Cannizzaro rearrangement, which is responsible for further degradation of certain degradants possessing an α-keto-aldehyde functionality on the steroid D-ring.[78] For example, betamethasone enol aldehyde is a dehydration degradant of betamethasone which has two regioisomers (refer to Section 4.1.1). Both the *E*- and *Z*-enol aldehyde can be rehydrated to form two

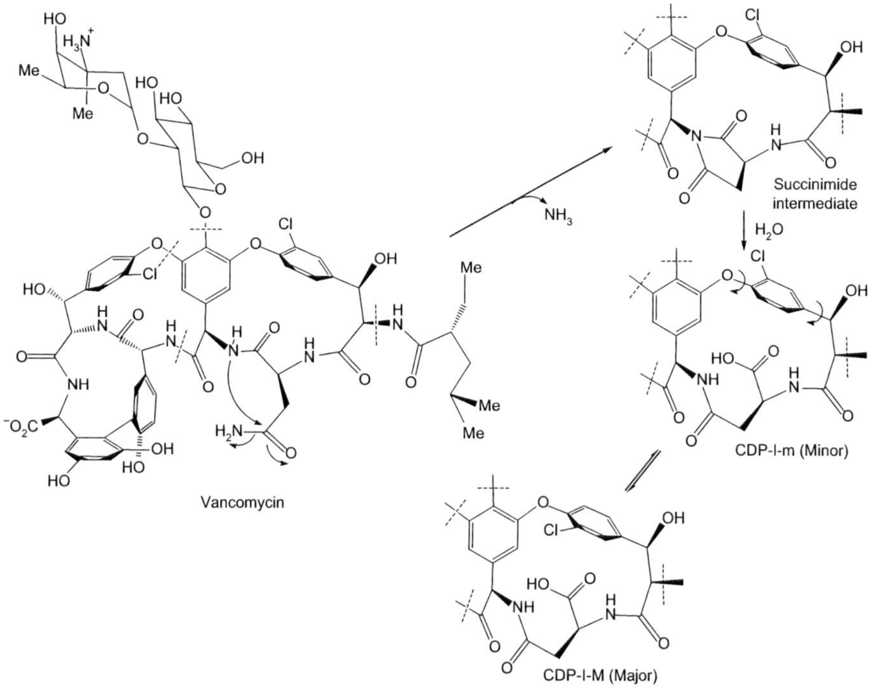

Scheme 4.39

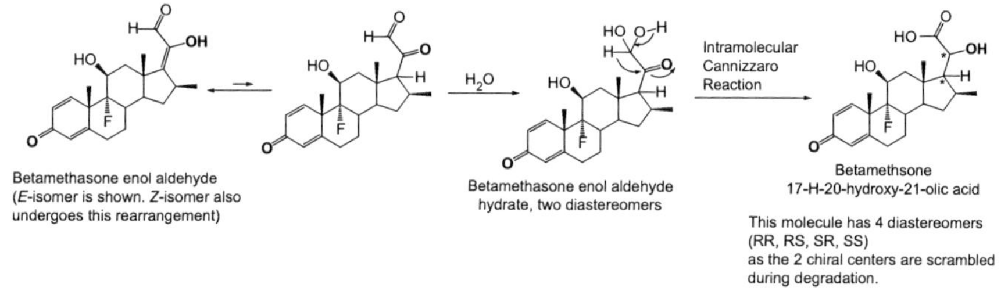

Scheme 4.40

enol aldehyde hydrates.[75] The latter can undergo Cannizzaro rearrangement to produce four additional isomeric degradants of betamethasone.[71,75] This degradation pathway is summarized in Scheme 4.40.

4.6 Cyclization

4.6.1 Formation of Diketopiperazine (DKP)

A fairly large number of small molecule drugs are dipeptides or dipeptide analogs. If the *N*-terminal of these dipeptides is not protected, the amino group

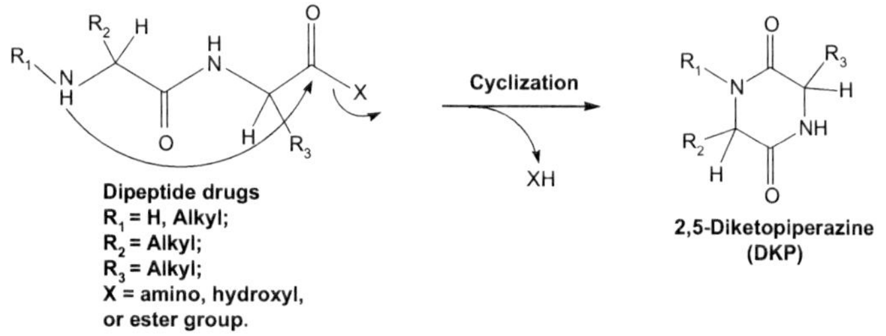

Scheme 4.41

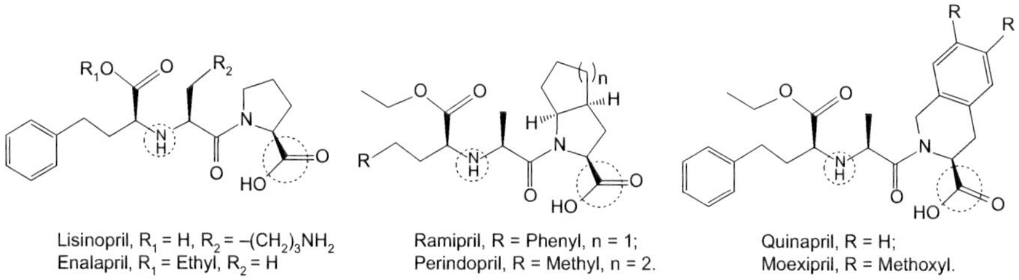

Figure 4.5   Structures of exemplary ACE inhibitors that undergo DKP cyclization. The groups involved in cyclization are highlighted in dotted circles.

can react with the carbonyl group of the *C*-terminal, resulting in the formation of a diketopiperazine (DKP) ring (Scheme 4.41). Dipeptide drugs containing a *C*-terminal proline residue are particularly susceptible to this cyclization,[79] which may be due to the fact that the proline residue predisposes the conformation of the dipeptides in a way favorable to the cyclization.

Quite a few angiotensin-converting-enzyme (ACE) inhibitors such as lisinopril, enalapril, ramipril, perindopril, quinapril,[80] and moexipril,[81] are dipeptide analogs containing an unprotected *N*-terminal secondary amino group and a *C*-terminal proline residue or its analog (Figure 4.5). Hence, DKP cyclization is a common and significant degradation pathway for these drugs. DKP cyclization can be catalyzed by both acid and base.[82,83]

4.6.2   Other Cyclization Reactions

Denagliptin, a dipeptidyl peptidase-4 (DPP-4) inhibitor that was being developed for type 2 diabetes, contains a primary amino and a cyano group. During pharmaceutical development regarding the stability of the drug substance and its experimental formulations, it was found that the amino group attacks the cyano group of the same molecule to produce a cyclized amidine (Scheme 4.42).[84]

The initial (3*S*,7*S*,8a*S*)-amidine degradant formed undergoes epimerization at the 8a-position, probably *via* an imine–enamine tautomerization, producing

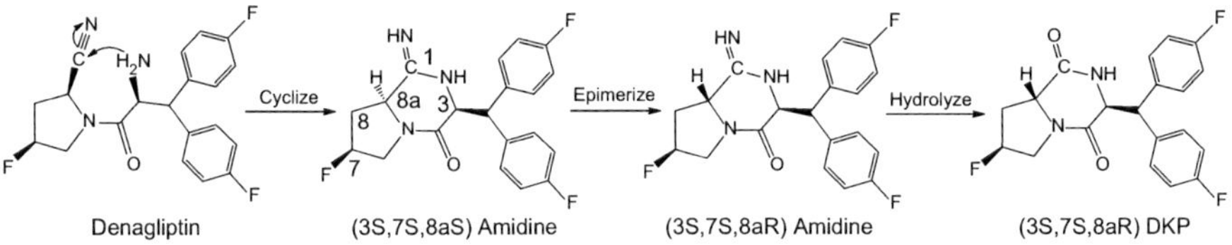

Scheme 4.42

the second amidine degradant (the $3S,7S,8aR$-isomer). The latter can be hydrolyzed to yield the DKP degradant.

4.7 Dimerization/Oligomerization

In this section, degradation *via* dimerization is broadly defined as the chemical association of two molecules of a drug substance, which can include several types such as $M + M = 2M$, $M + M = 2M - m$, and $M + X + M = M\text{-}X\text{-}M$, whereby M is the molecular formula of the drug substance, m is a fragment eliminated during the dimerization, and X is a linker that typically originates from an excipient or its impurities.

Drug degradation *via* dimerization or oligomerization can proceed by many different mechanisms. Such degradation is more likely in liquid-formulated drug products with high dose concentrations. A number of cases discussed in previous chapters and/or sections involve dimeric or oligomeric degradation: for example, the morphine dimer (pseudomorphine) formed by oxidative coupling of the drug substance (Section 3.59), the ziprasidone dimer produced by oxidation and subsequent aldol condensation (Section 4.4.1), and the captopril dimer formed by oxidative coupling of the two thiol groups (Section 3.5.6). Likewise, peptide/protein drugs containing cysteine residues can dimerize *via* the same oxidative coupling pathway like that of captopril to form dimers and oligomers. This topic will be discussed in Chapter 7.

Losartan potassium is a potent, first-in-class angiotensin II receptor antagonist used clinically for the treatment of hypertension. It contains a tetrazole ring, which is an isostere of the carboxylic group. The negative charge on the tetrazole ring is nucleophilic and can localize on each one of the five atoms of the ring. As a result, the tetrazole ring can attack the hydroxymethyl group attached to the imidazole ring of another losartan molecule *via* nucleophilic substitution, resulting in the formation of two main, dimeric degradants. It is apparent that the hydroxyl group should leave as water which is facilitated under acidic conditions (Scheme 4.43).[85,86]

In all β-lactam antibiotics such as those in the penicillin, cephalosporin, and carbapenem families, the lactam bond is in a constrained four-membered ring and thus prone to nucleophilic attack. If the attacking nucleophile is water, hydrolytic degradants are formed, discussed in Chapter 2, Hydrolytic Degradation. If the attacking nucleophile is an amino group from another molecule of the antibiotic, which is referred to as intermolecular aminolysis, dimeric and

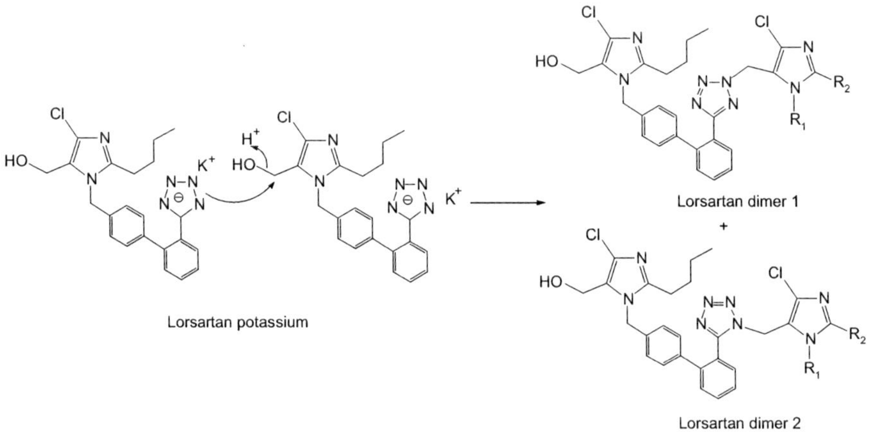

Scheme 4.43

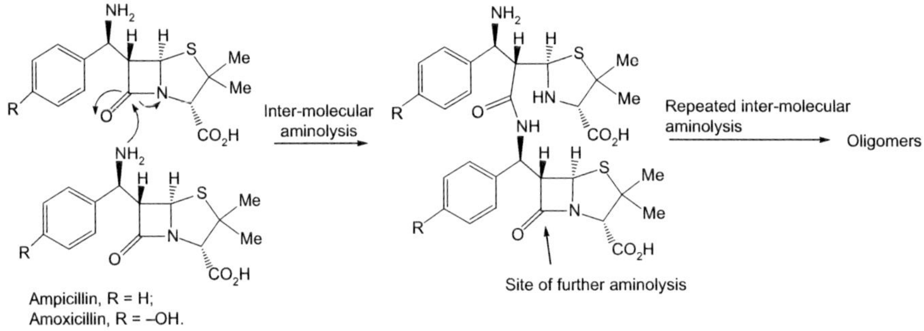

Ampicillin, R = H;
Amoxicillin, R = –OH.

Scheme 4.44

oligomeric degradants can be produced, in particular in formulations with high concentrations of antibiotics. For example, ampicillin and amoxicillin are two antibiotics in the penicillin family, both of which contain a primary amino group. This amino group can attack the four-membered lactam ring of another molecule, resulting in the formation of dimers and oligomers (Scheme 4.44).[87–89]

Ertapenem, a synthetic broad-spectrum β-lactam antibiotic in the carbapenem family, contains a secondary amino group and hence, it also degrades *via* aminolytic dimerization. Once the four-membered lactam ring is opened during the aminolysis, the enamine moiety, unmasked from the previously fused five-membered ring, can tautomerize to produce the corresponding imine, resulting in two interchangeable dimeric degradants (dimers I and II in Scheme 4.45).[90]

Furthermore, the benzoic acid moiety of ertapenem was also found to be capable of opening the lactam ring, producing an anhydride intermediate which rearranges *via* acyl migration to ertapenem dimer III with a more stable benzoyl amide linkage. Finally, another two dimeric degradants (dimer–H$_2$O a and dimer–H$_2$O b) can also be formed through intermolecular amide linkages

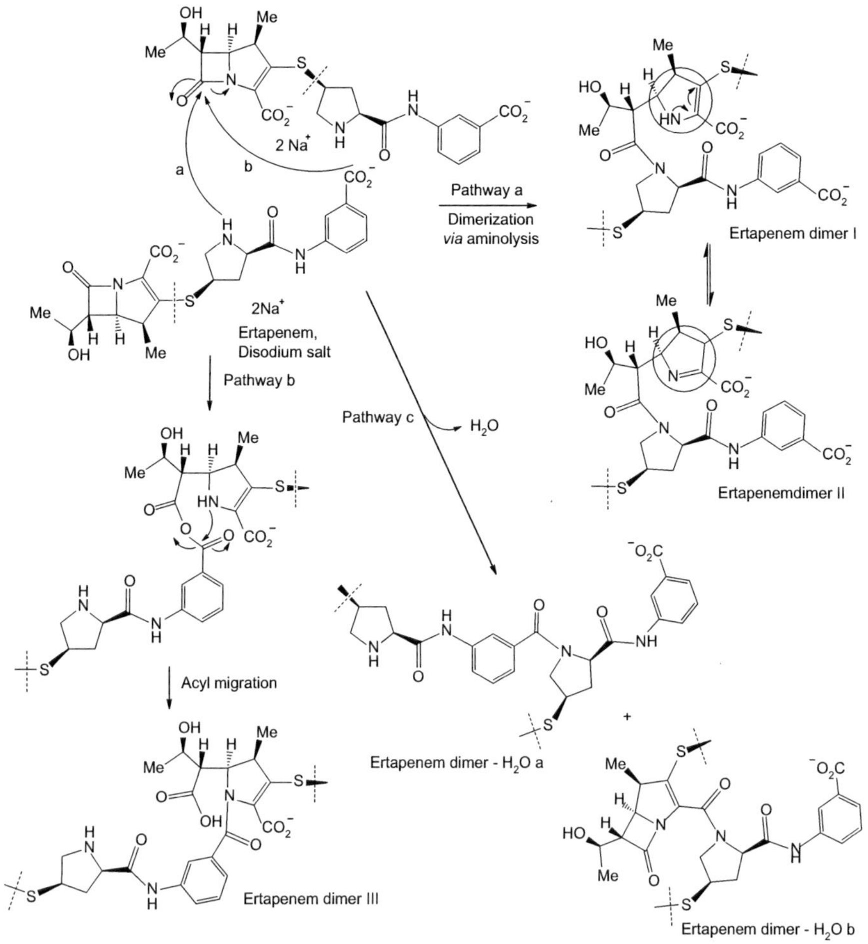

Scheme 4.45

between the secondary amino group and the two carboxylic groups from another ertapenem molecule.

Biapenem, another antibiotic in the carbapenem family, does not contain an amino group unlike the antibiotics discussed above. Hence, it cannot undergo a direct dimerization described above. Nevertheless, the carboxyl group on the 4-membered lactam ring is capable of attacking the lactam ring of another bia-penem molecule, under acidic or basic conditions and elevated temperature, to produce a dimeric anhydride intermediate. The latter can rearrange *via* acyl migration to dimer A.[91] Up to this point, this degradation pathway is analo-gous to pathway b in the degradation of ertapenem (Scheme 4.45) in terms of the reaction between a carboxyl and the lactam ring. The carboxyl group on the five membered ring in dimer A can now attack the remaining lactam ring triggering another sequence of ring opening followed by acyl migration, which

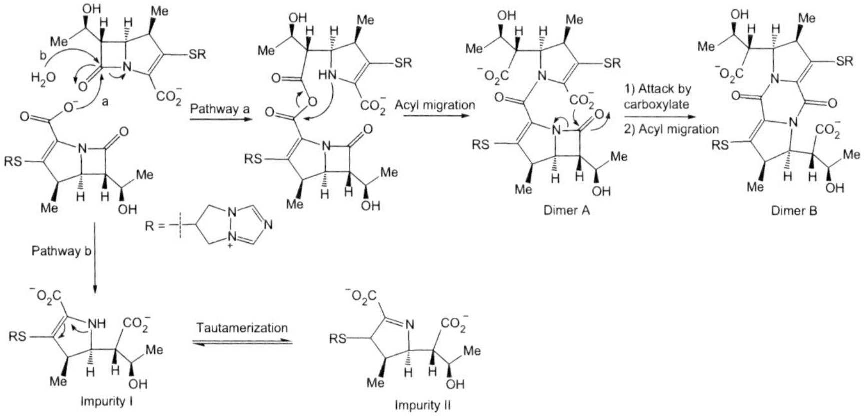

Scheme 4.46

ultimately yields isomeric dimer B containing a fused DKP ring. Separately, hydrolysis of the lactam ring of biapenem gives rise to another set of isomeric degradants, Impurity I and II. All the above degradation pathways of biapenem are summarized in Scheme 4.46. The nomenclature of the degradants of biapenem is the same as in the original publication.[91]

Imipenem, the first antibiotic in the carbapenem family which was discussed in Chapter 2, Hydrolytic Degradation, also degrades to form a similar DKP dimer *via* the same mechanism as shown above.[92]

The above cases of degradation *via* dimerization, either belong to type $M + M = 2M$ (like ampicillin, amoxicillin, ertapenem, biapenem, and imipenem) or type $M + M = 2M - m$ (like losartan). Sometimes, a dimeric degradant can be formed *via* a linker X as in type $M + X + M = M\text{-}X\text{-}M$, and formaldehyde frequently plays such a linker role.

Formaldehyde is a synthon employed in the synthesis of certain drug substances. For example, hydrochlorothiazide, a common diuretic drug, is made by the reaction of 5-chloro-2,4-disulfamylaniline with formaldehyde. Nevertheless, formaldehyde can also cause undesirable dimerization of hydrochlorothiazide during this synthesis. Additionally, it is known that hydrochlorothiazide decomposes *via* a retro-synthetic pathway to regenerate formaldehyde. Hence, the hydrochlorothiazide dimer may be a degradant as well. Two studies of this impurity were published independently by Franolic *et al.*[93] and Fang *et al.*[94] at approximately the same time. The structure of this impurity was unequivocally identified by the use of 2D NMR spectroscopy.[94] The process and degradation chemistry of hydrochlorothiazide discussed above is summarized in Scheme 4.47.

In the above pathway, the key steps are the sequential formation of two electrophiles, that is, an imine and iminium cation, followed by respective nucleophilic attacks. In the first key step, the imine intermediate is attacked by the intramolecular sulfonylamido group. In the second key step, the hydrochlorothiazide formed condenses with another molecule of formaldehyde to

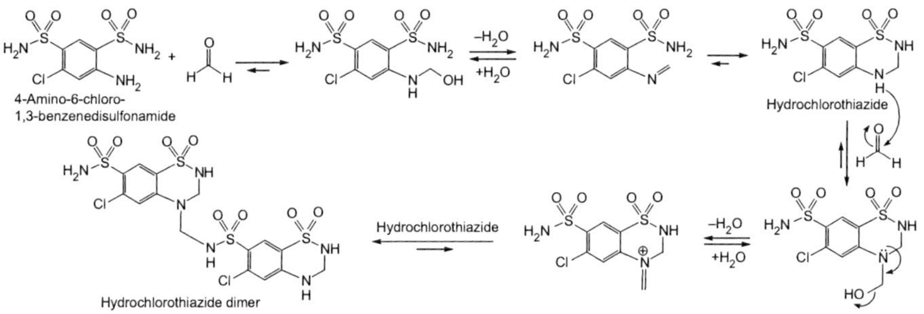

Scheme 4.47

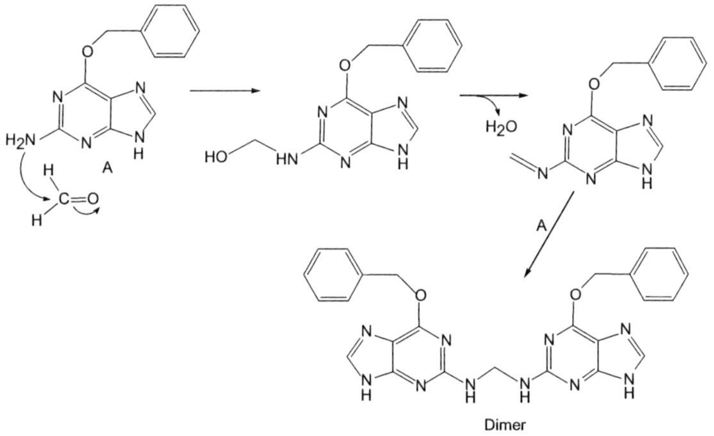

Dimer

Scheme 4.48

produce the iminium intermediate. Attack on the latter by the unsubstituted sulfonylamido group of hydrochlorothiazide yields the dimer.

Formaldehyde is also an impurity present in a number of pharmaceutical excipients, like PEG and polysorbate, which are prone to autooxidation during which process various oxidative degradants, including formaldehyde, can be produced.[95] Hence, in drug products formulated with these excipients, degradation of the APIs *via* formaldehyde/methylene-bridged dimerization may be possible. For example, a stability study of an experimental formulation of O^6-benzylguanine (NSC-637037), a drug candidate that demonstrated anti-cancer potential, in aqueous PEG 400 showed that a methylene-bridged dimer is the main degradant.[96] The degradation is most likely to proceed through the imine intermediate, followed by nucleophilic attack by a second molecule of O^6-benzylguanine as illustrated in Scheme 4.48.

In drug entities containing activated aromatic rings that can react with formaldehyde *via* electrophilic aromatic substitution, formaldehyde-bridged dimerization is also possible. For instance, an indolocarbazole derivative (Scheme 4.49) based on rebeccamycin, a potent inhibitor of topoisomerase I in

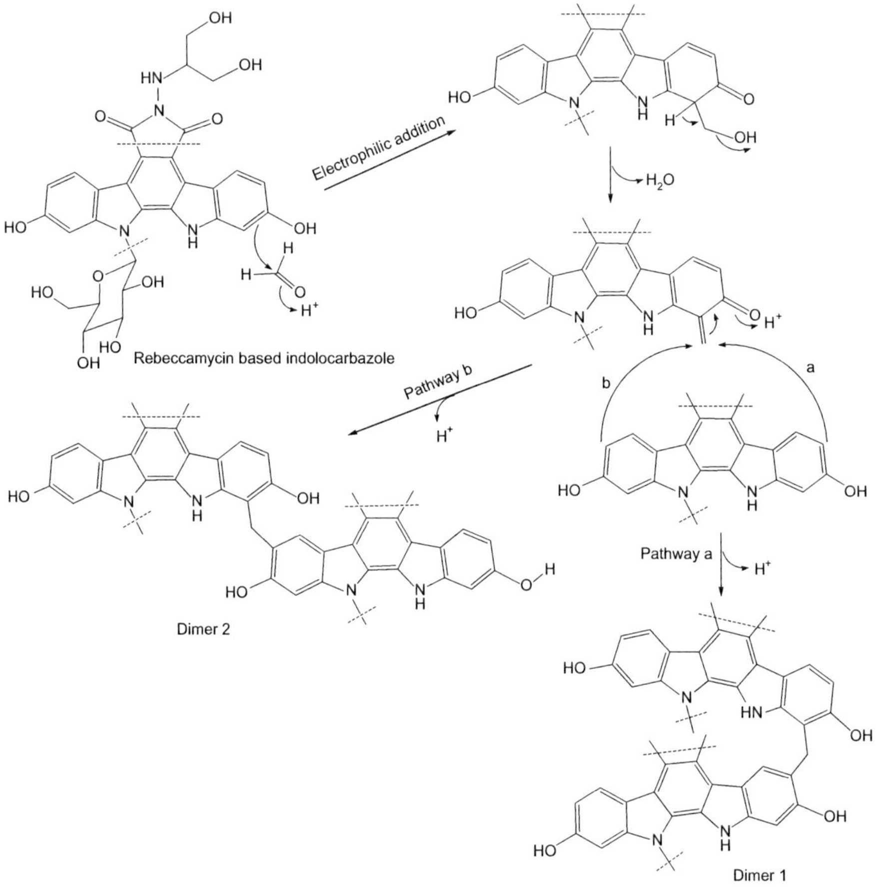

Scheme 4.49

a phase III clinical trial for anti-cancer treatment, contains two fused 6-hydroxyindole moieties. In experimental formulations containing high concentrations of the drug candidate in aqueous PEG, two dimeric degradants formed *via* a methylene linkage between the respective 5-hydroxyindole moieties were observed and fully characterized by liquid chromatography-mass spectroscopy (LC-MS), LC-MS/MS, and NMR analysis (1D and 2D).[97] The formation of the two dimeric degradants should proceed through the Lederer–Manasse hydroxylalkylation mechanism, which is shown in Scheme 4.49.

4.8   Miscellaneous Degradation Mechanisms

4.8.1   Diels–Alder Reaction

Although the Diels–Alder reaction is widely utilized in synthetic organic chemistry, drug degradation *via* Diels–Alder does not appear to be common.

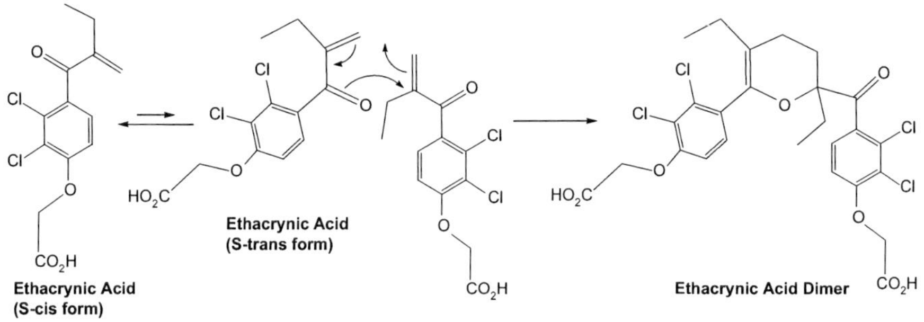

**Ethacrynic Acid
(S-cis form)**

**Ethacrynic Acid
(S-trans form)**

Ethacrynic Acid Dimer

Scheme 4.50

One rare case is ethacrynic acid, a diuretic drug whose degradation behavior is mostly attributable to its α,β-unsaturated carbonyl functionality, discussed in Section 4.3. The same functionality makes ethacrynic acid susceptible to degradation *via* the $4+2$ Diels–Alder addition mechanism (Scheme 4.50),[98] which gives another dimeric degradant that is different from the one mentioned in Section 4.3.

4.8.2 Degradation *via* Reduction or Disproportionation

As we discussed quite extensively in Chapter 3, Oxidative Degradation, drug degradation *via* various oxidative pathways is one of the two most frequently observed events in drug stability (or instability; the other being hydrolytic degradation). This is understandable since in the vast majority cases, the ultimate oxidizing agent is molecular oxygen. However, drug degradation *via* reduction is rare, if not impossible, owing to the lack of reducing agents that are capable of reductively degrading the drug substances. Hence the current author was quite surprised to come across a paper that reported the reductive degradation of rabeprazole sodium in a simulated intestinal fluid (a pH 6.8, 50 mM phosphate buffer).[99] Upon careful review of the experimental evidence presented, it appears that the "reductive degradation" of rabeprazole, a proton-pump inhibitor containing a sulfoxide moiety, may be a case of degradation *via* disproportionation. In such a case, when one molecule of rabeprazole is reduced to the reductive degradant, rabeprazole thioether, as observed by the original authors, another molecule of rabeprazole is most likely to be oxidized to form rabeprazole sulfone. In the chromatograms presented in the paper, it can be seen that when the late-eluting peak corresponding to rabeprazole thioether started to occur and increased over time, two early eluting peaks (one major and one minor; both immediately after the void volume), occurred and also increased in their peak areas. The sum of the peak areas apparently increased in sync with the thioether peak area. It seems that these peaks could be rabeprazole sulfone and a related oxidative degradant, which might be overlooked during the original study because they eluted immediately after the void volume in the high-performance liquid chromatography (HPLC) analysis.

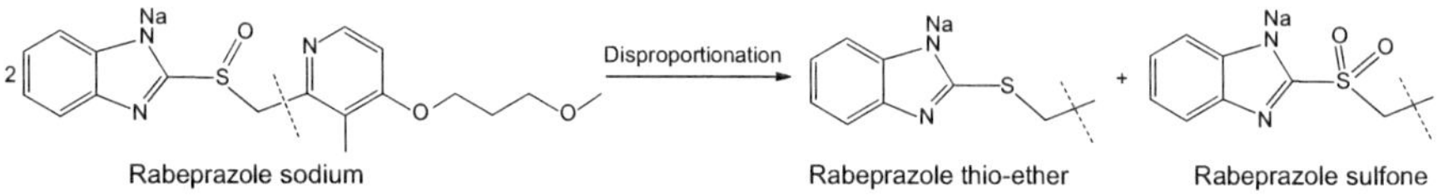

Scheme 4.51

Furthermore, since the thioether degradant elutes much later than rabeprazole, which is a sulfoxide, it is quite possible that the sulfone degradant is more polar than rabeprazole and hence elutes very closely to the void volume under the HPLC conditions employed. The likely scenario of rabeprazole degradation *via* disproportionation is shown in Scheme 4.51.

References

1. V. R. Mattox, *J. Am. Chem. Soc.*, 1952, **74**, 4340.
2. M. L. Lewbart and V. R. Mattox, *J. Org. Chem.*, 1964, **29**, 513.
3. T. Hidaka, S. Huruumi, S. Tamaki, M. Shiraishi and H. Minato, *Yakugaku Zasshi*, 1980, **100**, 72.
4. M. Li, B. Chen, M. Lin, T.-M. Chan, X. Fu and A. Rustum, *Tetrahedron Lett.*, 2007, **48**, 3901.
5. B. Chen, M. Li, M. Lin, G. Tumambac and A. Rustum, *Steroids*, 2009, **74**, 30.
6. M. Li, B. Chen, M. Lin and A. Rustum, *Am. Pharm. Rev.*, 2008, **1**, 98.
7. M. Shou, W. A. Galinada, Y.-C. Wei, Q. Tang, R. J. Markovich and A. M. Rustum, *J. Pharm. Biomed. Anal.*, 2009, **50**, 356.
8. P. Kaur, G. Wilmer, Y.-C. Wei and A. M. Rustum, *Chromatographia*, 2010, **71**, 805.
9. R. M. Bianchini, P. M. Castellano and T. S. Kaufman, *J. Pharm. Biomed. Anal.*, 2008, **48**, 1151.
10. P. J. Atkins, T. O. Herbert and N. B. Jones, *Int. J. Pharm.*, 1986, **30**, 199.
11. T. Cachet, G. Van den Mooter, R. Hauchecorne, C. Vinckier and J. Hoogmartens, *Int. J. Pharm.*, 1989, **55**, 59.
12. P. Alam, P. C. Buxton, J. A. Parkinson and J. Barber, *J. Chem. Soc. Perkin Trans. 2*, 1995, 1163.
13. A. Hassanzadeh, M. Helliwellb and J. Barber, *Org. Biomol. Chem.*, 2006, **4**, 1014.
14. J. Barber, J. I. Gyi, L. Lian, G. A. Morris, D. A. Pye and J. K. Sutherland, *J. Chem. Soc. Perkin Trans.*, 1991, **2**, 1489.
15. E. F. Fiese and S. H. Steffen, *J. Antimicrob. Chemother.*, 1990, **25**(Suppl. A), 39.
16. X. W. Teng, D. C. Cutler and N. M. Davies, *Int. J. Pharm.*, 2003, **259**, 129.
17. S. Sahasranaman, M. Issar, G. Tóth, G. Horváth and G. Hochhaus, *Pharmazie*, 2004, **59**, 367.
18. I. Nikcevic, P. Sajonz, M. Li, R. Markovich, A. Rustum, Challenges in the Analytical Method Development for Drug Product Containing a Steroid

Active Pharmaceutical Ingredient, presentation at *Pittcon 2011*, Session number 600-4.

19. J. B. Stenlake, R. D. Waigh and G. H. Dewar, *Eur J. Med. Chem.*, 1981, **16**, 515.
20. R. M. Welch, A. Brown, J. Ravitch and R. Dahl, *Clin. Pharmacol. Ther.*, 1995, **58**, 132.
21. G. R. Humphrey, R. A. Miller, P. J. Pye, K. Rossen, R. A. Reamer, A. Maliakal, S. S. Ceglia, E. J. J. Grabowski, R. P. Volante and P. J. Reider, *J. Am. Chem. Soc.*, 1999, **121**, 11261.
22. Ö. Almarsson, R. A. Seburg, D. Godshall, E. W. Tsai and M. J. Kaufman, *Tetrahedron*, 2000, **56**, 6877.
23. Z. Zhao, X.-Z. Qin and R. A. Reed, *J. Pharm. Biomed. Anal.*, 2002, **29**, 173.
24. S. G. Jivani and V. G. Stella, *J. Pharm. Sci.*, 1985, **74**, 1274.
25. R. K. Palsmeier, D. M. Radzik and C. E. Lunte, *Pharm. Res.*, 1992, **9**, 933.
26. X. Zhang and C. S. Foote, *J. Am. Chem. Soc.*, 1993, **115**, 8867.
27. M. Li, B. Conrad, R. G. Maus, S. M. Pitzenberger, R. Subramanian, X. Fang, J. A. Kinzer and H. J. Perpall, *Tetrahedron Lett.*, 2005, **46**, 3533.
28. Y. L. Lee, J. Padula and H. Lee, *J. Pharm. Sci.*, 1988, **77**, 81.
29. A. S. Kalgutkar, I. Gardner, R. S. Obach, C. L. Shaffer, E. Callegari, K. R. Henne, A. E. Mutlib, D. K. Dalvie, J. S. Lee, Y. Nakai, J. P. O'Donnell, J. Boer and S. P. Harriman, *Curr. Drug Metab.*, 2005, **6**, 161.
30. M. Burg and N. Green, *Kidney Int.*, 1973, **4**, 301.
31. D. A. Koechel, *Ann. Rev. Pharmacol. Toxicol.*, 1981, **21**, 265.
32. D. E. Duggan and R. M. Noll, *Arch. Biochem. Biophys.*, 1965, **109**, 388.
33. D. A. Koechel and E. J. Cafruny, *J. Med. Chem.*, 1973, **16**, 1147.
34. D. A. Koechel and E. J. Cafruny, *J. Pharmacol. Exp. Ther.*, 1975, **192**, 179.
35. R. J. Yarwood, W. D. Moore and J. H. Collett, *J. Pharm. Sci.*, 1985, **74**, 220.
36. A. Marin, A. Espada, P. Vidal and C. Barbas, *Anal. Chem.*, 2005, **77**, 471.
37. J. Wong, L. Wiseman, S. Al-Mamoon, T. Cooper, L.-K. Zhang and T.-M. Chan, *Anal. Chem.*, 2006, **78**, 7891.
38. L. Ding, X. Wang, Z. Yang and Y. Chen, *J. Pharm. Biomed. Anal.*, 2008, **46**, 282.
39. G. Orgován, K. Tihanyi and B. Noszál, *J. Pharm. Biomed. Anal.*, 2009, **50**, 718.
40. C.-Y. Liang, Y. Yang, M. A. Khadim, G. S. Banker and V. Kumar, *J. Pharm. Sci.*, 1995, **84**, 1141.
41. B. J. Fish and G. P. R. Carr, *J. Chromatogr.*, 1986, **353**, 39.
42. C. A. Janicki and C. Y. Ko, *Anal. Profiles Drug Subst.*, 1980, **9**, 341.
43. S. Fors, in *The Maillard Reaction in Foods and Nutrition*, ed. G.R. Waller and M.S. Feather, ACS Symposium Series, Vol. 215, American Chemical Society, Chapter 12, pp. 185–286.
44. M. Jun, Y. Shao, C.-T. Ho, U. Koetter and S. Lech, *J. Agric. Food Chem.*, 2003, **51**, 6340.
45. J. Hong, J. C. Shah and M. D. McGonagle, *J. Pharm. Sci.*, 2011, **100**, 2703.
46. D. E. Guttman and F. D. Meister, *J. Am. Pharm. Assoc. Sci. Ed.*, 1958, **47**, 773.
47. J. Hansen and H. Bundgaard, *Int. J. Pharm.*, 1980, **6**, 307.

48. H. S. Wendler in *Molecular Rearrangements, Part Two*, ed. P. Mayo, Interscience, New York, 1967, pp. 1067–1075.
49. K. Florey in *Analytical Profiles of Drug Substances*, Vol. 12, ed. K. Florey, Academic Press, New York, 1983, p. 277–324.
50. M. Li, B. Chen, S. Monteiro and A. M. Rustum, *Tetrahedron Lett.*, 2009, **50**, 4575.
51. H.-Y. Young, C.-T. Chiang, Y.-L. Huang, F. P. Pan and G.-L. Chen, *J. Food Drug Anal.*, 2002, **10**, 149.
52. N. Fukutsu, T. Kawasaki, K. Saito and H. Nakazawa, *J. Chromatogr. A*, 2006, **1129**, 153.
53. INVEGA®, Scientific discussion, European Medicines Agency; http://www.ema.europa.eu/docs/en_GB/document_library/EPAR_-_Scientific_Discussion/human/000746/WC500034928.pdf.
54. C. Danel, N. Azaroual, A. Brunel, D. Lannoy, P. Odou, B. Décaudin, G. Vermeersch, J.-P. Bonte and C. Vaccher, *Tetrahedron: Asymmetry*, 2009, **20**, 1125.
55. D. Stepensky, M. Chorny, Z. Dabour and I. Schumacher, *J. Pharm. Sci.*, 2004, **93**, 969.
56. R. J. Strife, I. Jardine and M. Colvin, *J. Chromatogr.*, 1980, **182**, 211.
57. J. H. Beijnen, J. J. M. Holthuis, H. G. Kerkdijk, O.A.G.J. van der Houwen, A. C. A. Paalman, A. Bult and W. J. M. Underberg, *Int. J. Pharm.*, 1988, **41**, 169.
58. C. Dugave and L. Demange, *Chem. Rev.*, 2003, **103**, 2475.
59. S. Zhang, J. Xing and D. Zhong, *J. Pharm. Sci.*, 2004, **93**, 1300.
60. W. H. Okamura, M. M. Midland, M. W. Hammond, N. Abd Rahman, M. C. Dormanen, I. Nemere and A. W. Norman, *J. Steroid Biochem. Mol. Biol.*, 1995, **53**, 603.
61. M. Brandl, X. Y. Wu, Y. Liu, J. Pease, M. Holper, E. Hooijmaaijer, Y. Lu and P. Wu, *J. Pharm. Sci.*, 2003, **92**, 1981.
62. N. Hashimoto and K. Hirano, *J. Pharm. Sci.*, 1998, **87**, 1091.
63. A. Rüegger, M. Kuhn, H. Lichti, H.-R Loosli, R. Huguenin, C. Quiquerez and A. Von Wartburg, *Rifai. Helv. Chim. Acta*, 1976, **59**, 1075.
64. G. J. Friis and H. Bundgaard, *Int. J. Pharm.*, 1992, **82**, 79.
65. X. Liu, J. Zhang, L. Song, B. C. Lynn and T. G. Burke, *J. Pharm. Biomed. Anal.*, 2004, **35**, 1113.
66. L. Ruzicka and H. F. Meldahi, *Helv. Chim Acta*, 1938, **21**, 1760.
67. D. N. Kirk and M. P. Hartshorn, *Steroid Reaction Mechanisms*, Elsevier, Amsterdam, 1969, pp. 294–301.
68. N. L. Wendler, in *Molecular Rearrangements*, ed. P. de Mayo, Interscience, New York, 1964, Vol. 2, pp. 1099–1101; pp. 1114–1121.
69. D. N. Kirk and C. R. McHugh, *J. Chem. Soc. Perkin Trans 1*, 1978, **1**, 73.
70. D. Dekker and J. H. Beijnen, *Pharm. Weekbl., Sci. Ed.*, 1980, **2**, 112.
71. D. Dekker and J. H. Beijnen, *Acta Pharm. Suec.*, 1981, **18**, 185.
72. A. Liguori, F. Perri and C. Siciliano, *Steroids*, 2006, **71**, 1091.
73. E. J. Delaney, R. G. Sherrill, V. Palaniswamy, T. C. Sedergran and S. P. Taylor, *Steroids*, 1994, **59**, 196.

74. J. H. Beijnen and D. Dekker, *Pharm. Weekbl., Sci. Ed.*, 1984, **6**, 1.
75. M. Li, X. Wang, B. Chen, T.-M. Chan and A. Rustum, *J. Pharm. Sci.*, 2009, **98**, 894.
76. J. C. Rotschafer, K. Crossley, D. E. Zaske, K. Mead, R. J. Sawchuck and L. D. Solem, *Antimicrob. Agents Chemother.*, 1982, **22**, 391.
77. C. M. Harris, H. Kopecka and T. M. Harris, *J. Am. Chem. Soc.*, 1983, **105**, 6915.
78. M. L. Lewbart and V. R. Mattox, *J. Org. Chem.*, 1963, **28**, 1779.
79. S. M. Steinberg and J. L. Bada, *J. Org. Chem.*, 1983, **48**, 2295.
80. B. N. Roy, G. P. Singh, H. M. Godbole and S. P. Nehate, *Indian J. Pharm. Sci.*, 2009, **71**, 395.
81. R. G. Strickley, G. C. Visor, L.-H. Lin and L. Gu, *Pharm. Res.*, 1989, **6**, 971.
82. N. F. Sepetov, M. A. Krymsky, M. V. Ovchinnikov, Z. D. Bespalova, O. L. Isakova, M. Soueek and M. Lebl, *Pept. Res.*, 1991, **4**, 308.
83. M. Beyermann, M. Bienert, H. Niedrich, L. A. Carpino and D. Sadat-Aalace, *J. Org. Chem.*, 1990, **55**, 721.
84. B. K. Josh, B. Ramsey, B. Johnson, D. E. Patterson, J. Powers, K. L. Facchine, M. Osterhout, M. P. Leblanc, R. Bryant-Mills, R. C. B. Copley and S. L. Sides, *J. Pharm. Sci.*, 2010, **99**, 3030.
85. K. E. McCarthy, Q. Wang, E. W. Tsai, R. E. Gilbert, D. P. Ip and M. A. Brooks, *J. Pharm. Biomed. Anal.*, 1998, **17**, 671.
86. Z. Zhao, Q. Wang, E. Tsai, X. Qin and D. Ip, *J. Pharm. Biomed. Anal.*, 1999, **20**, 129.
87. H. Bundgaard, *Acta Pharm. Suec.*, 1976, **13**, 9.
88. H. Bundgaard, *Acta Pharm Suec.*, 1977, **14**, 47.
89. C. Y. Lu and C. H. Feng, *J. Sep. Sci.*, 2007, **30**, 329.
90. P. Sajonz, T. K. Natishan, Y. Wu, J. M. Williams, B. Pipik, L. DiMichele, T. Novak, S. Pitzenberger, D. Dubost and Ö. Almarsson, *J. Liq. Chromatogr. Relat. Technol.*, 2001, **24**, 2999.
91. M. Xia, T.-J. Hang, F. Zhang, X.-M. Li and X.-Y. Xu, *J. Pharm. Biomed. Anal.*, 2009, **49**, 937.
92. R. W. Ratcliffe, K. J. Wildonger, L. Di Michele, A. W. Douglas, R. Hajdu, R. T. Goegelman, J. P. Springer and J. Hirshfield, *J. Org. Chem.*, 1989, **54**, 653.
93. J. D. Franolic, G. J. Lehr, T. L. Barry and G. Petzinger, *J. Pharm. Biomed. Anal.*, 2001, **26**, 651.
94. X. Fang, R. T. Bibart, S. Mayr, W. Yin, P. A. Harmon, J. F. McCafferty, R. J. Tyrrell and R. A. Reed, *J. Pharm. Sci.*, 2001, **90**, 1800.
95. K. C. Waterman, W. B. Arikpo, M. B. Fergione, T. W. Graul, B. A. Johnson, B. C. MacDonald, M. C. Roy and R. J. Timpano, *J. Pharm. Sci.*, 2008, **97**, 1499.
96. D. S. Bindra, T. D. Williams and V. J. Stella, *Pharm. Res.*, 1994, **11**, 1060.
97. Y. Sato, D. Breslin, H. Kitada, W. Minagawa, T. Nomoto, X.-Z. Qin and S. B. Karki, *Int. J. Pharm.*, 2010, **390**, 128.
98. R. J. Yarwood, A. J. Phillips, N. A. Dickinson and J. H. Collett, *Drug Dev. Ind. Pharm.*, 1983, **9**, 35.
99. S. Rena, M.-J. Park, H. Sah and B.-J. Lee, *Int. J. Pharm.*, 2008, **350**, 197.

CHAPTER 5

Drug–Excipient Interactions and Adduct Formation

A great number of cases in drug degradation in formulated formats involve drug–excipient interactions. This chapter discusses some representative cases in which an excipient, excipient- or packaging-related impurity is incorporated into the structure of a degradant (adduct formation) or causes the formation of a degradant.

5.1 Degradation Caused by Direct Interaction between Drugs and Excipients

5.1.1 Degradation *via* the Maillard Reaction

The Maillard reaction is an oversimplified classification of the chemical interaction of primary and secondary amines with reducing sugars, which can be mono- or polysaccharides. Numerous chemical reactions are involved in the Maillard reaction. The first reaction begins between the amino group and the glycosidic hydroxyl group at the reducing end of the sugar to form *N*-glycosides (glycosylamines). The *N*-glycosides can then undergo the Amadori rearrangement to produce α-aminoketoses. Numerous other parallel or consecutive reactions can occur, which ultimately lead to the formation of brown pigments as well as numerous volatile compounds. This phenomenon, also referred to as the "browning" reaction, was first described by Louis-Camille Maillard in the 1910s.[1,2] The Maillard reaction is very significant in food chemistry as the process produces flavoring aroma and taste as well as toxins in cooked food.[3] Since a great number of drug products contain primary and secondary amino functionalities as well as reducing mono-, di-, and polysaccharides (such as glucose, lactose, and starch) as excipients, drug degradation *via* the Maillard reaction pathways is frequently observed. Some commonly described reactions that have

RSC Drug Discovery Series No. 29
Organic Chemistry of Drug Degradation
By Min Li
© Min Li 2012
Published by the Royal Society of Chemistry, www.rsc.org

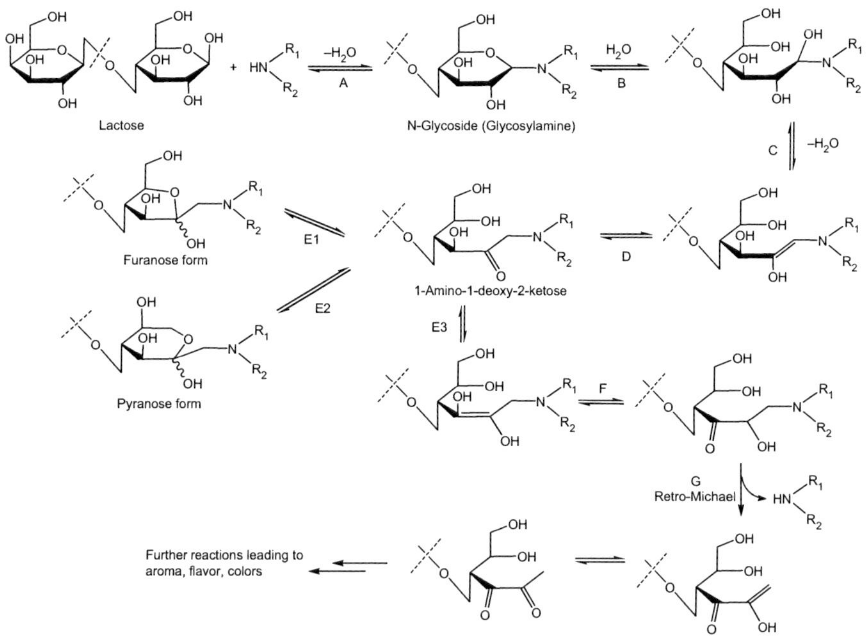

Lactose   N-Glycoside (Glycosylamine)

Furanose form   1-Amino-1-deoxy-2-ketose

Pyranose form

G
Retro-Michael

Further reactions leading to aroma, flavor, colors

Scheme 5.1

pharmaceutical and food chemistry significance in the early stage of the Maillard reaction are shown in Scheme 5.1, with lactose as an exemplary reducing sugar.

The *N*-glycoside formed in step A can readily undergo the Amadori rearrangement (steps B through D) to give the 1-amino-1-deoxy-2-ketose product, which is usually present predominantly in two cyclized forms, that is, the furanose and pyranose forms, especially in solution. The ketose can also proceed by further enolization with the 1′-hydroxyl group, resulting in the formation of the β-aminoketo compound. This compound is susceptible to retro-Michael addition, in particular in food chemistry where high temperatures are typically utilized during food preparation (such as baking and grilling), yielding the α-diketo product. According to Hartings,[4] compounds containing α-diketo moieties are the intermediates that ultimately lead to the formation of molecules that impart aroma, flavor, taste, and colors to cooked food. Among a large multitude of volatile products of the Maillard reaction are a sleuth of furans such as 2-furaldehyde and 5-(hydroxymethyl)-2-furaldehyde (5-HMF).[5,6] Other cyclized products connected through an N–C bond (such as pyrrole derivatives, with amines reintroduced into the reaction) or C–C bond (such as cyclopentenone derivatives) are also observed.[7–9]

In a long term stability study of pregabalin capsules, an anticonvulsant drug used for treating neuropathic pain, a total of seven major degradants were observed along with the pregabalin lactam degradant. The seven major degradants were found to be the Maillard reaction products formed between the active pharmaceutical ingredient (API) and respectively, lactose (a main

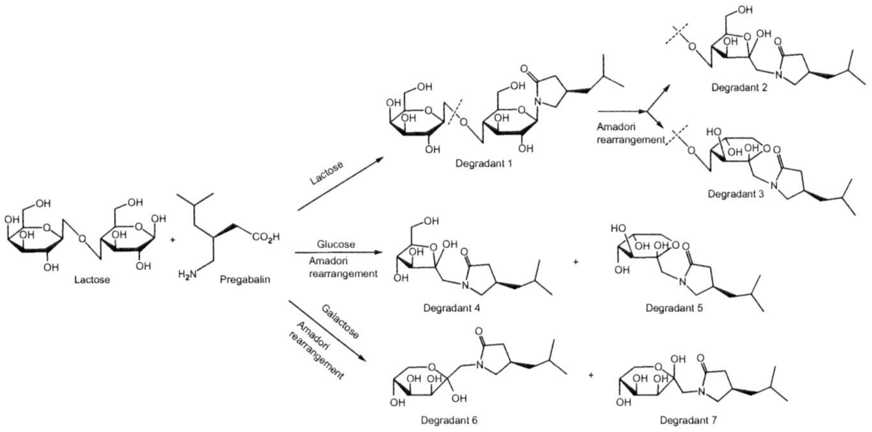

Scheme 5.2

excipient) and lactose components (galactose and glucose), all of which are in the lactam form indicating facile cyclization to form the five-membered lactam ring (Scheme 5.2).[10]

Degradants 4 and 5 can be direct degradants formed between glucose, present as either an impurity or degradant of lactose, and the API; alternatively they can be hydrolytic degradants of degradants 2 and 3, respectively. Degradants 6 and 7 must be direct degradants formed between galactose, present as either an impurity or degradant of lactose, and the API. In this case, the seven degradants were isolated and analyzed by NMR spectroscopy. Hence, the structures were determined between the possible isomers resulting from the Maillard degradation process. In other cases, isolation of individual isomers may not be possible owing to facile equilibration between these isomers, in particular the aminoketose and its two cyclized forms. For example, during a study of several commercial formulations of fluoxetine (a classic antidepressant which is a secondary amine) in the presence of lactose, both *N*-glycoside and Amadori-rearranged degradants were observed.[11] However, the structures of the Amadori-rearranged degradants were established by reduction of the ketone group in the Amadori-rearranged product to prevent the formation of the two cyclized forms. On the other hand, under accelerated stability conditions (40 °C/75% RH), the Amadori-rearranged product was found to undergo oxidative degradation, resulting in the formation of *N*-formylfluoxetine along with a number of volatile degradants. The overall degradation chemistry is illustrated in Scheme 5.3.

The formation of *N*-formylfluoxetine takes place by oxidative degradation of the Amadori-rearranged product. The exact mechanism of the oxidation is not clear. Nevertheless, it is known that the products generated during the Amadori rearrangement can be susceptible to autooxidation.[12]

Other examples of degradation *via* the Maillard reaction include lactose incompatibility with amlodipine, which contains a primary amine group,[13,14]

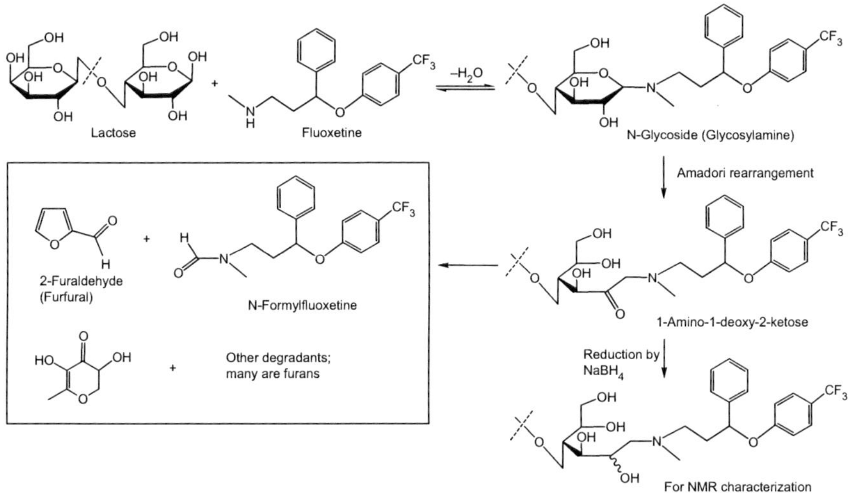

Scheme 5.3

and ceronapril.[15] Maillard reaction between hydrochlorothiazide, a diuretic containing both primary and secondary sulfonamide moieties, and lactose was also reported.[16]

5.1.2 Drug–Excipient Interaction *via* Ester and Amide Linkage Formation

Citric acid is a widely used excipient in the capacity of an acid, buffer, and/or chelating agent in both solid and liquid formulations. Since citric acid can be present in equilibrium with its two anhydride forms in solution,[17] formation of amide, imide, and ester degradation products could be an issue, particularly in liquid formulations. In an experimental enema formulation of 5-aminosalicylic acid, a drug used in the treatment of chronic inflammatory bowel diseases, in the presence of citric acid, three isomeric degradants were observed under both stressed and ambient storage conditions.[18] The degradants were the results of ester and amide formation between the API and citric acid, presumably proceeding through the anhydride intermediates (Scheme 5.4).[19]

Another example of degradation due to amide formation can be found in the case of oxytocin, which is a nonapeptide hormone used clinically to induce labor and prevent postpartum hemorrhage. In an effort to develop a formulation that does not require refrigeration or "cold chain" for storage, Poole *et al.* found two degradants resulting from the formation of amides between the *N*-terminal amino group of cysteine and citric acid. A minority of the amides apparently underwent dehydration to produce the two corresponding imide degradants (Figure 5.1).[20]

When the formulation of an amine drug contains inorganic carbonate salts, unstable carbamate (carbon dioxide adduct) may be formed. A commercial

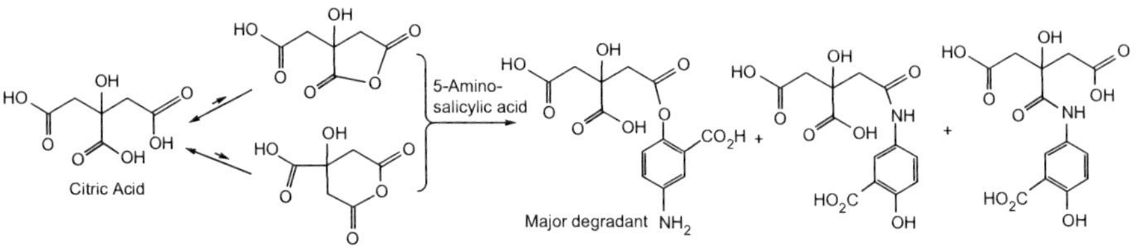

Scheme 5.4

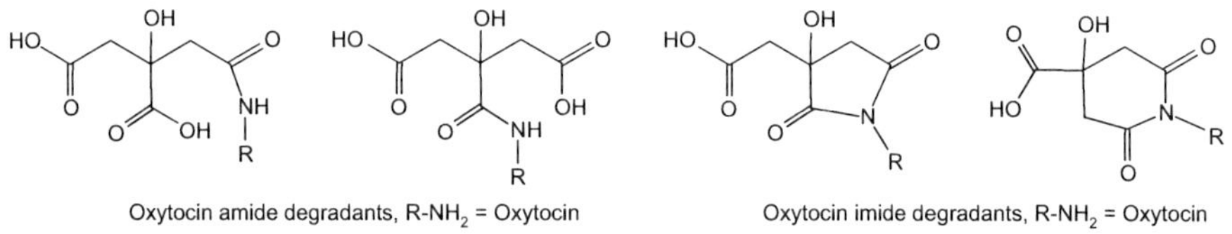

Oxytocin amide degradants, R-NH₂ = Oxytocin   Oxytocin imide degradants, R-NH₂ = Oxytocin

Figure 5.1   Oxytocin degradants.

formulation of meropenem, a 1-β-methylcarbapenem antibiotic, consisted of the crystalline API and sodium carbonate. Upon reconstitution into solution for injection, part of the API was found to form a covalent carbon dioxide adduct based on various NMR and mass spectrometric experiments, in conjunction with the use of ^{13}C-labelled sodium carbonate.[21] As discussed in Chapter 2, these types of carbamate is not stable and can only be present where there is an overwhelming presence of carbonate salts in the product formulation or in its solution. Once diluted in solutions devoid of carbonate, they quickly decarboxylate to reform the drug substances.

5.1.3   Drug–Excipient Interaction *via* Transesterification

Transesterification is analogous to hydrolysis of esters in that an alcohol plays the role of water molecule in a nucleophilic ester hydrolysis. Hence, drugs possessing an ester or alcohol functionality can undergo transesterification if the formulation contains excipients with a hydroxyl or ester functionality. For example, four major degradants were seen in an experimental tablet formulation of vitamin D_3 under stressed conditions (60 °C/ambient humidity for 7 months). These degradants were found to be the octanoate and decanoate esters of vitamin D_3 and pre-vitamin D_3, respectively. Apparently, they resulted from transesterification with the octanoate and decanoate components of triglycerides in the formulation. (Scheme 5.5).[22]

5.1.4   Degradation Caused by Magnesium Stearate

Magnesium stearate is a very common excipient used as a lubricant in pharmaceutical solid dosage form. Magnesium stearate can have an impact on drug

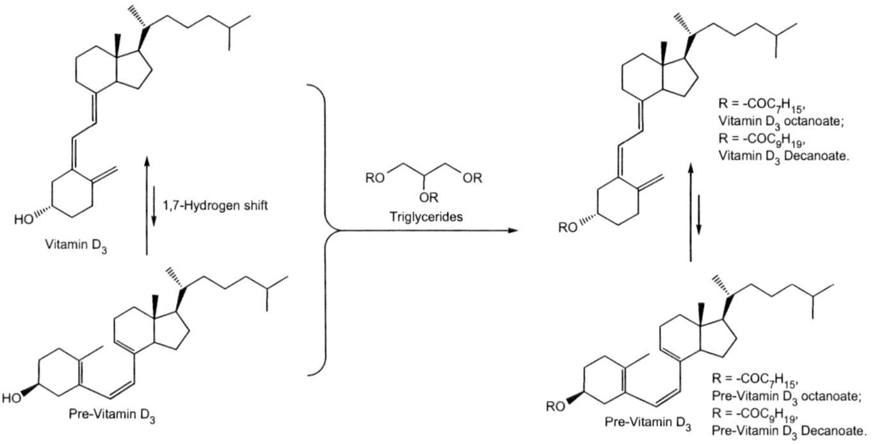

Scheme 5.5

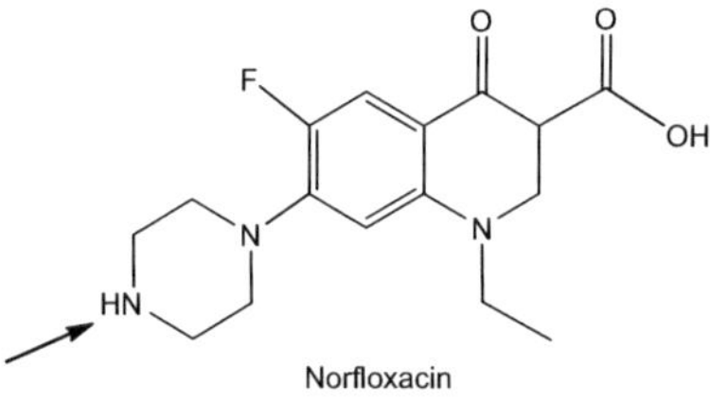

Figure 5.2 Norfloxacin. The arrow indicates the site of stearoyl amide linkage formation.

degradation behavior in at least three ways. First, it is a basic lubricant that can induce a microenvironment pH change in a solid dosage formulation. As such, it can alter the degradation behavior of drugs that are susceptible to hydrolytic degradation.[15] Second, the stearate portion of this excipient can react with primary and secondary amine drugs to produce the corresponding amide degradants. A stearoyl amide degradant was observed in a tablet formulation of norfloxacin (Figure 5.2) under stressed conditions.[23]

Third, the magnesium ion can promote drug degradation. For instance, fosinopril sodium is a diester pro-drug of an angiotensin converting enzyme (ACE) inhibitor. A prototype tablet formulation of the drug contained magnesium stearate as the lubricant. The magnesium ion was found to cause degradation of the drug molecule *via* a novel rearrangement mechanism followed by hydrolysis (Scheme 5.6).[24]

The role of magnesium ion in the degradation was verified by performing a stress study in which magnesium acetate was used. Kinetic analysis indicated that the degradation is a bimolecular reaction. This finding that the reaction order for magnesium is one (rather than a catalytic role for the metal ion) was attributed to the binding of the metal ion by the three degradants, which

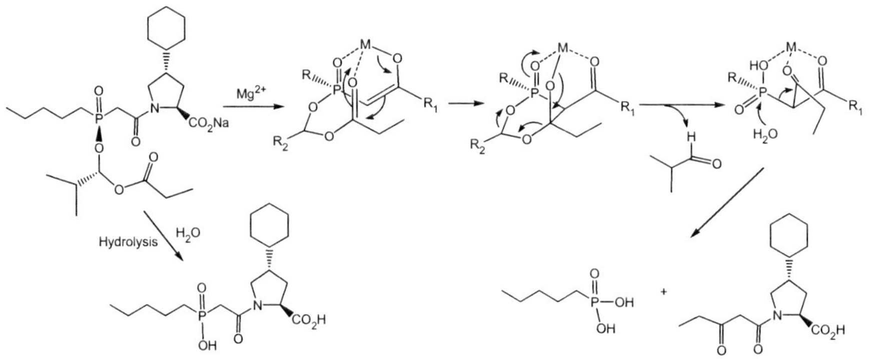

Scheme 5.6

prevents the metal ion from recomplexing with the drug substance. Because of this self-limiting effect, the formation of the degradants in the novel rearrangement pathway was limited and leveled off over time. In the meantime, formation of the straight hydrolytic degradant increased continuously, presumably catalyzed by the acidic degradants.

5.1.5 Degradation Caused by Interaction between API and Counter Ions and between Two APIs

Various organic acids are commonly employed in appropriate salt forms of amine-containing drugs. Maleic acid and its regio-isomer fumaric acid are among the most utilized organic acids in salt selection.[25] Both are α,β-conjugated diacids that are Michael acceptors. Hence, when they are used as counterions for basic drugs, degradation of primary and secondary amine drug substances can occur by Michael-type addition between the amines and the conjugated acids. For example, as discussed in Section 4.3, an adduct was formed between phenylephrine, a secondary amine, and maleate through Michael-type addition during pharmaceutical development of drug products for the common cold that contain phenylephrine HCl and dexbromphenir-amine maleate or chlorpheniramine maleate.[26]

An example of drug degradation caused by reaction between two APIs can be found in tablet formulations of phenylephrine HCl and aspirin.[27] The main degradant was formed *via trans*-acylation from aspirin to the secondary amine group of phenylephrine (Scheme 5.7). Under elevated temperatures, acylation of both the phenolic and hydroxyl groups occurred.

Another example of drug degradation caused by the interaction between the API and its counter ion involves an experimental drug containing a 2-aminopyridine moiety.[28] This drug was found to produce a formaldehyde-bridged dimer among other degradants, many of which are oxidative degradants; formaldehyde was determined to be a degradant of the tris(hy-droxymethyl)aminomethane (TRIS) salt of the drug substance.[29] A plausible

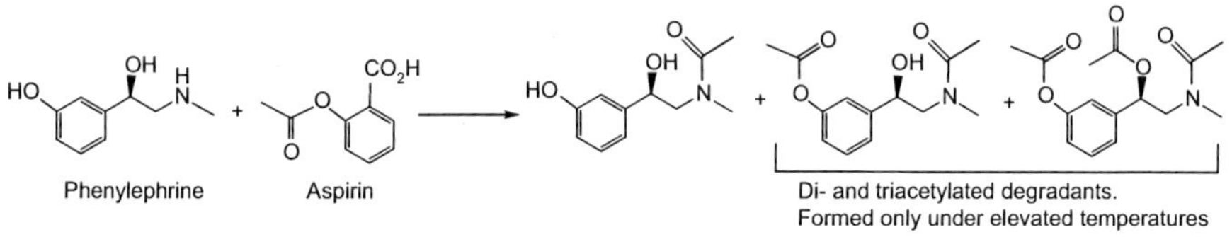

Phenylephrine   Aspirin   Di- and triacetylated degradants.
                                        Formed only under elevated temperatures

Scheme 5.7

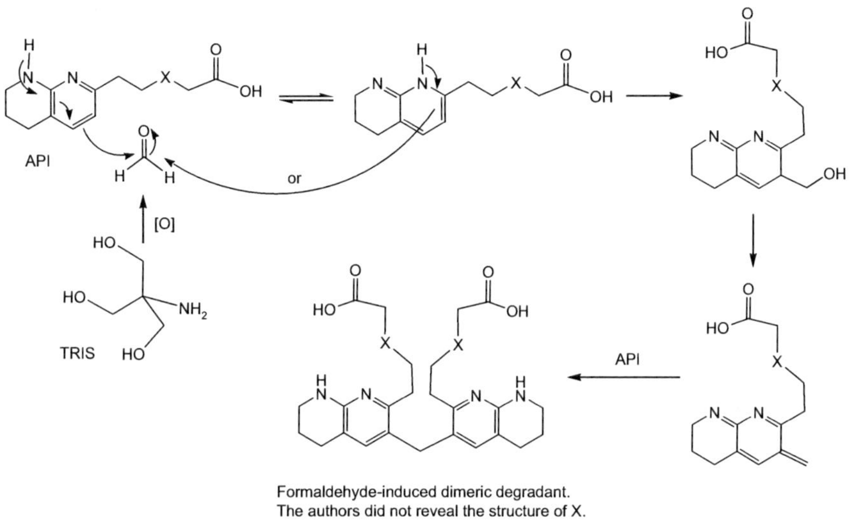

Formaldehyde-induced dimeric degradant.
The authors did not reveal the structure of X.

Scheme 5.8

mechanism for the formation of this dimeric degradant is shown in Scheme 5.8. This mechanism is slightly different from the one shown by the original authors in that the possibility of direct nucleophilic attack on formaldehyde by the 2-aminopyridine moiety (rather than its imine tautomer) is also presented here. In either case, the nucleophile can be viewed as an enamine or its equivalent.

5.1.6   Other Cases of Drug–Excipient Interactions

Sodium bisulfite and sodium metabisulfite are frequently used as preservatives as well as antioxidants especially in liquid formulations. In solution, the two forms are interconvertible *via* dehydration/hydration with the equilibrium favoring the bisulfite form. As discussed in Section 4.5.3, a liquid formulation of epinephrine was found to undergo metabisulfite-catalyzed racemization *via* the intermediacy of a sulfite adduct.[30] Similar degradation pathways were also observed in liquid formulations of phenylephrine, which is structurally very similar to epinephrine.[31] In other cases, the bisulfite or metasulfite salts induced drug degradation *via* their nucleophilic capability. For example, it was reported that sodium bisulfite caused hydrolysis of aspirin.[32]

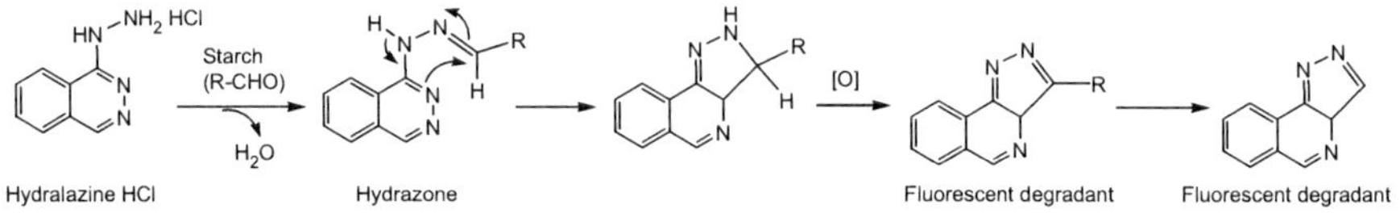

Scheme 5.9

In a stability study of hydralazine HCl tablets, an antihypertension drug, the potency of the API decreased significantly over time.[33] The lost potency was attributed to interaction between the API, which contains a strong nucleophilic hydrazine group, and the reducing sugar units in starch, a component of the formulation. The initial degradation product is a hydrozone resulting from nucleophilic attack and subsequent dehydration. Although structurally similar to imines (Schiff bases), formation of hydrozones is usually irreversible in contrast to the formation of imines, probably because hydrozones are stabilized by a few resonance forms. In this case, the hydrozone formed can undergo cyclization and then autooxidation to produce a fluorescent tricyclic degradant.[34] A small percentage of the latter degradant further degraded, cleaving itself from the starch molecule. The degradation pathway is summarized in Scheme 5.9.

5.2 Degradation Caused by Impurity of Excipients

Many excipients, in particular polymeric excipients, contain residual monomers, oligomers and other impurities and/or degradants. These impurities and degradants can react with drug substances to cause drug degradation.

5.2.1 Degradation Caused by Hydrogen Peroxide, Formaldehyde, and Formic Acid

Quite a few excipients such as polyethylene glycol (PEG), polysorbate, and povidone (polyvinylpyrrolidone) contain impurities resulting from autooxidation. Hydrogen peroxide, formaldehyde, and formic acid are the most frequently observed impurities caused by autooxidation (refer to Chapter 3 for the formation mechanism of these impurities).[35–38] Oxidative degradation caused by hydrogen peroxide, such as *N*-oxidation, has been discussed in detail in Chapter 3, Oxidative Degradation, while some of the degradation pathways caused by formaldehyde, such as dimerization, have been discussed in Chapter 4. In this sub-section, we will discuss a couple more examples of degradation caused by formaldehyde and formic acid.

In an osmotic tablet formulation of the first smoking cessation drug, varenicline, *N*-methylation and *N*-formylation degradants were found to be formed by the reaction of the drug with formaldehyde and formic acid, respectively.[39] The formation of the former is apparently through direct condensation with formic acid, while the formation of the latter should be *via* the Eschweiler–Clarke reaction (Scheme 5.10).

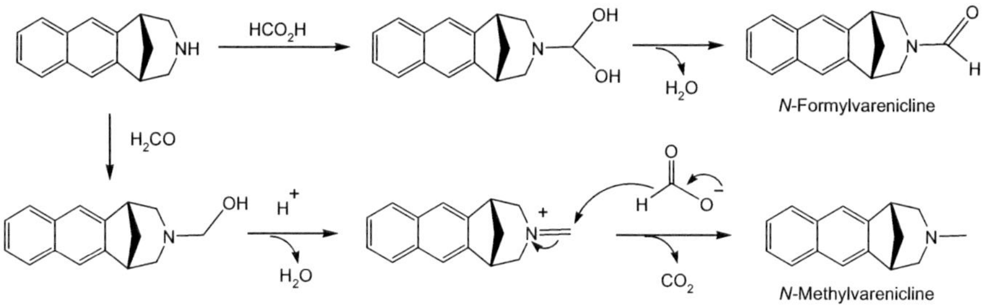

Scheme 5.10

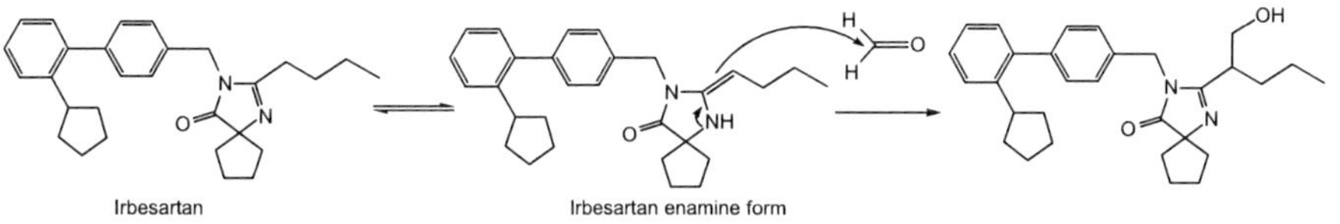

Scheme 5.11

The impurities, that is, formaldehyde and formic acid, come from the PEG coating of the control-released tablets and the formation of the two degradants could be minimized by reducing the mobility of these two impurities. Such measures include increasing PEG phase compatibility with cellulose acetate, another key component of the tablet coating, and reducing the moisture content.

A similar case caused by formaldehyde migrating from the coating to the core of the tablet can be found in a low dose film-coated formulation of irbesartan, a non-peptide angiotensin II receptor antagonist for the treatment of hypertension (Scheme 5.11).[40]

There are numerous additional cases of drug degradation that are caused by formaldehyde, many of which induced the formation of the dimers of the corresponding drug substances connected through a methylene bridge (from formaldehyde), in addition to the initially formed formyl degradants. Some of these dimerization cases have been discussed in Section 4.7. Formaldehyde can also cause crosslinking of both hard and soft shell gelatin capsules,[41,42] resulting in depressed dissolution of the active pharmaceutical ingredients.

5.2.2 Degradation Caused by Residual Impurities in Polymeric Excipients

In an enteric polymer-coated formulation of duloxetine hydrochloride containing the enteric polymers hydroxypropyl methylcellulose acetate succinate (HPMCAS) and hydroxypropyl methylcellulose phthalate (HPMCP), the API was found to react with residual succinic acid and phthalic acid to form

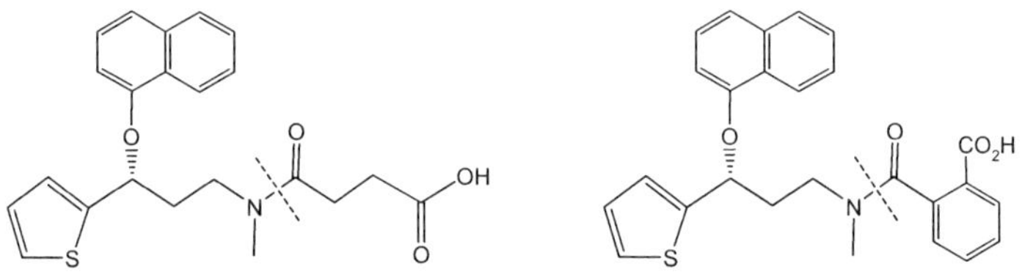

Duloxetine-succinic acid adduct   Duloxetine-phthalic acid adduct

Figure 5.3   Duloxetine succinic and phthalic acid adducts.

degradants possessing, respectively, succinamide and phthalamide linkages (Figure 5.3).[43]

The formation of the two degradants was apparently caused by migration of the residual impurities in the enteric coating into the core tablet; increasing the thickness of the physical barrier separating the coating and core tablet minimized the formation of the degradants.

5.3   Degradation Caused by Degradants of Excipients

Antioxidants are species that are more easily oxidized than the drug substances they are intended to protect from oxidative degradation in pharmaceutical formulations. Consequently, antioxidants are preferentially oxidized. Usually, the oxidized forms of antioxidants are not reactive toward drug substances. In some limited cases, however, they can interact with drug substances. During the development of a petrolatum-based topical ointment formulation of an antifungal drug, miconazole nitrate, a 1:1 adduct between the API and 2,6-di-*tert*-butyl-4-methylphenol (BHT) was formed under accelerated stability conditions (40 °C/75% RH).[44] The structure of this adduct was elucidated using tandem liquid chromatography-mass spectrometry (LC-MS^n), organic synthesis, stress studies, and various 1D and 2D nuclear magnetic resonance (NMR) (including both H-^{13}C and H-^{15}N experiments). The linkage between the two components of the adduct is the C–N bond formed between one of the imidazole nitrogens and the methyl (now a methylene) group of BHT. The authors postulated a quinone methide intermediate, resulting from the immediately oxidized form of BHT (BHT phenolic radical) *via* disproportionation.[45,46] The quinone methide is a rather electrophilic species which can be readily intercepted by the nucleophilic imidazole moiety of miconazole (Scheme 5.12).

As mentioned in Section 5.1.6, metabisulfite and bisulfite salts are used as preservatives as well as antioxidants. However, in commercial formulations of propofol, a widely utilized intravenous sedative for the induction and maintenance of anesthesia and sedation, sodium metabisulfite was found to promote the oxidation of the drug substance.[47–49] The two major oxidative degradants are propofol dimer and propofol dimer quinone. The quinone degradant is a yellow compound caused by the extended conjugation. According to a study by Baker *et al.*,[50] the formation of these two degradants was mediated through sulfite anion

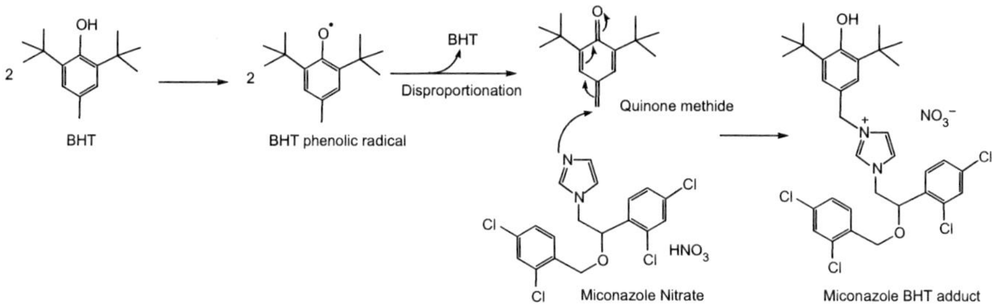

Scheme 5.12

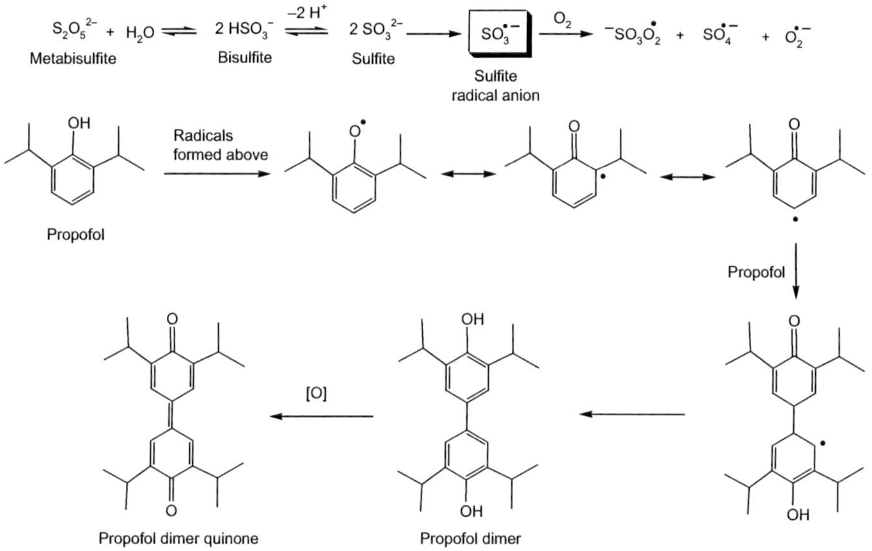

Scheme 5.13

radical, the presence of which was detected by electron spin resonance (ESR) in commercial propofol formulations containing sodium metabisulfite. The sulfite anion radical, which is produced from sulfite by a one-electron oxidation process, can readily react with molecular oxygen to give sulfite peroxyl radical ($^-SO_3O_2{}^\bullet$), sulfate radical ($SO_4{}^{-\bullet}$), and to a lesser extent, superoxide radical anion. These species, as well as sulfite anion radical itself, can directly or indirectly trigger the formation of propofol dimer and propofol dimer quinone (Scheme 5.13).

5.4 Degradation Caused by Impurities from Packaging Materials

In a lyophilized formulation of an experimental drug for stroke treatment, BMS-204352, a formaldehyde adduct of the API was produced in stability

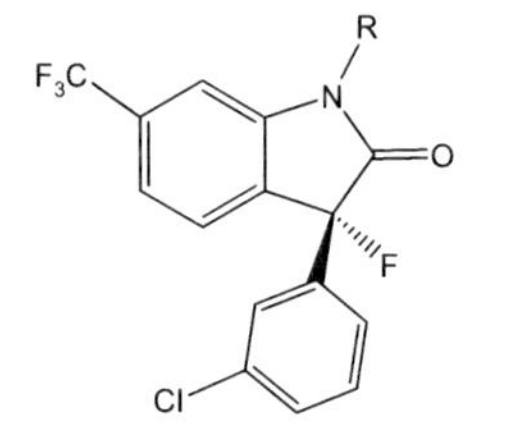

BMS-204352, R = H;
BMS-204352 formaldehyde adduct,
R = -CH₂OH

Figure 5.4   BMS-204352.

samples (Figure 5.4). Although the formulation contains polysorbate 80, in which formaldehyde could be an impurity, the authors attributed the source of formaldehyde as a leachable from the rubber stopper of the formulation vial, based on a stopper study and information from the stopper vender that formaldehyde was used as a reinforcing agent in the manufacture of the stopper.[51]

In a stability study of an orange-red amine hydrochloride drug substance stored in plastic antistatic liners, partial discoloration of the drug substance was observed.[52] The discolored drug substance became yellow, which is the color of the neutral form of the drug. Knowing that the antistatic agents in the plastic liners are *N,N*-bis(2-hydroxyethyl)alkylamines (alkyl = C_{12}–C_{18}), it was apparent that the discoloration was caused by leaching of the antistatic agents, which neutralized the hydrochloride salt of the drug into its free base. The infrared (IR) spectrum of the discolored material lacks the characteristic hydrogen-bonded amine (NH^+) stretch at ~2,400 cm^{-1}, which is consistent with the free base form of the drug substance.

References

1. L.-C. Maillard, *Compt. Rend.*, 1912, **154**, 66.
2. L.-C. Maillard, *Ann. Chim.*, 1916, **9**, 258.
3. H.-D. Belitz, W. Grosch, and P. Schieberle, *Food Chemistry*, Translated by M.M. Burghagen from the 5th German edition, Springer, Berlin, 2004, p. 268.
4. M. Hartings, *C&E News*, American Chemical Society, Nov. 21, 2011, p. 36.
5. S. Fors, in *The Maillard Reaction in Foods and Nutrition*, ed. G. R. Waller and M. S. Feather, ACS Symposium Series, Vol. 215, American Chemical Society, Chapter 12, pp. 185–286.
6. M. Jun, Y. Shao, C.-T. Ho, U. Koetter and S. Lech, *J. Agicr. Food Chem.*, 2003, **51**, 6340.
7. T. Hofmann, *J. Agric. Food Chem.*, 1998, **46**, 3918.
8. S. Estendorfer, F. Ledl and T. Severin, *Tetrahedron*, 1990, **46**, 5617.
9. B. Kramholler, F. Ledl, H. Lerche and T. Severin, *Z. Lebensm.-Unters. Forsch.*, 1992, **194**, 431.

10. M. J. Lovdahl, T. R. Hurley, B. Tobias and S. R. Priebe, *J. Pharm. Biomed. Anal.*, 2002, **28**, 917.

11. D. D. Wirth, S. W. Baertschi, R. A. Johnson, S. R. Maple, M. S. Miller, D. K. Hallenbeck and S. M. Gregg, *J. Pharm. Sci.*, 1998, **87**, 31.

12. V. V. Mossine, M. Linetsky, G. V. Glinsky, B. J. Ortwerth and M. S. Feather, *Chem. Res. Toxicol.*, 1999, **12**, 230.

13. A. Abdoh, M. M. Al-Omari, A. A. Badwan and A. M. Y. Jaber, *Pharm. Dev. Tech.*, 2004, **9**, 15.

14. T. Murakami, N. Fukutsu, J. Kondo, T. Kawasaki and F. Kusu, *J. Chromatogr. A*, 2008, **1181**, 67.

15. A. T. M. Serajuddin, A. B. Thakur, R. N. Ghoshal, M. G. Fakes, S. A. Ranadive, K. R. Morris and S. A. Varia, *J. Pharm. Sci.*, 1999, **88**, 696.

16. P. A. Harmon, W. Yin, W. E. Bowen, R. J. Tyrrell and R. A. Reed, *J. Pharm. Sci.*, 2000, **89**, 920.

17. T. Higuchi, T. Miki, A. C. Shah and A. K. Herd, *J. Am. Chem. Soc.*, 1963, **85**, 3655.

18. J. Larsena, D. Staerkb, C. Cornett, S. H. Hansen and J. W. Jaroszewski, *J. Pharm. Biomed. Anal.*, 2009, **49**, 839.

19. T. Higuchi, H. Uno, S. O. Eriksson and J. J. Windheuser, *J. Pharm. Sci.*, 1964, **53**, 280.

20. R. A. Poole, P. T. Kasper and W. Jiskoot, *J. Pharm. Sci.*, 2011, **100**, 3018.

21. Ö. Almarsson, M. J. Kaufman, J. D. Stong, Y. Wu, S. M. Mayr, M. A. Petrich and J. M. Williams, *J. Pharm. Sci.*, 1998, **87**, 663.

22. J. M. Ballard, L. Zhu, E. D. Nelson and R. A. Seburg, *J. Pharm. Biomed. Anal.*, 2007, **43**, 142.

23. C. Mazuel, in *Analytical Profiles of Drug Substances*, ed. K. Florey, Vol. 20, Academic Press, New York, 1991, p. 557.

24. A. B. Thakur, K. Morris, J. A. Grosso, K. Himes, J. K. Thottathil, R. L. Jerzewski, D. A. Wadke and J. T. Carstensen, *Pharm. Res.*, 1993, **10**, 800.

25. P. L. Gould, *Int. J. Pharm.*, 1986, **33**, 201.

26. J. Wong, L. Wiseman, S. Al-Mamoon, T. Cooper, L.-K. Zhang and T.-M. Chan, *Anal. Chem.*, 2006, **78**, 7891.

27. A. E. Troup and H. Mitchner, *J. Pharm. Sci.*, 1964, **53**, 375.

28. Y. Wu, T.-L. Hwang, K. Algayer, W. Xu, H. Wang, A. Procopio, L. DeBusi, C.-Y. Yang and B. Matuszewska, *J. Pharm. Biomed. Anal.*, 2003, **33**, 999.

29. Y. Song, R. L. Schowen, R. T. Borchardt and E. M. Topp, *J. Pharm. Sci.*, 2001, **90**, 1198.

30. D. Stepensky, M. Chorny, Z. Dabour and I. Schumacher, *J. Pharm. Sci.*, 2004, **93**, 969.

31. V. J. Stella, *J Parenter. Sci. Technol.*, 1986, **40**, 142.

32. J. W. Munson, A. Hussain and R. Bilous, *J. Pharm. Sci.*, 1977, **66**, 1775.

33. T. Lessen and D.-C.(D.) Zhao, *J. Pharm. Sci.*, 1996, **85**, 326.

34. M.A.E. Shaban, R. S. Ali and S. M. El-Badry, *Carbohydr. Res.*, 1981, **95**, 51.

35. J. W. McGinity, T. R. Patel and A. H. Naqvi, *Drug Dev. Commun.*, 1976, **2**, 505.

36. T. Huang, M. E. Garceau and P. Gao, *J. Pharm. Biomed. Anal.*, 2003, **31**, 1203.
37. W. R. Wasylaschuk, P. A. Harmon, G. Wagner, A. B. Harman, A. C. Templeton, H. Xu and R. A. Reed, *J. Pharm. Sci.*, 2007, **96**, 106.
38. K. J. Hartauer, G. N. Arbuthnot, S. W. Baertschi, R. A. Johnson, W. D. Luke, N. G. Pearson, E. C. Rickard, C. A. Tingle, P. K. S. Tsang and R. E. Wiens, *Pharm. Dev. Technol.*, 2000, **5**, 303.
39. K. C. Waterman, W. B. Arikpo, M. B. Fergione, T. W. Graul, B. A. Johnson, B. C. MacDonald, M. C. Roy and R. J. Timpano, *J. Pharm. Sci.*, 2008, **97**, 1499.
40. G. Wang, J. D. Fiske, S. P. Jennings, F. P. Tomasella, V. A. Palaniswamy and K. L. Ray, *Pharm. Dev. Technol.*, 2008, **13**, 393.
41. G. A. Digenis, T. B. Gold and V. P. Shah, *J. Pharm. Sci.*, 1994, **83**, 915.
42. C. M. Ofner III, Y.-E. Zhang, V. C. Jobeck and B. J. Bowman, *J. Pharm. Sci.*, 2001, **90**, 79.
43. P. J. Jansen, P. L. Oren, C. A. Kemp, S. R. Maple and S. W. Baertschi, *J. Pharm. Sci.*, 1998, **87**, 81.
44. F. Zhang and M. Nunes, *J. Pharm. Sci.*, 2004, **93**, 300.
45. R. H. Bauer and G. M. Coppinger, *Tetrahedron*, 1963, **19**, 1201.
46. N. Zhang, S. Kawakami, M. Higaki and V. T. Wee, *J. Am. Oil. Chem. Soc.*, 1997, **74**, 781.
47. M. T. Baker, *Am. J. Anesthesiol.*, 2000, **27**(Suppl.), 19.
48. M. T. Baker, *Am. J. Health-Syst. Pharm.*, 2001, **58**, 1042.
49. D. Mirejovsky, *Am. J. Health-Syst. Pharm.*, 2001, **58**, 1046.
50. M. T. Baker, M. S. Gregerson, S. Marc, S. M. Martin and G. R. Buettner, *Crit. Care Med.*, 2003, **31**, 787.
51. M. N. Nassar, V. V. Nesarikar, R. Lozano, Y. Huang and V. Palaniswamy, *Pharm. Dev. Technol.*, 2005, **10**, 227.
52. M. D. Argentine and P. J. Jansen in *Pharmaceutical Stability Testing to Support Global Markets, Chapter 18*, ed. K. Huynh-Ba, Springer, New York, 2010.

CHAPTER 6
Photochemical Degradation

6.1 Overview

Photochemical degradation of drugs occurs during the manufacture and storage of drug substances and drug products when they are exposed to ambient lighting, which in most cases is from fluorescent lamps. Additionally, after a photosensitive drug is ingested into a patient, photochemical degradation of the drug can occur when the subject is exposed to sunlight, fluorescent, and/or incandescent light. This process is responsible for phototoxicity and photoallergic reactions of drugs. These three light sources emit in the UV-A wavelengths between 320 and 400 nm, while the first two also emit in the UV-B wavelengths between 290 and 320 nm. A great number of organic drugs absorb in the UV-A and UV-B regions to various degrees. Emissions in the shorter wavelengths of the UV-B region are largely absorbed by the skin, while those in the longer wavelengths of the UV-A region are capable of penetrating the skin. Hence, emission in the UV-A region is mostly responsible for phototoxicity and photoallergic reactions of drugs.[1]

Photochemical reactions are chemical transformations that take place at electronically excited states of molecules. Since the electronic properties in excited states can be dramatically different from those in the ground state, photochemical degradation usually displays unique mechanisms and pathways. In order for a drug molecule to be photochemically reactive, it has to meet at least the following two requirements: first, it should contain a chromophore with reasonable absorption coefficiency at wavelengths comparable to the light sources. Second, it should contain functional groups or moieties that are reactive at electronically excited states, such as carbonyl group and several types of double bonds. Otherwise, the excited state can go back to the ground state *via* light emissions or physical relaxation processes such as collision with other molecular entities.

RSC Drug Discovery Series No. 29
Organic Chemistry of Drug Degradation
By Min Li
© Min Li 2012
Published by the Royal Society of Chemistry, www.rsc.org

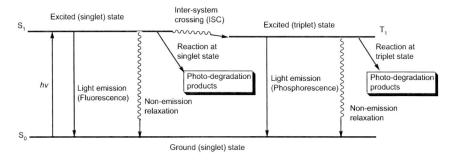

Figure 6.1   Schematic diagram showing the events after photoactivation of a molecule.

The vast majority of organic molecules in the ground state are singlet species (S_0). A schematic presentation of the events following initial photoactivation of a molecule is shown in Figure 6.1. When irradiated by light, an electron in the ground state molecular orbit of the molecule is activated to electronically excited states while keeping its electron spin state. The higher excited states can quickly deactivate to the lowest excited singlet state, S_1. S_1 will lose its energy and decay to S_0, *via* a number of deactivation events: first, it can go back to the ground state through multi-stage vibrational relaxation by colliding with other molecules (non-radiative relaxation). Second, it can emit light (fluorescence). Third, the singlet S_1 state can cross over to the triplet state, T_1, by reversing the spin of the excited electron. This process, called intersystem crossing (ISC), is made possible by the overlap between the vibrational levels of the two excited states. T_1 can then decay to S_0 either by emitting light (phosphorescence) or by colliding with other molecules. All of the above processes give back the original molecule, that is, no (photo)chemical reaction takes place. On the other hand, chemical transformation can happen at both the S_1 and T_1 states; in such cases, photochemical reactions occur and the efficiency of the reactions can be measured by quantum yield, Φ, which is defined as $\Phi = $ No. of reacting molecules/ no. of photons absorbed.

In the photochemical processes discussed above, the role of oxygen has not been discussed. This subject will be discussed in Section 6.3.

Photochemical degradation is not desirable for regular drugs. On the contrary, drugs used in photodynamic therapies need to be photochemically reactive. Different types of photoactive functional groups have different photochemical reactions; those that are significant in photodegradation of drugs will be discussed in the following sections.

6.2   Non-oxidative Photochemical Degradation

Some photodegradation pathways have non-photodegradation counterparts in terms of degradation products observed. For example, hydrolytic degradation of a drug can occur under both photochemical and non-photochemical conditions. Likewise, isomerization of drugs containing the oxime ether moiety can also take place under both types of conditions. Although the transition states may be different between the photo and non-photodegradation pathways, they can share similar degradation products. On the other hand, there are a number of photochemically unique degradation pathways. The non-oxidative photochemical degradation of a number of representative photolabile drugs will be discussed in this section.

6.2.1 Photodecarboxylation: Photodegradation of Drugs Containing a 2-Arylpropionic Acid Moiety

Quite a few non-steroidal anti-inflammatory drugs (NSAIDs) contain a 2-arylpropionic acid or arylacetic acid moiety. In the aryl portion of these drug molecules, many contain chromophores that have significant absorption at UV-A and UV-B regions. These chromophores include benzophenone or benzophenone-like (the so-called heterocyclic benzophenone) moieties (*e.g.*, ketoprofen, suprofen, and tiaprotinic acid), naphthalene (naproxene), and indole or fused indole rings (indomethacin and carprofen).

Ketoprofen is a simple 2-propionic acid derivative of benzophenone, known for its phototoxicity. Benzophenone is one of the most widely studied molecules for its photochemistry and photophysics. Hence, the photodegradation chemistry of ketoprofen has been much more thoroughly studied than others, thanks in part to knowledge of benzophenone photochemistry and photophysics. The ionized form of the drug molecule undergoes rapid decarboxylation upon photolysis. In aqueous buffers under anaerobic condition, Borsarelli *et al.* found that decarboxylation takes place within 10 ns of the excitation of ketoprofen anion with 355 nm laser pulses.[2] Such a short time scale seems to suggest that the decarboxylation occurs from either a singlet or a very short-lived triplet state. Additional evidence obtained by Cosa *et al.* appears to favor a singlet state preceding the decarboxylation.[3]

The photodecarboxylation of ketoprofen is pH dependent and efficient decarboxylation occurs at pH greater than the pK_a of ketoprofen, which is 4.7. The decarboxylation step is highly efficient with a quantum yield (Φ) of ~ 0.75.[4] The decarboxylation results in a carbanion intermediate; at high pH, the carbanion quickly decays to the ground state within a nanosecond timescale, while at low pH, the carbanion is first converted to a long-lived bi-radical, which eventually also decays to the same ground state photodegradant within a timescale of 10 µs. The photodegradation of ketoprofen in aqueous buffers under anaerobic condition is illustrated in Scheme 6.1.

In the presence of oxygen, the decarboxylated product, 3-ethylbenzophenone, is still observed as the major degradant, in addition to a few minor oxidative degradants.[4,5] The mechanism leading to the formation of the latter degradants will be discussed in Section 6.3.

Other 2-arylpropionic acid NSAIDs that undergo decarboxylation at singlet state include benoxaprofen[6,7] and indoprofen;[8] analogous degradants were

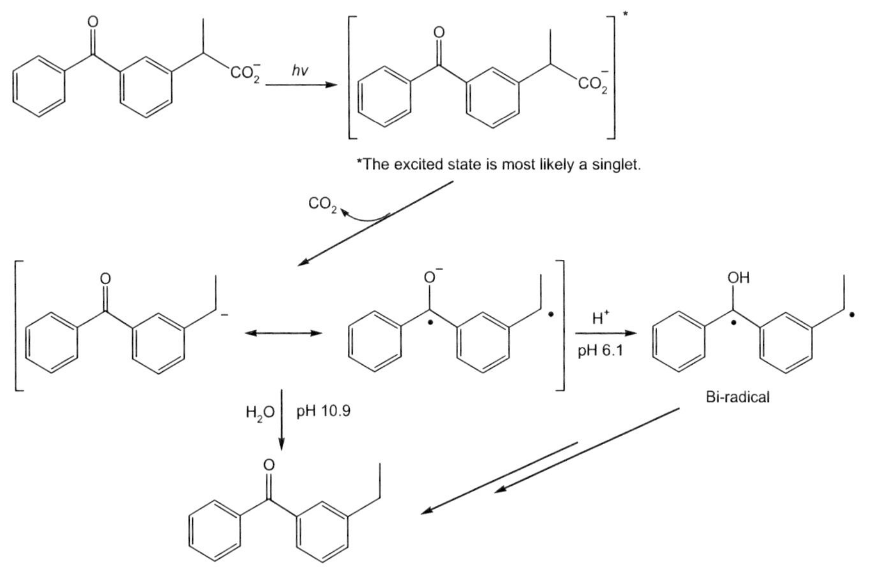

*The excited state is most likely a singlet.

Scheme 6.1

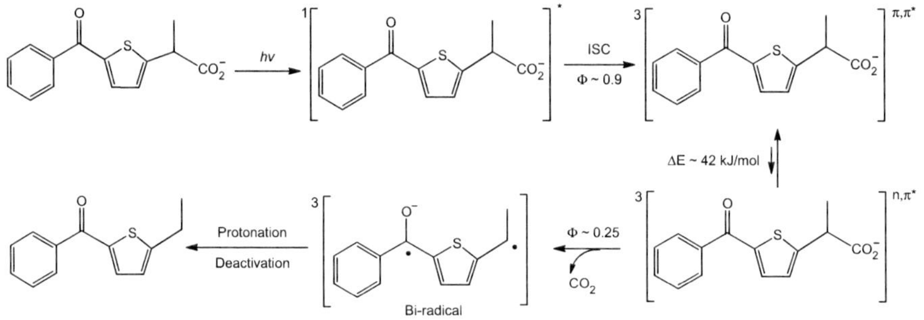

Scheme 6.2

formed. On the other hand, many 2-propionic acid derivatives of benzophenone or analogous NSAIDs undergo photochemical decarboxylation in the triplet state. Examples in this category include tiaprofenic acid, suprofen, and ketorolac. Tiaprofenic acid is said to be the most phototoxic NSAID, and it exerts its phototoxic effect through photosensitization.[9] After excitation to its lowest singlet state, tiaprofenic acid undergoes an efficient intersystem crossing to its lowest triplet state, which is a π,π^* triplet, with a Φ_{ISC} of ~ 0.9. This triplet state is interconvertible with the n,π^* triplet state which is only about 42 kJ mol^{-1} ($\sim 10\,\text{kcal mol}^{-1}$) more energetic than the former. Decarboxylation then takes place at the n,π^* state, giving rise to the decarboxylated product *via* a triplet bi-radical intermediate with a quantum yield of ~ 0.25.[10] The photodecarboxylation mechanism of tiaprofenic acid is illustrated in Scheme 6.2.

The decarboxylated product still contains the same chromophore and in the presence of oxygen, the excited triplet states of tiaprofenic acid and its photodegradant can be quenched by molecular oxygen to produce singlet oxygen. The singlet oxygen together with superoxide anion radical and the subsequently formed hydrogen peroxide contribute collectively to the phototoxicity of the drug. This topic will be discussed in detail in Section 6.3.

Suprofen[11,12] and ketorolac[13] both undergo the same or a similar photodecarboxylation pathway to that shown in Scheme 6.2; obviously, the respective quantum yields and energetics of their n,π^* and π,π^* triplets may be different in each case.

Similar to the NSAIDs containing benzophenone and heterocyclic benzophenone moieties, the photodegradation of naproxen in an aqueous environment is mainly decarboxylation under anaerobic condition.[14] However, its decarboxylation mechanism (Scheme 6.3) differs from those of the former drugs.

As illustrated in Scheme 6.3, the singlet excited state of naproxen, formed upon irradiation, ejects an electron into the aqueous medium to produce a solvated electron and a carboxyl radical.[15] The latter then undergoes decarboxylation to give rise to a carbon-centered radical. Under anaerobic conditions, the major degradant is 2-ethyl-6-methoxynaphthalene, formed by abstracting a hydrogen from a hydrogen donor. In the presence of oxygen, two photooxidative degradants, 2-(1-hydroxyethyl)-6-methoxynaphthalene and 2-acetyl-6-methoxynaphthalene, predominate. The overall quantum yields for the decarboxylated degradation products were reported to be 0.012 and 0.001, respectively, under aerobic and anaerobic conditions.[15] These results indicate that the overall degradation due to photodecarboxylation is much more significant under aerobic conditions than under anaerobic ones. A fairly significant portion ($\Phi = 0.28$)[16] of the singlet state molecules cross over to the triplet state, which are then quenched by ground state oxygen to generate

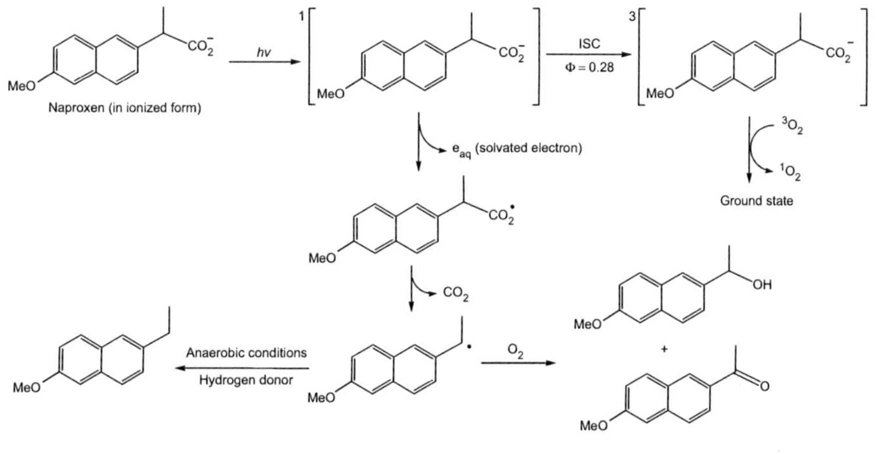

Scheme 6.3

singlet oxygen. The high efficiency of the latter step contributes to the photo-toxicity of naproxen.[14]

Carprofen is another NSAID that has a 2-arylpropionic acid moiety; the aryl part of the molecule is a chlorocarbazole ring. In contrast to the majority of the NSAIDs in the 2-arylpropionic acid family, photodecarboxylation is only a minor degradation pathway in the photodegradation of carprofen, which occurs in its singlet state.[17] Its major photodegradation pathways are poly-merization and dechloronation dependent upon the availability of hydrogen donors.[17,18] These two dominant pathways will be discussed later.

6.2.2   Photoisomerization

6.2.2.1   *Cis–Trans Isomerization Around Carbon–Carbon, Carbon–Heteroatom, and Heteroatom–Heteroatom Double Bonds*

One of the photochemical reactions of carbon–carbon, some carbon–hetero-atom, and heteroatom – heteroatom double bonds is *cis–trans* isomerization which takes place around the double bonds. The non-conjugated double bonds are practically photochemically inactive owing to their lack of absorption in the UV-A and UV-B regions. Stilbene and related compounds are α, β-substituted olefins by aromatic substituents. As such, they usually have reasonable absorption in the UV-A and UV-B regions and, hence, are readily susceptible to *cis–trans* isomerization. One good example is the photoisomerization of montelukast,[19] which contains a stilbene-like chromophore. When exposing a montelukast solution under ambient laboratory lighting, it was found that the majority of the drug was converted to its *cis*- or Z-isomer in four days (Scheme 6.4).[20]

During the pharmaceutical development of an avermectin analog, L-648,548, Wang *et al.* observed two unknown degradants in an animal health formulation, in addition to the two known oxidative degradants, 5-oxo and 8a-oxo degra-dants.[21] The first unknown degradant was identified as a dehydration product of the 5-oxo degradant. The dehydration resulted in the formation of a substituted phenol ring, which is now conjugated with the butadiene moiety of the macro-cyclic ring. A majority of the conjugated phenol degradant quickly isomerized to the (8,9)-Z-isomer, that is, phenol degradant 2, upon exposure to visible light. The overall degradation pathways of L-648,548 are summarized in Scheme 6.5.

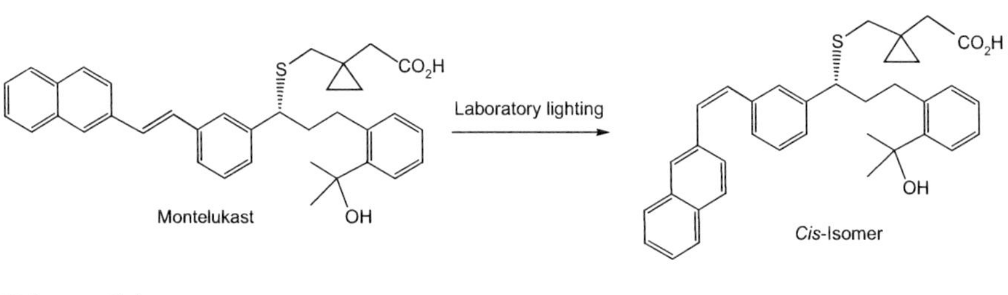

Scheme 6.4

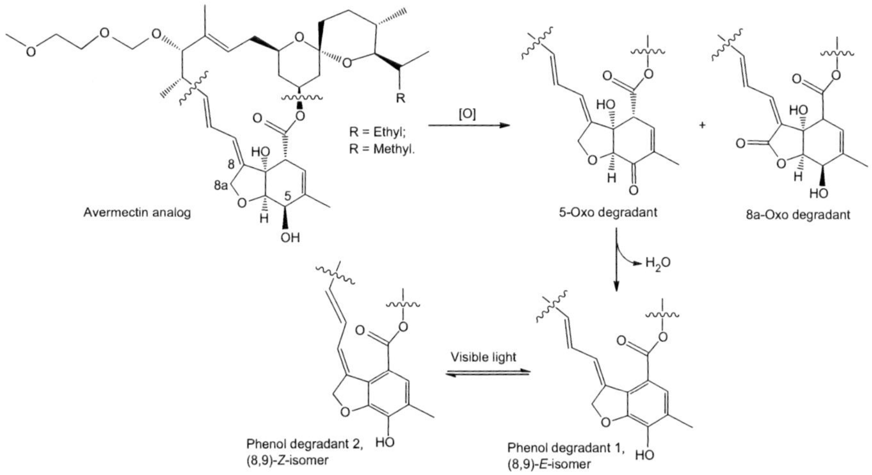

Scheme 6.5

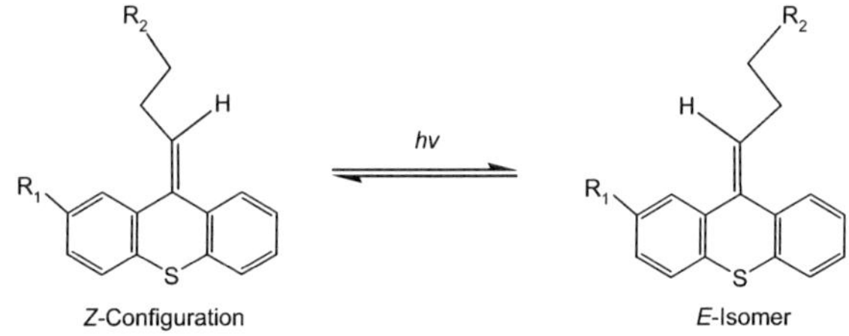

Chlorprothixene, R_1 = -Cl, R_2 = -N(Me)$_2$;
Zuclopenthixol, R_1 = -Cl, R_2 = 4-(2-Hydroxyethyl)piperizinyl;
Flupenthixol, R_1 = -CF$_3$, R_2 = 4-(2-Hydroxyethyl)piperizinyl;
Thiothixene, R_1 = -S(O$_2$)N(Me)$_2$, R_2 = 4-(2-Hydroxyethyl)piperizinyl.

Scheme 6.6

The carbon–carbon double bond in a number of thioxanthine antipsychotic drugs, like chlorprothixene, zuclopenthixol, flupenthixol, and thiotixene, is conjugated to the tricyclic aromatic ring. As such these drug molecules are also subject to photoisomerization around the double bond until an equilibrium is reached between the *Z*- and *E*-isomers (Scheme 6.6).[22] The *Z*-forms of these drugs are more active than their *E*-isomers. Hence, the ratio between the two isomers needs to be controlled in order to maintain the proper potency of the drugs.

Retinoic acid (tretinoin) contains an extensively conjugated system consisting of five conjugated carbon–carbon double bonds and a carboxyl group. It is a yellow to light orange color compound that has very strong absorption in UV-A region with a maximum absorption at ∼350 nm.[23,24] It is therefore not surprising that irradiation in UV-A region was found to be the major culprit in

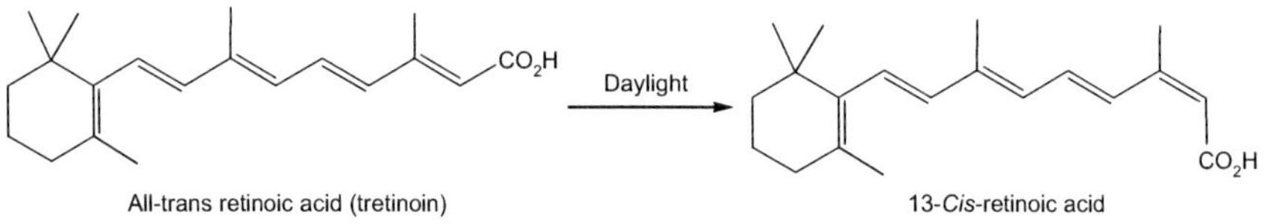

All-trans retinoic acid (tretinoin)   13-*Cis*-retinoic acid

Scheme 6.7

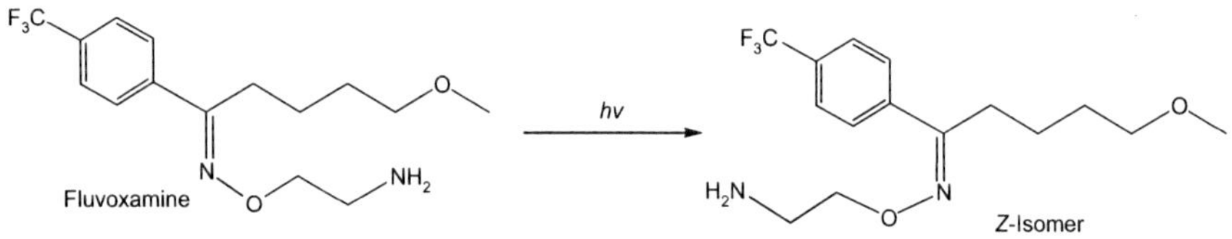

Scheme 6.8

its photodegradation giving several positional isomers,[25] among which 13-*cis*-retinoic acid (isotretinoin) is the major one (Scheme 6.7).[24] The minor isomeric degradants are believed to result from rotation around other double bonds but their exact structures were not determined.

   Drugs containing oxime or oxime ether moieties are susceptible to isomerization around the carbon–nitrogen double bond. The difference between oxime and stilbene in terms of their isomerization behavior is that oxime can also isomerize under non-photochemical conditions, in particular under catalysis by acid or base.[26] Fluvoxamine, a selective serotonin reuptake inhibitor, is an oxime ether which can be considered to be a condensation product between a valerophenone and an *O*-alkylhydroxamine. The conjugated oxime chromophore has UV absorption up to 300 nm. Under irradiation by lamps covering either the UV-A or UV-B region, fluvoxamine is converted to the Z-isomer (Scheme 6.8).[27,28]

   A number of cephalosporin antibiotics, for example, cefotaxime, contain the oxime ether moiety and hence, can also undergo the same isomerization as long as their chromophores absorb adequately in the UV-A and UV-B regions.[29]

6.2.2.2   Photoisomerization of Drugs Containing a 2,5-Cyclodienone Ring

A large number of corticosteroids such as prednisolone,[30] betamethasone,[31] dexamethasone,[32] and their related esters[33,35] have the pregna-1,4-diene-3,20-dione core structure, of which the 1,4-diene-3-one A ring (*i.e.* cross-conjugated 2,5-cyclohexadienone) is particularly photoreactive. Under irradiation by UV-A and UV-B light sources, the so-called "lumi" isomeric degradant was formed quite efficiently, as evidenced by their quantum yields being generally in the range 0.1–0.3.[36] This photoisomerization should take place in the triplet excited state according to a study by Ricci *et al.*[37] The initially formed lumi degradant can isomerize again to form a "photolumi" degradant upon further

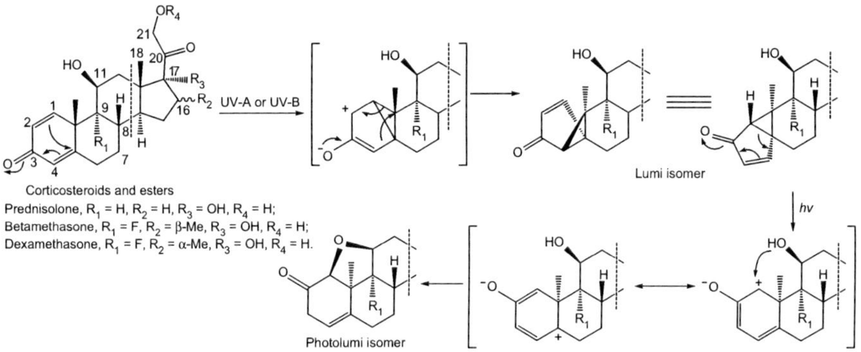

Scheme 6.9

irradiation.[35] This photodegradation process, which was proposed *via* a dipolar intermediate,[32] is illustrated in Scheme 6.9.

It should be pointed out that in the original photodegradation study of betamethasone performed by Hidaka *et al.*[31] the orientation of the C_{10} methyl group in the lumi degradant was incorrectly assigned to the β-position. As shown in Scheme 6.9, the orientation of C_{10} methyl is inverted during attack by the 3,4-double bond. Unfortunately, this error was propagated in some of the literature[34] citing the original work.

When these steroids are irradiated with UV-B light over prolonged periods, a minor photodegradant due to the Norris type I photochemistry at the C_{20} ketone can also be formed. Beclomethasone-17,21-dipropionate, which contains a chlorine at the 9α-position, does not undergo the "lumi" rearrangement, which was attributed to the quenching of the cyclodienone triplet state by chlorine (heavy atom effect).[36] Instead, its photodegradation is solely due to the C_{20} ketone photochemistry. Photodegradation caused by a ketone group will be discussed in Section 6.2.8.

6.2.2.3 Photo-Fries Rearrangement

The photo-Fries rearrangement is an intramolecular rearrangement of the acyl group in phenyl esters and related compounds to the *ortho-* and/or *para*-position of the phenyl ring.[38–41] Benorylate is a co-drug which is made by linking acetaminophen and aspirin through an ester bond. Photodegradation of the drug caused the formation of several degradants, among which the isomeric degradant, formed by a photo-Fries rearrangement, was the major one.[42] The initially formed isomer is a benzophenone derivative which readily underwent transacylation to produce a second isomeric degradant as illustrated in Scheme 6.10.

In the mechanism of the photo-Fries rearrangement, it appears that phenoxyl and acyl radicals are generated from the excited singlet state within a solvent-caged complex, according to the mechanistic study of photo-Fries

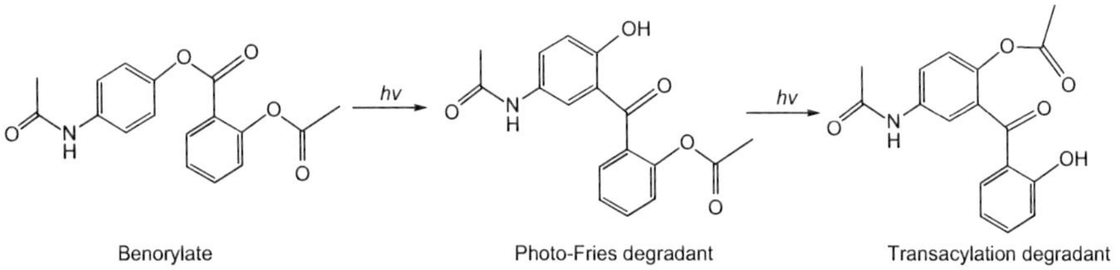

Benorylate   Photo-Fries degradant   Transacylation degradant

Scheme 6.10

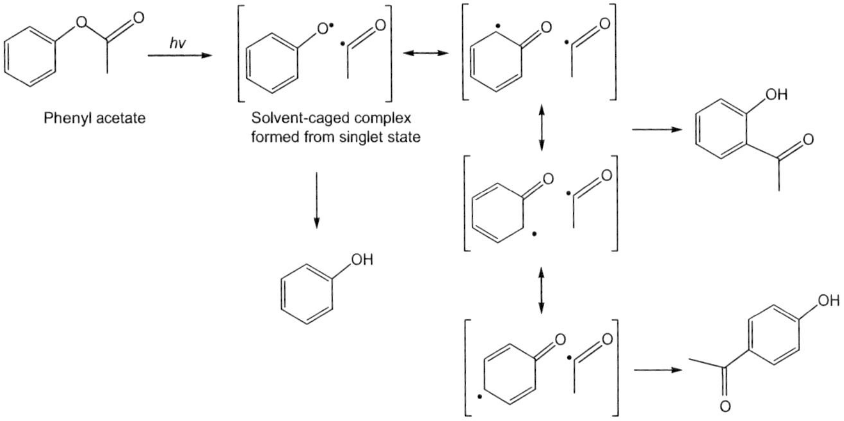

Phenyl acetate   Solvent-caged complex
                            formed from singlet state

Scheme 6.11

rearrangement of phenyl acetate performed by Kalmus and Hercules.[43] The phenoxyl radical has a resonance contribution from the three cyclohexadienone forms and recombination of the acyl radical with these forms should give rise to photorearrangement products. Some of phenoxyl radical can escape from the solvent cage to produce phenol. The mechanism is shown in Scheme 6.11.

6.2.3   Aromatization of 1,4-Dihydropyridine Class of Drugs

There are quite a few drugs, for example, amlodipine, isradipine, nifedipine, and nisoldipine, in the 1,4-dihydropyridine family that are calcium channel blockers, used in the treatment of hypertension. The chromophore of these drug molecules, that is, the substituted 1,4-dihydropyridine ring, is susceptible to photochemical degradation resulting in the formation of an aromatized pyridine ring. Fasani *et al.* performed a mechanistic study of the aromatization degradation of amlodipine.[44] They found that this degradation process, which can be regarded as an oxidation process, is independent of oxygen, as the photodegradation reaction remains the same under both aerobic and anaerobic conditions. Furthermore, the photodegradation was essentially not influenced by other experimental conditions, such as solvents, which led the authors to

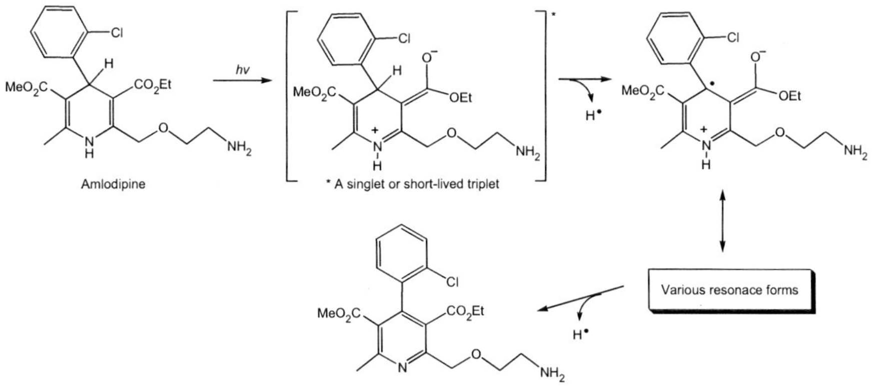

Scheme 6.12

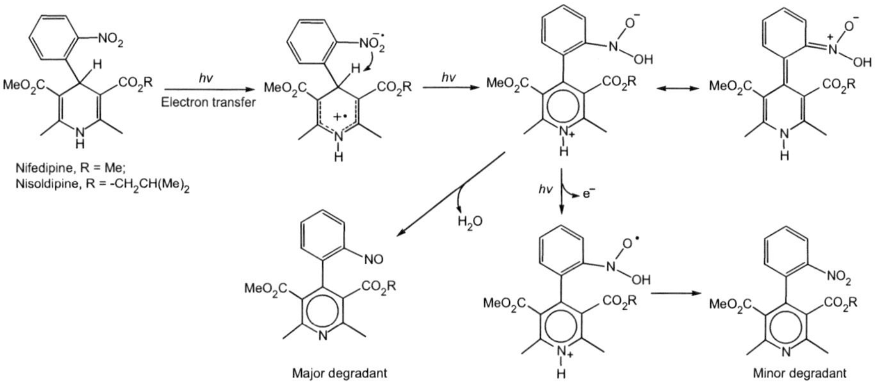

Scheme 6.13

conclude that the photodegradation process must occur from an excited singlet or a very short-lived triplet. Lastly, the authors estimated that the quantum yield (Φ) of this photodegradation is ~ 0.001 when the photolysis was conducted under UV-A irradiation. Despite its low quantum yield, amlodipine, as well as other drugs in this family, is still considered a photolabile drug, owing to its strong absorption coefficiency, ε, in the UV-A region. As pointed out by the authors, it is the product of Φ times ε, which is relevant practically in the photodegradation of drugs. The mechanism for aromatization of amlodipine by Fasani *et al.* is shown in Scheme 6.12.

The key step in the above mechanism is the loss of the hydrogen radical at the 4-position of the 1,4-dihydropyridine ring; the resulting radical can be stabilized by various resonance forms. This mechanism should be applicable to other drugs in the 1,4-dihydropyridine family, except for nifedipine and nisoldipine which contain a 2′-nitro group on the phenyl ring. The presence of the 2′-nitro group leads to a different mechanism (Scheme 6.13) for the photoaromatization according to a mechanistic study of the drugs and related compounds carried out by Fasani *et al.*[45] Upon irradiation, the 1,4-dihydropyridine moiety is activated in the lowest excited singlet state, which is then immediately quenched by the 2′-nitrophenyl group through electron transfer from the former to the latter. This event should result in the formation of an intramolecular radical ion pair. The radical ion pair would readily abstract the 4-hydrogen and the zwitterion formed can dehydrate to produce the major 2′-nitrosopyridine photodegradant. A minor portion of the radical ion pair would give rise to the 2′-nitropyridine photodegradant *via* a radical cation by ejecting an electron. The evidence supporting this mechanism includes the following observations. First, the analogous 4-methyl-1,4-dihydropyridine compound has a blue fluorescence, while nifedipine and nisoldipine do not fluoresce. Second, the redox potentials of 1,4-dihydropyridine and 2′-nitrophenyl moieties indicate that the former tends to give away an electron, while the latter tends to accept an electron. Third, the key zwitterion intermediate and solvated electron are observed in experiments performed in solid matrix at low temperature.

As illustrated by the above mechanism, the presence of the 2′-nitro group on the phenyl ring greatly facilitates the photodegradation of the two drugs, displaying a quantum yield of ~ 0.3.[45]

6.2.4 Dehalogenation of Aryl Halides

Photochemical dehalogenation is a widely observed photochemical degradation pathway for aryl halides. Since a key step in this process is the breakage of the C–X bond, in general, the ease of photodehalogenation correlates inversely with the bond dissociation energies (BDE) of the C–X bond involved. Using phenyl halides as the model compounds for aryl halides, the BDE of the C–X bonds are 67.2, 82.6, 97.6, and 101 $kcal\,mol^{-1}$,[46,47] respectively, from iodo- to fluorobenzene. Hence, aryliodines are expected to be most susceptible to dehalogenation, followed respectively by arylbromines, arylchlorines, and arylfluorines. Drug molecules containing the four different halogens have all been reported to undergo photochemical dehalogenation. For example, under irradiation at 313 nm in a phosphate buffer, diflunisal was found to undergo defluorination as part of an overall photochemical $S_{RN}1$ degradation process,[48] in which a diflunisal dimer was formed (Scheme 6.14).[49]

Under anaerobic conditions, the diflunisal dimer was the only photodegradant formed. In the presence of oxygen, a number of unidentified, presumably oxidative photodegradants were also produced in addition to the dimer. Studies with various scavengers/quenchers, such as butylated hydroxyanisole (BHA) and reduced glutathione (GSH) (radical scavengers), superoxide dismutase (SOD) (superoxide anion scavenger), mannitol (hydroxyl radical scavenger) and sodium azide (singlet oxygen quencher), suggest that superoxide anion, hydroxyl radical and singlet oxygen were involved in the formation of oxidative photodegradants. Another point that should be noted is

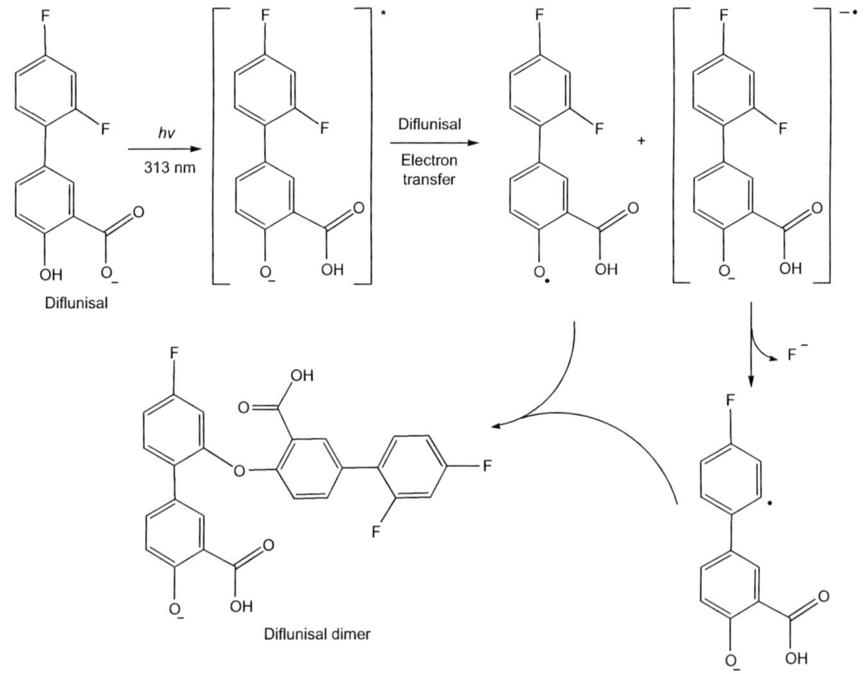

Scheme 6.14

that defluorination occurs at the 2-fluoro position (*ortho* to the 3-carboxyl-4-hydroxyphenyl group). The 2-fluoro group should have more steric hindrance compared to the 4-fluoro group and as such the former would be expected to be more reactive in the excited state. This is consistent with the observation made by Grimshaw and de Silva that *ortho*-halogen substituents of aryl halides are in general more photolabile.[50]

Carprofen is an aryl-2-propionic acid NSAID with chlorocarbazole as the aryl portion. As mentioned previously, decarboxylation is a minor degradation pathway for carprofen, which occurs from its excited singlet state. A significant portion of the singlet converts to the triplet state, which was proposed to form an excimer with a ground state carprofen molecule.[14] The excimer would produce a radical anion and a radical cation through an internal charge transfer. The radical anion can then be dechlorinated by releasing a chloride ion to produce a carbazole-based radical. Meanwhile, the radical cation can produce a chlorocarbazole-based radical by releasing a proton. In the presence of hydrogen donors (such as organic solvents), both radicals can abstract a hydrogen, respectively, to form deschlorocarprofen or give back carprofen. On the hand, the radicals can initiate polymerization of carprofen in the absence of hydrogen donors.[17] The complete photodegradation pathways are illustrated in Scheme 6.15.

The two case studies presented above were both proposed to involve radical anion intermediates from which dehalogenation occurs. For other arylhalide

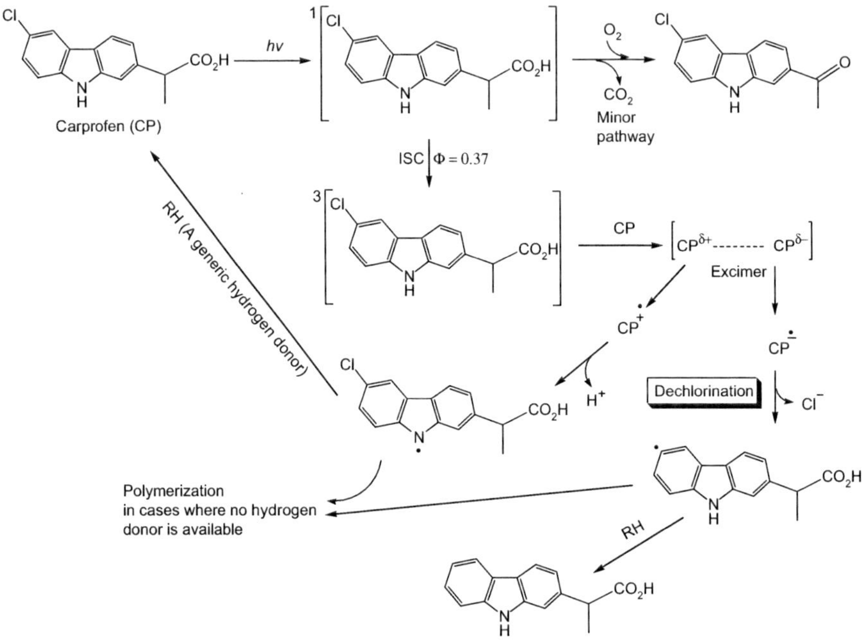

Scheme 6.15

drugs, such as chlorpromazine, a phenothiazine tranquilizer which is well known for its phototoxicity, homolytic cleavage of the carbon–chlorine bond without a charge transfer step was proposed (pathway a, Scheme 6.16)[51] for photodegradation under irradiation by sunlight.

In a separate study by Motten *et al.*, the corresponding carbon-based promazinyl radical, which would be expected from homolytic cleavage, was observed by ESR spin-trapping experiments.[52] According to a study by Chignell *et al.*, the promazinyl radical is most likely to be the species responsible for the phototoxicity of the drug.[53] When irradiated under either 254 or 347 nm, photoionization of chlorpromazine was observed (pathway b, Scheme 6.16),[54,55] during which process an electron was ejected from chlorpromazine to produce the solvated electron. The latter can react with a chlorpromazine molecule in the ground state to effect dechlorination *via* an anion radical intermediate.

Galli and Pau examined the impact on C–X BDE of the addition of one electron to the subject aryl halides[46] and found that the BDE values decrease dramatically in the corresponding aryl halides. For example, the BDE of the C–Br bond in bromobenzene is lowered by approximately 75% once it takes in an electron. Hence, it would be much more favorable for arylfluorines, aryl-chlorines, and arylbromides to proceed *via* an anion radical intermediate when electron transfer is possible. In the case of aryliodides, since they contain a low energy σ^* orbital, which is absent in other aryl halides, homolytic C–I bond

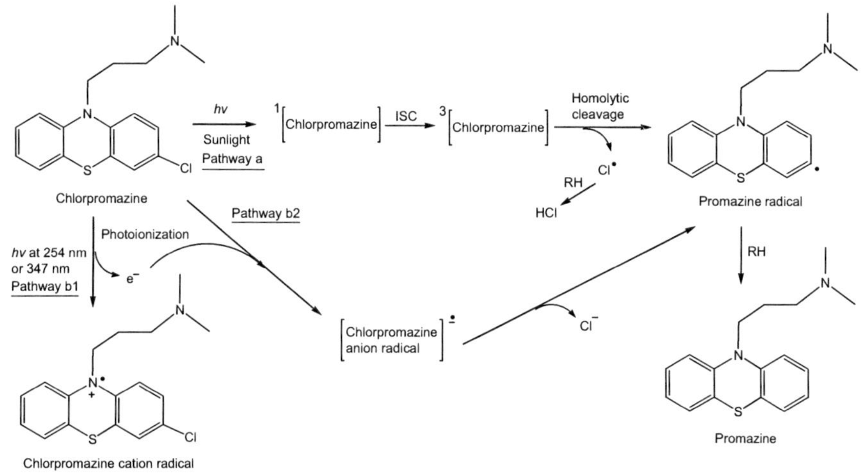

Scheme 6.16

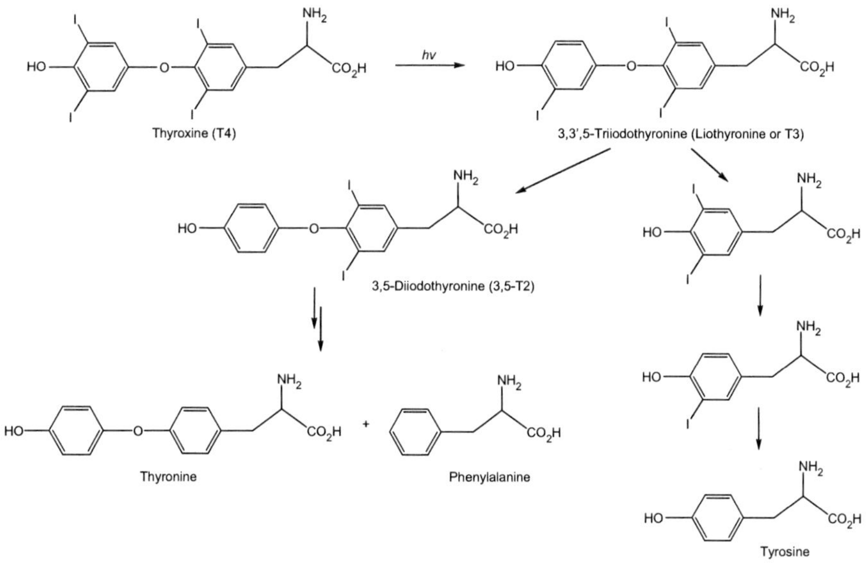

Scheme 6.17

cleavage without charge transfer would be possible from their $n\sigma^*$ or $\sigma\sigma^*$ excited states.[50] This homolytic C–I dissociation mechanism may explain the facile, consecutive deiodination observed with thyroxine (T4). Under irradiation by UV-A light, thyroxine quickly deiodinates to form 3,5,3'-triiodothyronine (T3).[56] Upon further photolysis, all the remaining three iodo groups are eventually cleaved, along with side chain breakage (Scheme 6.17).[57]

6.2.5   Cyclization in Polyaromatic Ring Systems

Photocyclization in molecules containing polyaromatic rings is a commonly observed phenomenon.[58] For aryl halides containing multiple aromatic rings, one likely event following dehalogenation is the cyclization of two adjacent rings. For examples, diclofenac and meclofenamic acid are two NSAIDs that contain a diphenylamine core structure. Upon irradiation, both drug molecules undergo dechlorination which is accompanied by the cyclization of the two substituted phenyl rings, resulting in tricyclic chlorocarbazole rings. Since the dehalogenation of aryl halides usually produces free radicals by proceeding through the anion radical intermediates discussed in previous sub-sections, it had been assumed that the cyclization was mediated *via* a free radical intermediate. Nevertheless, based on a study by Encinas *et al.*,[59] the cyclization does not involve a free radical intermediate (pathway a, Scheme 6.18). Instead, these researchers proposed that cyclization occurs at an excited triplet state through a 6π electrocyclization mechanism (pathway b, Scheme 6.18).

The readily formed chlorocarbazoles dechlorinate to yield the corresponding carbazole radicals. Diclofenac also undergoes facile decarboxylation and the calculated energy barriers for the decarboxylation and dechlorination of the excited deprotonated carbazole are only 4.5 and $6.2\,\mathrm{kcal\,mol^{-1}}$, respectively.[60] Photolysis of the diclofenac-derived chlorocarbazole has been demonstrated to cause cell toxicity much more efficiently than diclofenac itself.[61]

Stilbene and stilbenoid compounds are 1,2-diarylethylenes whose two aromatic rings are connected by a carbon–carbon double bond. As we have discussed before, stilbene and its analogs undergo E/Z isomerization upon irradiation. With further irradiation, the two aromatic rings of the Z-isomers can cyclize to form an additional ring. In the presence of oxygen, the latter ring is usually further oxidized to afford a new fused aromatic ring. This oxidative cyclization as illustrated in Scheme 6.19 is a well known photochemical process for stilbene and stilbenoids and synthetically it is the method of choice for making polycondensed arenes.[62,63] The key cyclization step is a spin allowed, conrotatory electrocyclization process involving the lowest unoccupied molecular orbit (LUMO) of six π electrons.

Quite a few drug molecules contain stilbene-like moieties and two good examples are the anticancer drug, tamoxifen and the cholesterol-lowering drug, atorvastatin (Figure 6.2).[64,65]

In the former case, further photolysis resulted in cleavage of the amino-alkoxyl substituent. In the latter case, the center pyrrole ring of the cyclized degradant was degraded by singlet oxygen, discussed in more detail in Section 6.3.2.

In other cases, cyclization of aromatic rings takes place through the substituents of the rings. Telmisartan, a relatively new angiotensin II receptor antagonist, is such an example. It possesses a 2-carboxylbiphenyl moiety. Irradiation of an acidic telmisartan solution in a photostability chamber with both UV and visible lamps for 13 days caused the formation of a photocyclized product in approximately 15% yield.[66] The authors proposed a diradical

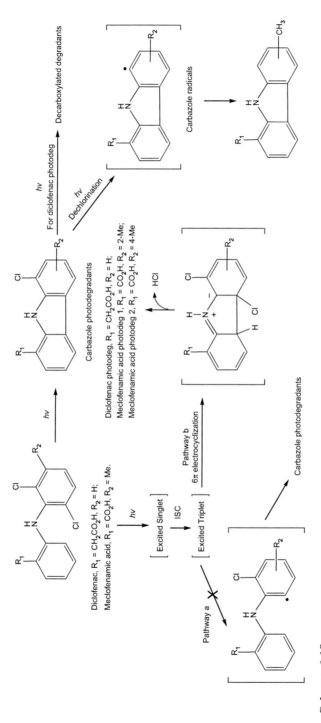

Scheme 6.18

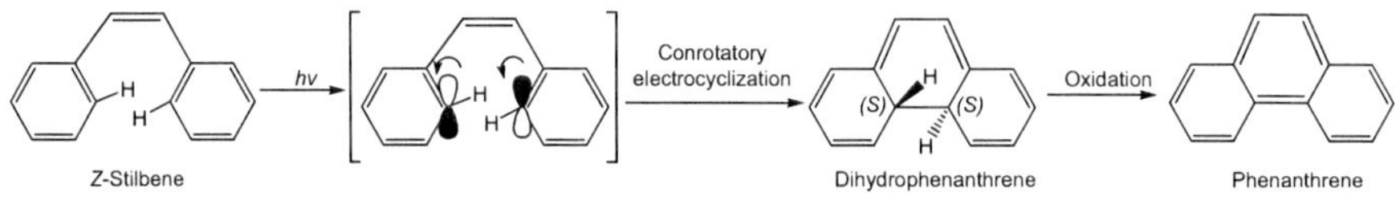

Scheme 6.19

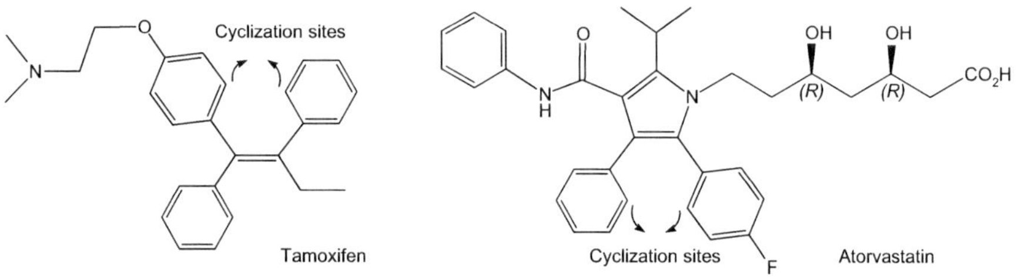

Figure 6.2 Anticancer drug, tamoxifen and cholesterol-lowering drug, atorvastatin.

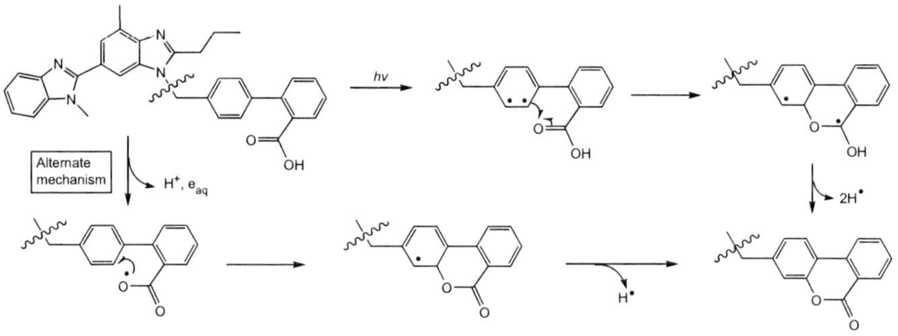

Scheme 6.20

mechanism mediated *via* the excitation of two π electrons to a non-bonding orbital (Scheme 6.20). Alternatively, a carboxyl radical could be generated through ejection of a proton and a solvated electron. The resulting oxygen-based radical would then attack the adjacent phenyl group to afford the new lactone ring.

Similar photocyclization of a 2-tetrazole-substituted biphenyl moiety can also be found in the photodegradation of irbesartan.[67]

6.2.6 Photochemical Elimination

Norfloxacin and ciprofloxacin are fluoroquinolone antibacterial drugs and their substituted quinolone rings are photolabile under both UV and sunlight. The two major photodegradation products result from the sequential destruction of the 7-piperazinyl substituent, as shown in Scheme 6.21.

The outcome of this photodegradation is equivalent to a photochemical elimination, although it is very likely to be mediated through a number of photooxidative degradants. In a photodegradation study of ofloxacin, oxidation occurred on the methyl and two methylene groups of the substituted piperazine ring.[68]

Clinafloxacin is another quinolone antibacterial drug that has different structural feature from norfloxacin and ciprofloxacin: it has a 3-aminopyrrolidinyl group rather than the piperazinyl group at the 7-position and an extra chloro group at the 8-position. Lovdahl and Priebe found that clinafloxacin incurs additional degradation because of the presence of these two substituents, in addition to a photoelimination pathway (pathway a, Scheme 6.22) similar to that observed in the cases of norfloxacin and ciprofloxacin.[69]

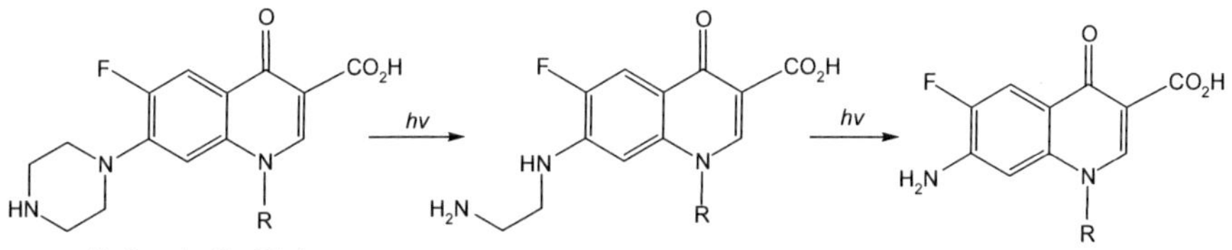

Norfloxacin, R = Ethyl;
Ciprofloxacin, R = Cyclopropyl.

Scheme 6.21

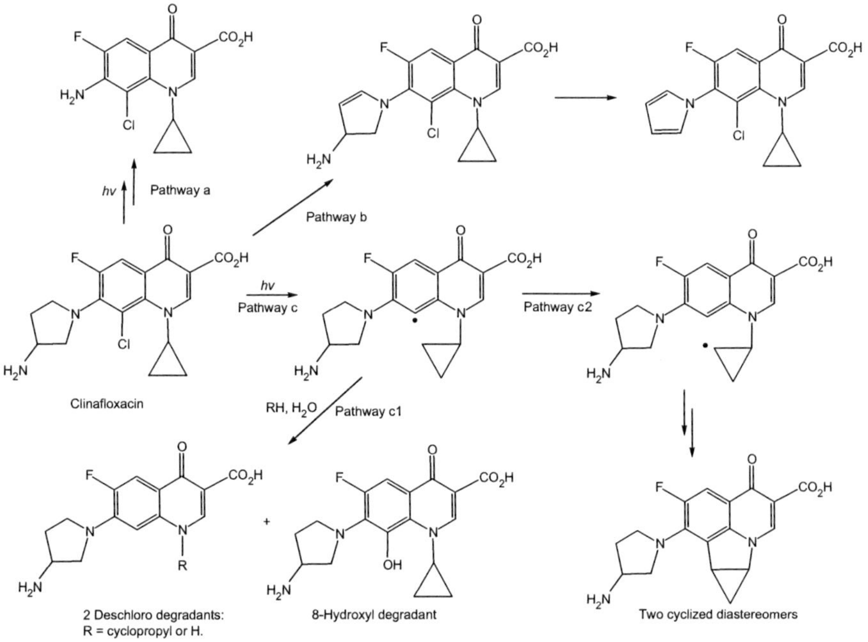

Clinafloxacin

2 Deschloro degradants:
R = cyclopropyl or H.

8-Hydroxyl degradant

Two cyclized diastereomers

Scheme 6.22

Additionally, the 7-(3-aminopyrrolidinyl) group was found to undergo apparent photooxidation which gave rise to the 7-dehydropyrrolidinyl and 7-pyrrolyl degradants, probably in a sequential fashion (pathway b). On the other hand, photodechlorination resulted in a number of degradants including two unusual cyclized degradants (pathway c). The researchers did not offer formation mechanisms for these degradation products. Based on what we have discussed so far in this chapter, it appears that dechlorination would be most likely proceed *via* a free radical pathway, from which the 8-hydroxyl and all other deschloro degradants were formed (pathways c1 and c2). Among approximately seven major degradants were two diastereomeric, cyclized degradants. The cyclization probably started from the abstraction of a hydrogen from the 2-position of cyclopropyl ring by the phenyl radical produced by dechlorination (pathway c2).

6.2.7 Photodimerization and Photopolymerization

$2 + 2$ photocyclization is a very common photodimerization pathway, which can occur in both solution and solid states, although the stereochemistry of the cyclized products may be different under the two different conditions. A good example can be found in the case of menadione, also known as vitamin K_3, whose photochemical degradation has been well studied.[70–73] When solid menadione was irradiated under sunlight, two *syn* dimers with head-to-head and head-to-tail configurations were formed (Scheme 6.23, pathway a). On the other hand, when a solution of menadione or a solid mixture dispersed over silica was irradiated under a "black light" UV lamp, which emits UV light in

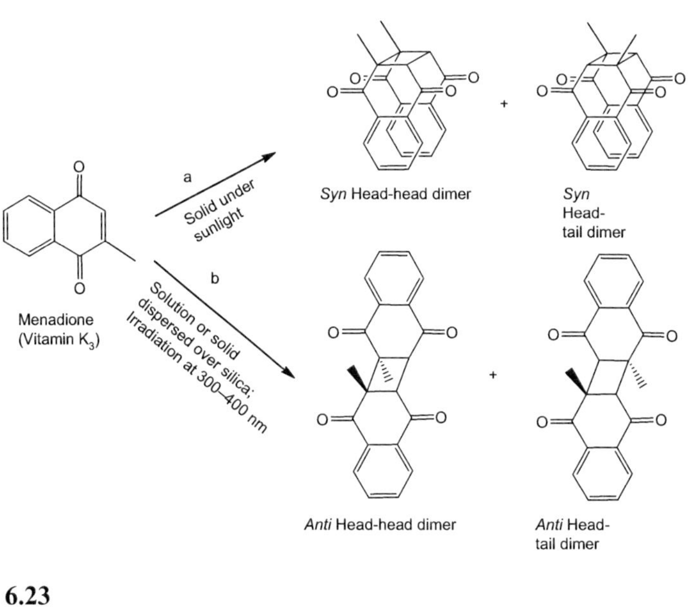

Syn Head-head dimer

Syn Head-tail dimer

Menadione (Vitamin K_3)

a
Solid under sunlight

b
Solution or solid dispersed over silica; Irradiation at 300–400 nm

Anti Head-head dimer

Anti Head-tail dimer

Scheme 6.23

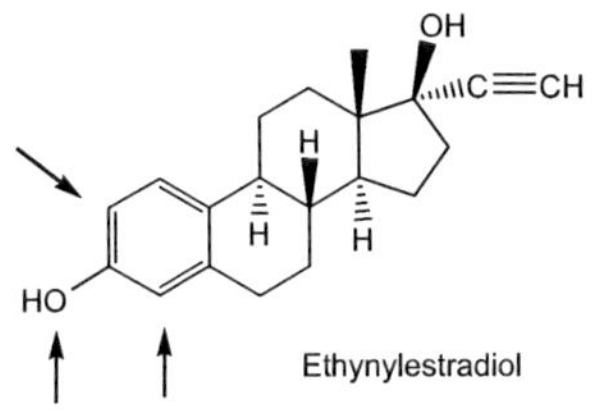

Ethynylestradiol

The arrows indicate the connecting points
for dimerization

Figure 6.3 Ethynylestradiol.

300 to 400 nm, two anti-dimers with head-to-head and head-to-tail configurations were produced (pathway b).[71]

Photochemical dimerization can also occur *via* many other mechanisms. For example, drugs containing phenolic moieties can readily undergo photodimerization between the photochemically generated carbon- and oxygen-based free radicals. Many of such radicals are the same as those produced by autooxidation. For example, photochemical stress of an ethynylestradiol solution in acetonitrile under xenon lamp irradiation (with a 310 nm cutoff filter) produced five regio-dimers (Figure 6.3), all of which were also observed in free radical-mediated oxidation of the drug.

A good example of drug degradation caused by photopolymerization is the photolysis of carprofen in phosphate buffers, which has been discussed in previous sections, particular in Section 6.2.4.[14] In such cases, the phosphate buffers are incapable of providing hydrogen to quench the various carprofen radicals generated from the photolysis. Consequently, polymerization of carprofen occurs.

6.2.8 Photochemistry of Ketones: Norris Type I and II Photoreactions

Non-conjugated ketones have a very weak UV absorption at ~ 280 nm due to the $n \rightarrow \pi^*$ transition, and their absorption coefficient (ε) values are typically no greater than 50.[74] Certain substitutions to the group, can shift the λ_{max} to ~ 300 nm. When irradiated by light sources containing this wavelength range, an isolated C_{20} keto group in a great number of corticosteroids, such as the ones discussed in Section 6.2.2.2, can undergo both Norris type I and II photoreactions. Norris type I photodegradation of three representative corticosteroids, flumethasone, triamcinolone acetonide, and fluocinolone acetonide, is shown in Scheme 6.24, in which the C_{17}–C_{20} bond, α to the C_{20} keto group, is cleaved. The C_{17}-based radicals give rise to different final degradation products depending upon different substituents on the C_{17} position: in flumethasone, loss of hydrogen from the 17-hydroxyl group readily affords flumethasone 17-ketone, while in triamcinolone acetonide and fluocinolone acetonide, the C_{17} radicals can only abstract a hydrogen to give 17-H degradants.

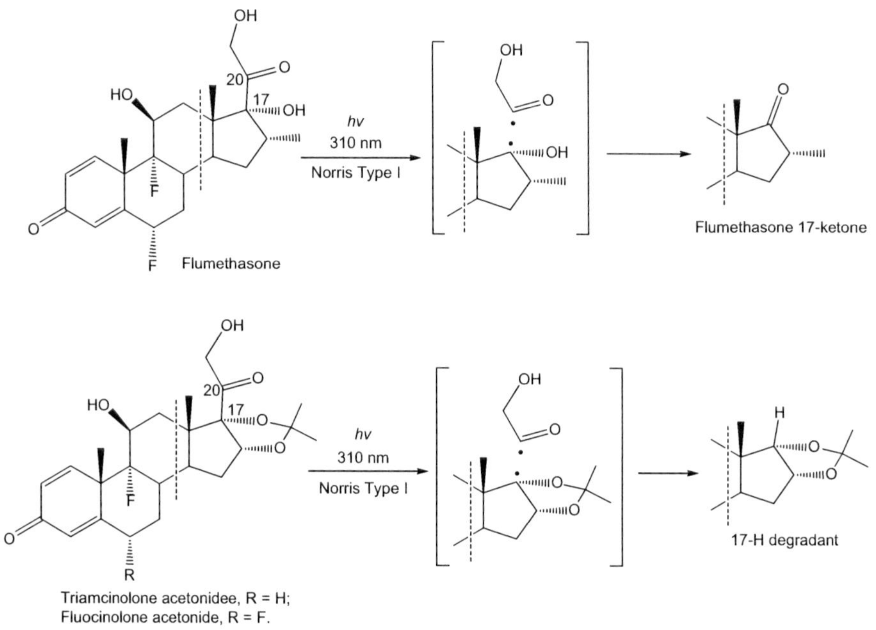

Scheme 6.24

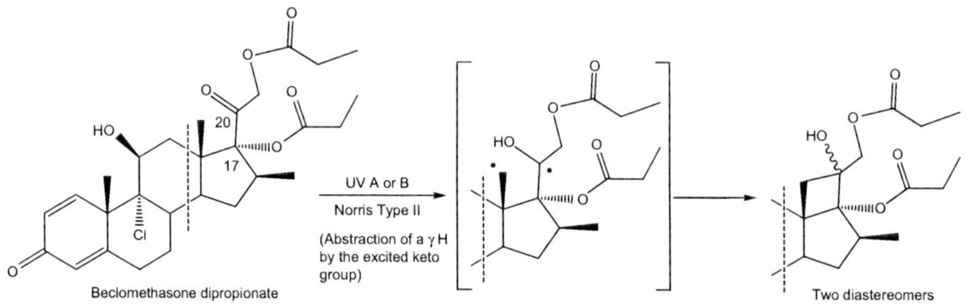

Scheme 6.25

In the case of beclomethasone-17,21-dipropionate, photochemistry of the A ring (isomerization; see Section 6.2.2.2) is completely suppressed by the 9-Cl group due to quenching of the triplet excited state of the A ring by the latter group. Consequently, the major photodegradants of beclomethasone-17,21-dipropionate are two diastereomers resulting from Norris type II photodegradation of the corticosteroid (Scheme 6.25).

The overall quantum yield of the C_{20} keto photochemistry is rather high, at ~ 0.3.[36] Nevertheless, the efficiency of formation of photodegradation products resulting from an isolated keto group (as we have discussed here) are quite low, in particular in cases where other photochemically more labile functional groups are present. This is due to the fact that the absorption of a keto group at 310 nm is very low, resulting in a low value of ε times Φ. As we mentioned previously in this chapter, the product of ε times Φ determines the ultimate photochemical degradation efficiency. Another conclusion that can be drawn from the discussion in this section is that the final product distribution is strongly influenced by the substituents that attach to or near the centers of photo-labile groups (such as the cross-conjugated cyclobutadienone and C_{20}-keto groups).

6.3 Oxidative Photochemical Degradation

In a real life scenario, photochemical degradation of a substrate almost always takes place in the presence of oxygen. Oxidative photochemical degradation occurs when oxygen gets involved in the process. As indicated previously, photochemical reaction is one of the deactivation events of an electronically excited state. The excited state (a sensitizer) can discharge an electron to molecular oxygen (to form superoxide anion radical initially) or abstract a hydrogen from the oxidation substrate itself. The oxidation substrate thus impacted typically gives rise to a carbon-based free radical, which can then react with $^3O_2$ to afford oxidative photodegradants such as peroxides, ketones/ aldehydes, and alcohols. On the other hand, the initially formed superoxide anion radical can produce hydrogen peroxide and then hydroxyl free radical (typically *via* Fenton chemistry). The latter two species, in particular hydroxyl free radical, are capable oxidants, both of which can cause various oxidative degradation pathways themselves. This photosensitized degradation process, involving initial free radical intermediate formation from the oxidation substrates and formation of a superoxide anion radical, hydrogen peroxide and ultimately hydroxyl free radical, is referred to as "type I photosensitized oxidation".[75] As discussed, type I photooxidation is similar to a regular auto-oxidation process in terms of degradation pathways and degradation product distribution, which can be very complex. On the other hand, the excited state can directly interact with molecular oxygen to produce singlet oxygen *via* energy transfer. This photosensitization process involving singlet oxygen is referred to as "type II photosensitized oxidation".[76]

A large number of photoactive compounds, including a sizable number of drugs, are capable of activating ground state molecular oxygen, which is a triplet species, *via* energy transfer from their photochemically excited triplet states to molecular oxygen. As a result, ground state molecular oxygen, $^3O_2$, is activated to become singlet oxygen, $^1O_2$.[75,76] The ground state oxygen, $^3O_2$, is also denoted by $O_2(^3\Sigma_g^-)$, while the lowest singlet oxygen is denoted by $O_2(^1\Delta_g)$. $^3\Sigma_g^-$ is the highest occupied molecular orbital of ground state oxygen containing two degenerate orbits, $^3\Sigma_{gx}^-$ and $^3\Sigma_{gy}^-$ with each one occupied by one electron. $^1\Delta_g$ is the lowest molecular orbit in singlet oxygen. For simplicity, we will use $^3O_2$ and $^1O_2$ in our discussion. The various type I and II photosensitized oxidative degradation pathways are summarized in Scheme 6.26.

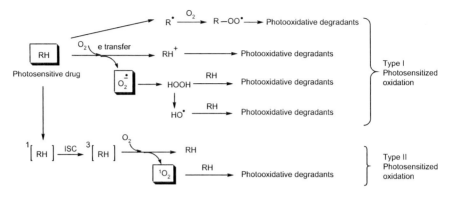

Schematic presentation of Type I and II photosensitized oxidation. Frequently, photosensitizer is the drug itself.

Scheme 6.26

Collectively, all these reactive oxygen species (ROS), superoxide anion radical, hydrogen peroxide, hydroxyl free radical, and singlet oxygen, contribute to a great number of oxidative photochemical events including photooxidative degradation and phototoxicity of drugs.

6.3.1 Type I Photosensitized Oxidation: Degradation *via* Radical Formation and Electron Transfer

As we have discussed in previous sections, free radicals are frequently generated during photochemical degradation events. One of the pathways for generating free radicals is through hydrogen abstraction *via* photosensitization. Regardless of their origin, these photochemically produced free radicals (typically triplet species) can readily react with molecular oxygen, also a triplet species, to form peroxide radicals. The latter can propagate and cause the formation of additional peroxides. The decomposition of peroxides can lead to the formation of hydroxyl radicals. In such cases, light acts as a free radical initiator and the subsequent process is similar to a typical free radical-mediated autooxidation process (for detailed pathways and discussion, see Section 3.2 of Chapter 3). Other outcomes of the decomposition of peroxides include the formation of the corresponding alcohols and ketones/aldehydes (refer to Section 3.5.8).

Indoprofen is a NSAID and its photochemical degradation behaves similarly to other NSAIDs in the arylpropionic acid family. In addition to undergoing straightforward decarboxylation, indoprofen can also transfer an electron from its carboxylate to molecular oxygen, during which process superoxide anion radical is generated. The carboxylate radical should readily decarboxylate. The resulting carbon-based free radical can then react with molecular oxygen to produce the corresponding peroxide and eventually the hydroxyl and keto degradants.[8] This oxidative photodegradation involving charge transfer and hydrogen abstraction is summarized in Scheme 6.27.

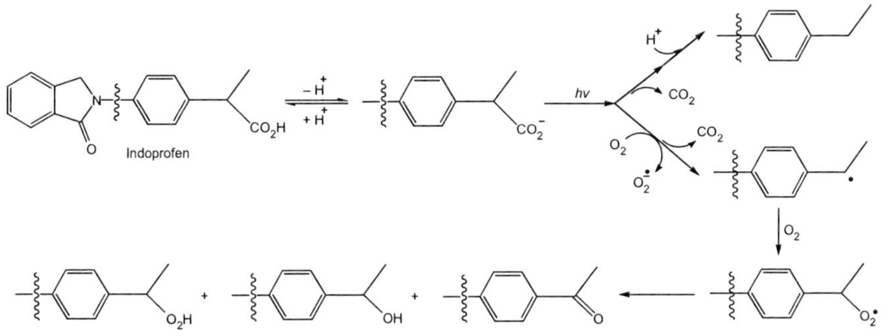

Scheme 6.27

Other NSAIDs in the arylpropionic acid family such as ketoprofen[14] and tiaprofenic acid[77] also give analogous peroxides and subsequent hydroxyl and keto degradants following photodecarboxylation. However, a notable difference between the photodegradation of the three drugs is that the photodegradation of indoprofen appears to occur only from its singlet excited state, while the photodegradation of ketoprofen and tiaprofenic acid appears to occur mostly from their triplet excited states. Consequently, the photolysis of ketoprofen and tiaprofenic acid would not only generate superoxide anion and hydrogen peroxide (evidence for the charge transfer mechanism), but also singlet oxygen (evidence for energy transfer from its triplet to molecular oxygen). The quantum yields for both the charge transfer and energy transfer processes appear to be quite high. For singlet generation, a yield of 0.39 was reported for ketoprofen[14] and 0.22 for tiaprofenic acid.[78] Singlet oxygen thus generated has unique reaction pathways, discussed in the following section.

6.3.2 Type II Photosensitized Oxidation: Degradation Caused by Singlet Oxygen

Singlet oxygen plays an important role in many natural photooxidative degradation processes.[79] It is also a key player in oxidative photodegradation of drugs. The photochemical generation of singlet oxygen takes place by quenching of the triplet state of a photoactive compound by molecular oxygen, as illustrated in Scheme 6.26 above. The efficiency of the triplet quenching can be very high, as we have seen with ketoprofen and tiaprofenic acid above. These photoactive molecules contain triplet energy levels that are greater than the energy level of singlet oxygen. For example, the triplet energy of ketoprofen is 62 kcal mol^{-1},[15,16] which is well above the energy levels of the two lowest electronically excited singlet states of molecular oxygen, $^1\Delta_g$ and $^1\Sigma_g^+$, which are 22.6 and 37.7 kcal mol^{-1}, respectively.[80] $O_2(^1\Delta_g)$ is a relatively long lived species in the absence of collisional deactivation, owing to the fact that its relaxation to the triplet ground state of molecular oxygen, $O_2(^1\Sigma_g^-)$ is spin-forbidden. On the other hand, $O_2(^1\Sigma_g^+)$ is a short lived species owing to its fast, spin-allowed transition to the lowest singlet state of $O_2(^1\Delta_g)$.[81] For this reason, $O_2(^1\Delta_g)$ plays a critical role in photochemical and photobiological processes,[79] including photodegradation and phototoxicity of drugs. In the following discussion, we will not specifically differentiate between the two electronic states of singlet (molecular) oxygen.

6.3.3   Degradation Pathways *via* Reaction with Singlet Oxygen

Singlet oxygen is a very reactive species and is capable of reacting with drugs containing various double bonds including conjugated systems with heteroatoms. A majority of these reactions can be categorized into three types, according to a comprehensive review by Frimer:[82] 1,2-addition (dioxetane formation), 1,3-addition ("ene" reaction), and 1,4-cycloaddition (a Diels–Alder reaction to form endoperoxide). Among these three types, the 1,4-cycloaddition is more frequently seen in the photodegradation of drugs. An example of 1,4-cycloaddition can be found in the photodegradation of losartan under ambient lighting.[83] The singlet oxygen generated *via* photosensitization adds onto the five-membered imidazole ring to form a bicyclic 2,5-endoperoxide intermediate (Scheme 6.28). The endoperoxide decomposes into an imide which can give further degradants *via* hydrolysis and isomerization.

Drug molecules containing analogous heterocyclic rings undergo similar photosensitized degradation by singlet oxygen. For example, an experimental drug containing a thiazole ring (Scheme 6.29) was found to degrade to an imide in a similar fashion.[84]

A good example for the 1,3-addition ("ene" reaction) can be found in the photosensitized oxidation of cholesterol (Scheme 6.30). Two sets of

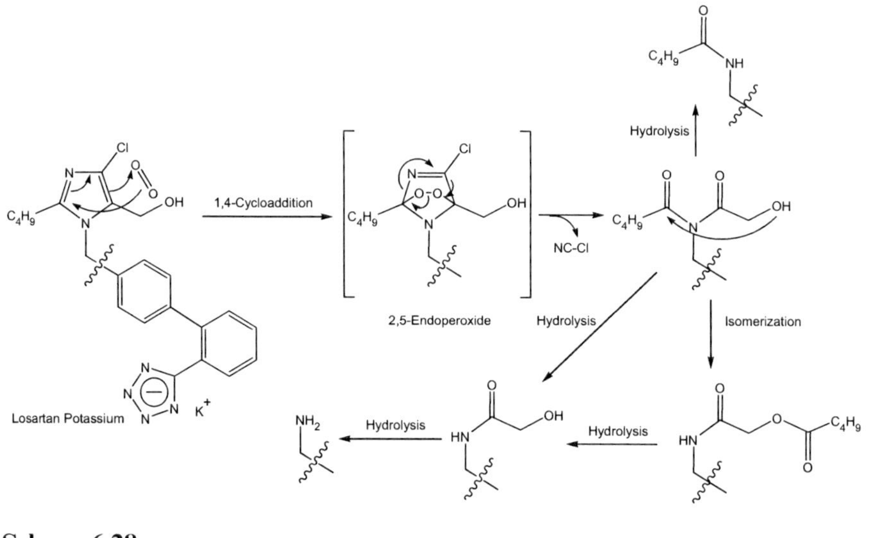

Scheme 6.28

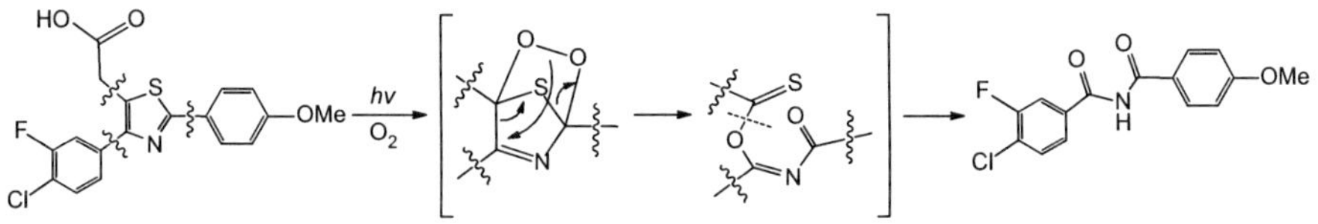

Scheme 6.29

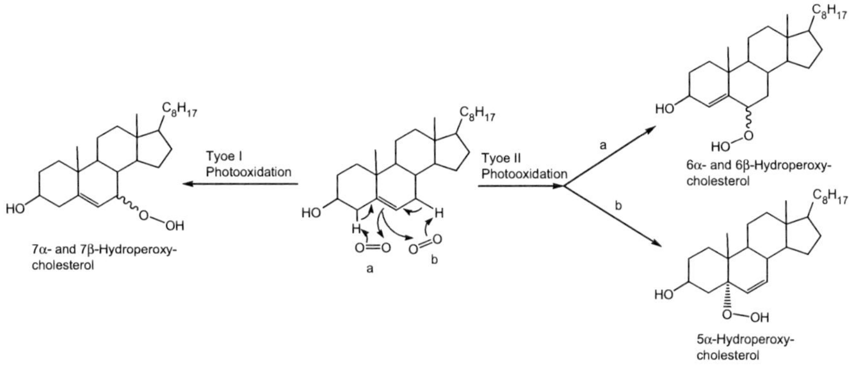

Scheme 6.30

regio-isomers, that is, 6- and 5-hydroperoxides are formed. There have been several proposed mechanisms for this reaction, including free radical, ionic, dioxetane, peroxirane, and "ene" mechanisms.[82] The schematic presentation above is close to the "ene" mechanism. It is worthwhile noting that the type I photooxidation produces 7-hydroperoxycholesterol.

Atorvastatin, the best selling cholesterol-lowering drug, contains a fully substituted pyrrole ring, which was found to undergo self-sensitized photo-oxidation in water under sunlight.[65] The major photodegradation occurred on the pyrrole ring and the experimental evidence strongly suggested that the photooxidation involves singlet oxygen. The formation of the major photo-degradants was proposed to proceed *via* attack by singlet oxygen onto the two double bonds of the pyrrole ring, respectively, to give the corresponding per-epoxides and then epoxides, which are followed by respective rearrangements (Scheme 6.31).

Singlet oxygen is capable of interacting with a sulfide to form a persulfoxide intermediate. The latter can either give back the original sulfide and ground state $^3O_2$, which represents the quenching of $^1O_2$ by the sulfide (Scheme 6.32, pathway a), or proceed with oxidation of the sulfide to produce sulfoxide, sulfone, and cleavage products *via* a number of pathways (pathways b to d). This mechanism is supported by both the experimental evidence and theoretical calculation.[85–87] It is noted that, according to this mechanism, the formation of sulfone does not necessarily have to go through the intermediacy of sulfoxide, as a direct formation pathway (pathway d2) exists.

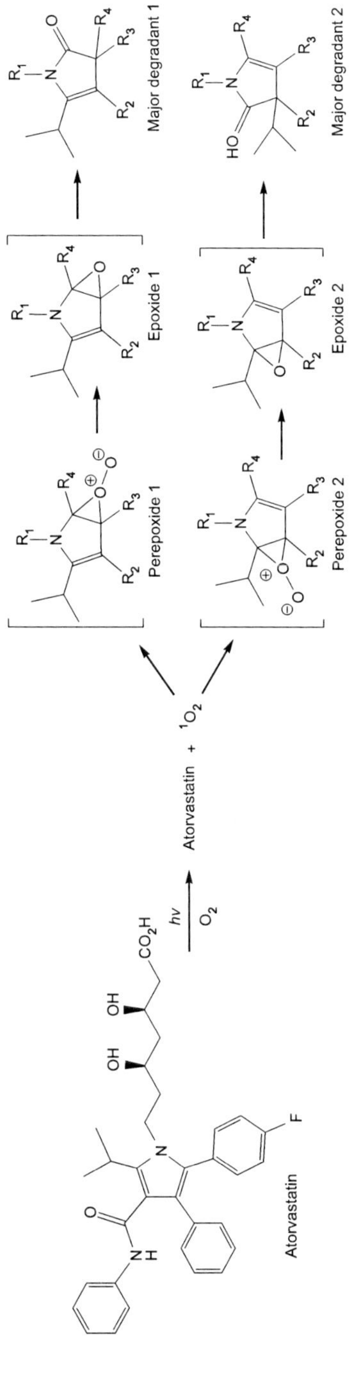
Atorvastatin
hv
O2
Atorvastatin + 1O2
Perepoxide 1
Epoxide 1
Major degradant 1
Perepoxide 2
Epoxide 2
Major degradant 2

Scheme 6.31

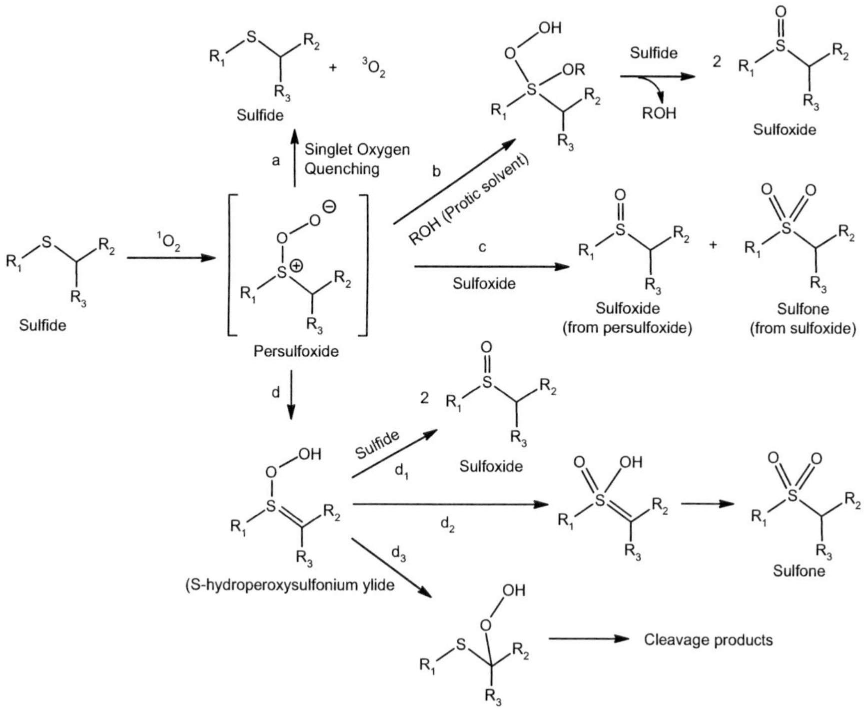

Scheme 6.32

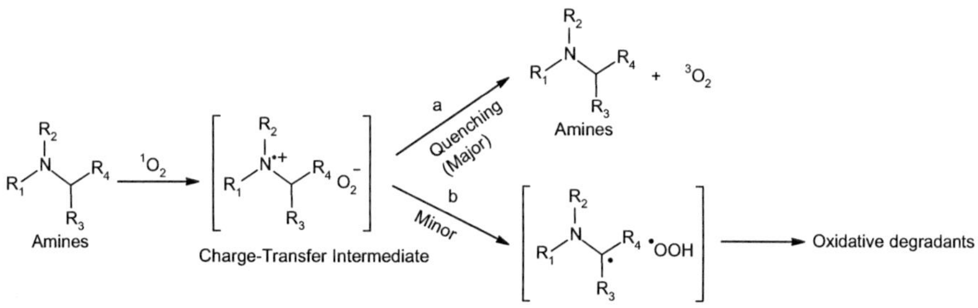

Scheme 6.33

It appears that the reaction of amines with singlet oxygen is quite sluggish. This is probably due to the fact that amines, in particular tertiary amines, are very good quenchers of singlet oxygen *via* a charge-transfer mechanism as illustrated in Scheme 6.33 (pathway a).[88,89] A single amine molecule can effectively quench numerous singlet oxygen molecules to ground state $^3O_2$.[79] Consequently, very little $^1O_2$ would have a chance to transform the amine chemically (pathway b).

Several alkaloid drugs have been found to be potent inhibitors of lipid peroxidation in a number of *in vitro* studies.[90–92] It is very likely that such a protection effect is attributable to the quenching of singlet oxygen by the alkaloids, as demonstrated in a separate study in which a number of plant alkaloids and related compounds were shown to be efficient singlet oxygen quenchers.[93]

References

1. G. Cosa, *Pure Appl. Chem.*, 2004, **76**, 263.
2. C. D. Borsarelli, S. E. Braslavsky, S. Sortino, G. Marconi and S. Monti, *Photochem. Photobiol.*, 2000, **72**, 163.
3. G. L. Cosa, L. Martinez and J. C. Scaiano, *Phys. Chem. Chem. Phys.*, 1999, **1**, 3533.
4. L. L. Constanzo, G. De Guidi, G. Condorelli, A. Cambria and M. Fama, *Photochem. Photobiol.*, 1989, **50**, 359.
5. F. Bosca, M. A. Miranda, G. Carganico and D. Mauleon, *Photochem. Photobiol.*, 1994, **60**, 96.
6. S. Navaratnam, B. J. Parsons and J. L. Hughes, *J. Photochem. Photobiol. A: Chem.*, 1993, **73**, 97.
7. D. Budac and P. Wan, *J. Photochem. Photobiol., A*, 1992, **67**, 135.
8. V. Lhiaubet-Vallet, J. Trzcionka, S. Encinas, M. A. Miranda and N. Chouini-Lalanne, *Photochem. Photobiol.*, 2003, **77**, 487.
9. E. Holzle, N. Neumann, B. Hausen, B. Przybilla, S. Schauder, H. Honigsmann, A. Bircher and G. Plewig, *J. Am. Acad. Dermatol.*, 1991, **25**, 59.
10. S. Encinas, M. A. Miranda, G. Marconi and S. Monti, *Photochem. Photobiol.*, 1998, **68**, 633.
11. J. V. Castell, M. J. Gomez-Lechon, C. Grassa, L. A. Martinez, M. A. Miranda and P. Tarrega, *Photochem. Photobiol.*, 1994, **59**, 35.
12. G. De Guidi, R. Chillemi, L. L. Costanzo, S. Giuffrida, S. Sortino and G. Condorelli, *J. Photochem. Photobiol., B*, 1994, **23**, 125.
13. L. Gu, H.-S. Chiang and D. Johnson, *Int. J. Pharm.*, 1988, **41**, 105.
14. F. Boscá, M. L. Marín and M. A. Miranda, *Photochem. Photobiol.*, 2001, **74**, 637.
15. D. E. Moore and P. P. Chappuis, *Photochem. Photobiol.*, 1988, **47**, 173.
16. L. J. Martinez and J. C. Scaiano, *Photochem. Photobiol.*, 1998, **68**, 646.
17. F. Bosca, S. Encinas, P. Heelis and M. A. Miranda, *Chem. Res. Toxicol.*, 1997, **10**, 820.
18. G. De Guidi, R. Chillemi, L. L. Costanzo, S. Giuffrida and G. Condorelli, *J. Photochem. Photobiol., B*, 1993, **17**, 239.
19. T. Radhakrishna, A. Narasaraju, M. Ramakrishna and A. Satyanarayana, *J. Pharm. Biomed. Anal.*, 2003, **31**, 359.
20. I. A. Alsarra, *Saudi Pharm., J.*, 2004, **12**, 136.

21. Q. Wang, J. D. Stong, P. Demontigny, J. M. Ballard, J. S. Murphy, J.-S. K. Shim and A. J. Faulkner, *J. Pharm. Sci.*, 1996, **85**, 446.
22. A. L. W. Lo and W. J. Irwin, *J. Pharm. Pharmacol.*, 1980, **32**, 25.
23. *Merck Index*, 13th edn, Merck & Co., Whitehouse Station, New Jersey, 2001.
24. M. Brisaert and J. Plaizier-Vercammen, *Int. J. Pharm.*, 2000, **199**, 49.
25. B. M. Tashtoush, E. L. Jacobson and M. K. Jacobson, *Int. J. Pharm.*, 2008, **352**, 123.
26. J. Szymanowski, *Hydroxyoximes and Copper Hydrometallurgy*, CRC Press, Boca Raton, USA, 1993.
27. G. Miolo, S. Caffieri, L. Levorato, M. Imbesi, P. Giusti, T. Uz, R. Manev and H. Manev, *Eur. J. Pharmacol.*, 2002, **450**, 223.
28. J.-W. Kwon and K. L. Armbrust, *J. Pharm. Biomed. Anal.*, 2005, **37**, 643.
29. D. A. Lerner, G. Bonnefond, H. Fabre, B. Mandrou and M. De Simeon Buochberg, *J. Pharm. Sci.*, 1988, **77**, 699.
30. J. R. Williams, R. H. Moore, R. Li and C. M. Weeks, *J. Org. Chem.*, 1980, **45**, 2324.
31. T. Hidaka, S. Huruumi, S. Tamaki, M. Shiraishi and H. Minato, *Yakugaku Zasshi*, 1980, **100**, 72.
32. O. T. Y. Fahmy, *Generation, isolation, characterization and analysis of some photolytic products of dexamethasone and related steroids.* Doctoral Thesis, University of Mississippi, 1997.
33. M. Lin, M. Li, A. V. Buevich, R. Osterman and A. M. Rustum, *J. Pharm. Biomed. Anal.*, 2009, **50**, 275.
34. G. Miolo, F. Gallocchio, L. Levorato, D. Dalzoppo, G. M. J. Beyersbergen van Henegouwen and S. Caffieri, *J. Photochem. Photobiol., B*, 2009, **96**, 75.
35. Y. Shirasaki, K. Inada, J. Inoue and M. Nakamura, *Steroids*, 2004, **69**, 23.
36. A. Ricci, E. Fasani, M. Mella and A. Albini, *J. Org. Chem.*, 2003, **68**, 4361.
37. A. Ricci, E. Fasani, M. Mella and A. Albini, *J. Org. Chem.*, 2001, **66**, 8086.
38. H. Kobsa, *J. Org. Chem.*, 1962, **27**, 2293.
39. J. C. Anderson and C. B. Reese, *Proc. Chem. Soc., London*, 1963, 1781.
40. R. A. Finnegan and J. J. Mattice, *Tetrahedron*, 1965, **21**, 1015.
41. D. P. Kelley and J. T. Pinhey, *Tetrahedron Lett.*, 1964, **46**, 3427.
42. J. V. Castell, M. J. Gomez-Lechon, V. Mirabet, M. A. Miranda and I. M. Morera, *J. Pharm. Sci.*, 1987, **76**, 374.
43. C. E. Kalmus and D. M. Hercules, *J. Am. Chem. Soc.*, 1974, **96**, 449.
44. E. Fasani, A. Albini and S. Gemme, *Int. J. Pharm.*, 2008, **352**, 197.
45. E. Fasani, D. Dondi, A. Ricci and A. Albini, *Photochem. Photobiol.*, 2006, **82**, 225.
46. C. Galli and T. Pau, *Tetrahedron*, 1998, **54**, 2893.
47. H. H. Cornehl, G. Hornung and H. Schwarz, *J. Am. Chem. Soc.*, 1996, **118**, 9960.
48. J. K. Kim and J. F. Burnett, *J. Am. Chem. Soc.*, 1970, **92**, 7463.
49. G. De Guidi, R. Chillemi, S. Giuffrida, G. Condorelli and M. Cambria Famà, *J. Photochem. Photobiol., B*, 1991, **10**, 221.
50. J. Grimshaw and A. P. de Silva, *Chem. Soc. Rev.*, 1981, **10**, 181.

51. F. W. Grant and J. Greene, *Toxicol. Appl. Pharmacol.*, 1972, **23**, 71.
52. A. G. Motten, G. R. Buettner and C. F. Chignell, *Photochem Photobiol.*, 1983, **42**, 9.
53. C. F. Chignell, A. G. Motten and G. R. Buettner, *Environ. Health Perspect.*, 1985, **64**, 103.
54. S. Navaratnam, B. J. Parsons, G. D. Phillips and A. K. Davies, *J. Chem. Soc., Faraday Trans. 1*, 1978, **74**, 1811.
55. T. Iwaoka and M. Kondo, *Bull. Chem. Soc. Jpn.*, 1974, **47**, 980.
56. B. Van Der Walt and H. J. Cahnmann, *Proc. Natl. Acad. Sci. U.S.A.*, 1982, **79**, 1492.
57. A. G. Kazemifard, D. E. Moore and A. Aghazadeh, *J. Pharm. Biomed. Anal.*, 2001, **25**, 697.
58. R. G. Harvey, *Org. Prep. Proced. Int.*, 1997, **29**, 243.
59. S. Encinas, F. Boscá and M. A. Miranda, *Photochem. Photobiol.*, 1998, **68**, 640.
60. K. A. K. Musa and L. A. Eriksson, *Phys. Chem. Chem. Phys.*, 2009, **11**, 4601.
61. S. Encinas, F. Bosca and M. A. Miranda, *Chem. Res. Toxicol.*, 1998, **11**, 946.
62. K. A. Muszkat, *Top. Curr. Chem.*, 1980, **88**, 89.
63. F. B. Mallory and C. W. Mallory, *Org. React.*, 1983, **30**, 1.
64. D. W. Mendenhall, A. Kobayashi, F. M. L. Shih, L. A. Sternson, T. Higuchi and C. Fabian, *Clin. Chem.*, 1978, **24**, 1518.
65. F. Cermola, M. DellaGreca, M. R. Iesce, S. Montanaro, L. Previtera and F. Temussi, *Tetrahedron*, 2006, **62**, 7390.
66. R. P. Shah and S. Singh, *J. Pharm. Biomed. Anal.*, 2010, **53**, 755.
67. R. P. Shah, A. Sahu and S. Singh, *J. Pharm. Biomed. Anal.*, 2010, **51**, 1037.
68. Y. Yoshida, E. Sato and R. Moroi, *Arzneim. Forsch.*, 1993, **43**, 601.
69. M. J. Lovdahl and S. R. Priebe, *J. Pharm. Biomed. Anal.*, 2000, **23**, 521.
70. Y. Asahi, *J. Pharm. Soc. Jpn.*, 1956, **76**, 373.
71. H. Werbin and E. T. Strom, *J. Am. Chem. Soc.*, 1968, **90**, 7296.
72. B. Marciniec and D. Witkowska, *Acta Pol. Pharm.*, 1988, **45**, 528.
73. A. Albini and E. Fasani, in *Drugs, Photochemistry and Photostability*, Royal Society of Chemistry, Cambridge, 1998, p. 44.
74. A. I. Scott, *Interpretation of the Ultraviolet Spectra of Natural Products*, Macmillan, New York, 1964, pp. 28–35.
75. M. J. Peak and J. G. Peak in *CRC Handbook of Organic Photochemistry and Photobiology*, ed. W. M. Horspool and P.-S. Song, CRC Press, Boca Raton, FL, 1995, pp. 1318–1325.
76. C. Foote, *Photochem. Photobiol.*, 1991, **54**, 659.
77. F. Bosca and M. A. Miranda, *J. Photochem. Photobiol., B*, 1998, **43**, 1.
78. D. de la Pena, C. Marti, S. Nonell, I. A. Martinez and M. A. Miranda, *Photochem. Photobiol.*, 1997, **69**, 828.
79. C. Schweitzer and R. Schmidt, *Chem. Rev.*, 2003, **103**, 1685.
80. J. M. Wessels and M. A. J. Rodgers, *J. Phys. Chem.*, 1995, **99**, 17586.
81. M. C. DeRosa and R. J. Crutchley, *Coord. Chem. Rev.*, 2002, **233–234**, 351.
82. A. A. Frimer, *Chem. Rev.*, 1979, **79**, 359.

83. R. A. Seburg, J. M. Ballard, T.-L. Hwang and C. M. Sullivan, *J. Pharm. Biomed. Anal.*, 2006, **42**, 411.
84. L. Wu, T. Y. Hong and F. G. Vogt, *J. Pharm. Biomed. Anal.*, 2007, **44**, 763.
85. C. Gu, C. S. Foote and M. L. Kacher, *J. Am. Chem. Soc.*, 1981, **103**, 5949.
86. F. Jensen, *J. Org. Chem.*, 1992, **57**, 6478.
87. F. Jensen, A. Greer and E. L. Clennan, *J. Am. Chem. Soc.*, 1998, **120**, 4439.
88. C. Quannes and T. Wilson, *J. Am. Chem. Soc.*, 1968, **90**, 6527.
89. W. F. Smith, Jr., *J. Am. Chem. Soc.*, 1972, **94**, 186.
90. C. Malvy, C. Paoletti, A. F. J. Searle and R. L. Willson, *Biochem. Biophys. Res. Commun.*, 1980, **95**, 734.
91. N. Shiriashi, T. Arima, K. Aono, B. Inouye, Y. Morimoto and K. Utsumi, *Physiol. Chem. Phys.*, 1980, **12**, 299.
92. K. Koreh, M. Seligman, E. S. Flamm and H. Demopolous, *Biochem. Biophys. Res. Commun.*, 1981, **102**, 1317.
93. R. A. Larson and K. A. Marley, *Phytochemistry*, 1984, **23**, 2351.

CHAPTER 7

Chemical Degradation of Biological Drugs

7.1 Overview

Biological drugs have become an increasingly important family of medications, especially since the millenium. According to a survey by About.com Pharma,[1] half of the top ten best-selling branded drugs in 2010 were biologics in terms of global sales. The vast majority of biological drugs are protein-based, as manifested by the fact that the top five biological drugs are all proteins: Enbrel, Remicade, Avastin, Rituxan, and Humira. Among the carbohydrate-based biological drugs, heparin and hyaluronan may be two of the very few that would fall into the scope of this chapter. Both heparin and hyaluronan (also called hyaluronic acid or hyaluronate) are polymeric, anionic glycosamino-glycans. On the other hand, there are a great number of small-molecule drugs that are carbohydrate-based or contain carbohydrate components, such as those in the aminoglycoside (e.g. tobramycin) and glycopeptide (e.g. vancomycin) antibiotics families. These small molecule drugs have been discussed in previous chapters and are outside the scope of this chapter.

To a much lesser degree, biological drugs are derived from DNA or RNA.[2] In the former category, the first of only two approved drugs is fomivirsen, which is a synthetic, modified 21-member oligonucleotide with phosphorothioate linkages replacing the usual phosphodiester ones. This drug, which was designed based on the concept of "antisense",[3] is used in the antiviral treatment of cytomegalovirus retinitis (CMV) in immune-compromised patients. For RNA-based therapies, the most promising drug candidates are a class of so-called "small interfering RNA", or siRNA, that work by the mechanism of RNA interference (RNAi).[4] There are a number of siRNA molecules in various stages of clinical trials[5,6] but none has been approved.

RSC Drug Discovery Series No. 29
Organic Chemistry of Drug Degradation
By Min Li
© Min Li 2012
Published by the Royal Society of Chemistry, www.rsc.org

Hence, due to the current landscape of biological drugs discussed above, the main emphasis of this chapter will be placed on protein drugs.

7.2 Chemical Degradation of Protein Drugs

The protein drugs that are discussed here include large peptide drugs. Small peptide drugs, in particular those based on di- and tripeptides, behave more or less like other small molecule drugs; some examples have been discussed in previous chapters. Proteins are made of amino acid residues that are connected by amide linkages, also called peptide bonds, forming the backbones of the proteins. Therefore, the degradation of protein drugs can be divided into two main categories: degradation of the backbones and the side chains of amino acid residues. Since this chapter is focused on the chemical degradation aspect, the degradation of protein drugs caused by physical changes, for example, irreversible conformation change and aggregation, will not be discussed. Readers who are interested in this topic are referred to a few review articles.[7–11]

7.2.1 Hydrolysis and Rearrangement of Peptide Backbone Caused by the Asp Residue

As mentioned in Chapter 2, Hydrolytic Degradation, the typical activation energy for hydrolysis of an amide bond is ~ 20 to 25 kcal mol^{-1} (see Table 2.1 of Chapter 2). Hence, in general, the amide linkage of a protein is quite stable with regard to hydrolytic degradation. Nevertheless, if a protein contains an Asp residue in its sequence, the peptide bonds before and after the Asp residue, that is, the preceding and succeeding peptide bonds, respectively, become much more susceptible to hydrolysis. It has been demonstrated that peptide bonds containing Asp hydrolyze at least 100 times faster than typical peptide bonds under acidic hydrolysis.[12] There have been a couple of mechanisms available to try to explain this phenomenon. The first mechanism, proposed by Inglis, attributed the lability of Asp–peptide bonds to the formation of a five-(pathway a of Scheme 7.1), or six-membered (pathway e) ring intermediate,[13] resulting from the attack of the succeeding and preceding peptide bonds by the side chain carboxyl group of Asp, respectively. According to this mechanism, both the preceding and succeeding peptide bonds would be labile towards hydrolytic degradation. Nevertheless, it appears that there has been no report regarding a particular weak X–Asp bond, although numerous reports have been made in the literature regarding weak Asp–X bonds.

In a comprehensive degradation study performed by Oliyai and Borchardt in which a model hexapeptide, Val-Tyr-Pro-Asp-Gly-Ala, was evaluated for its stability under different pHs, only the peptide fragment resulting from the succeeding peptide bond hydrolysis was observed.[14] Hence, the likelihood of pathway e being a viable degradation pathway is highly questionable. It appears that the formation of the proposed six-membered intermediate would be quite unfavorable thermodynamically compared to the formation of its

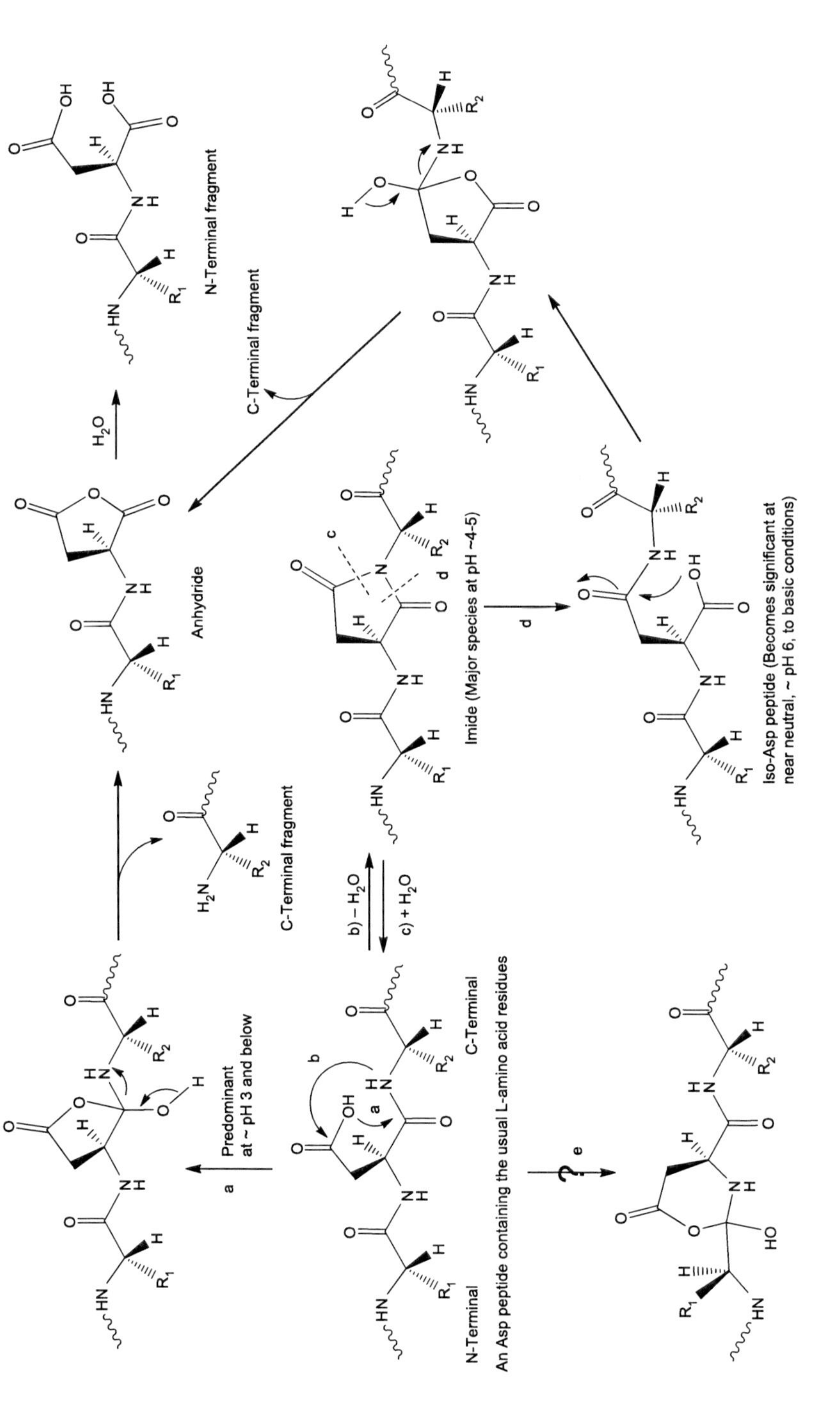
N-Terminal fragment
C-Terminal fragment
Anhydride
Predominant at ~ pH 3 and below
C-Terminal fragment
b) – H_2O
c) + H_2O
Imide (Major species at pH ~4-5)
Iso-Asp peptide (Becomes significant at near neutral, ~ pH 6, to basic conditions)
N-Terminal
C-Terminal
An Asp peptide containing the usual L-amino acid residues
? e

Scheme 7.1

five-membered counterpart.[15] This may also explain the fact that a Glu–X peptide bond is usually stable hydrolytically. Otherwise, such a bond would also be labile towards hydrolysis, since the Glu residue, a CH_2 homolog of Asp, could form a six-membered intermediate according to pathway a.

The pathways summarized in Scheme 7.1 are mainly based on the results obtained from the study by Oliyai and Borchardt,[14] along with the results reported by other research groups, many of which were discussed by these authors. Under a strongly acidic environment ($\sim$ pH 3 and below), the model hexapeptide predominantly undergoes hydrolysis *via* the five-membered ring intermediate (pathway a). With increased pH at $\sim$4 to 5, the formation of a succinimide intermediate becomes significant. At near neutral and basic pH, the succinimide readily converts to the iso-Asp peptide (pathway d). The latter may be susceptible to hydrolysis *via* the consecutive intermediacy of a five-membered ring, which is similar to that generated in pathway a, and the same anhydride intermediate.

It needs to be pointed out that the above mechanism is derived based on results from chemical hydrolysis studies of proteins and a model peptide. Under these conditions, the proteins/peptide are either in denatured state or in the case of the model peptide, do not have a well-defined three dimensional structure. Hence, the kinetic parameters obtained from the studies do not necessarily represent those from proteins that are in native state. Clarke has demonstrated that Asp residues in native proteins usually exist in conformations that do not favor the formation of the succinimide intermediate.[16] In other words, an Asp–X peptide bond in native proteins would be much more stable than the one in denatured proteins under mildly acidic pH ($\sim$4 to 5). If an Asp residue is followed by a Pro, the resulting Asp–Pro peptide bond becomes much more labile towards hydrolysis than other Asp–X bonds. A study with various model dipeptides indicated that Asp–Pro bonds were 8 to 20 times more susceptible to hydrolysis than other Asp–X or X–Asp bonds under acidic conditions.[17] The particular lability of the Asp–Pro bond was attributed to the greater basicity of the Pro amino group,[18] which is the only secondary amine among all common amino acids.

In the above discussion, there are two types of degradation of the protein backbone that are caused by the presence of an Asp residue: hydrolysis (also known as proteolysis) and rearrangement. In near neutral conditions, rearrangement also leads to the significant formation of iso-Asp peptides or proteins that are isomeric to the original Asp-peptides or proteins. In other words, the original Asp-proteins undergo isomerization under such conditions. Experimentally, it could be quite challenging to differentiate between the two isomers.[19] In the area of protein characterization by mass spectrometry, a high energy collision technique, electron capture dissociation (ECD),[20] is capable of generating differentiating fragmentation patterns between isomeric Asp- and iso-Asp-containing peptides, enabling the detection of protein degradation *via* the isomerization pathway. Since the rearrangement and subsequent isomerization require the formation of a succinimide intermediate, protein drugs containing Asp–Pro bonds do not undergo this pathway in positions where these bonds are present, because the secondary amine group of Pro cannot form the succinimide intermediate.

There have been numerous reports in the literature regarding hydrolytic degradation caused by the weak Asp–X linkage. Some examples are discussed here. Recombinant human interleukin II (rhIL-Il) is a multi-spectrum cytokine, which contains two Asp–Pro bonds. Under acidic stress conditions at 50 °C, Kenley and Warne found that the predominant degradation of the protein was peptide bond cleavage at the Asp_{133}–Pro_{134}, rather than the Asp_{12}–Pro_{13} bond.[21] A similar bias between multiple Asp–Pro bonds was also observed by the same research group during stress studies of recombinant human macrophage colony-stimulating factor.[22] In a formulation study of carbonic anhydrase, which contains two Asp–Pro bonds, with encapsulation in biodegradable poly(lactide-co-glycolide) (PLGA) polymer, $\sim 25\%$ of the enzyme molecules were found to experience chain cleavage after one week at pH 7.4 and 37 °C.[23]

7.2.2 Various Degradation Pathways Caused by Deamidation and Formation of Succinimide Intermediate

Protein deamidation refers to the degradation of the two amide side chains of Asn and Gln. Owing to the relatively high activation energy of an unactivated amide bond, mentioned earlier in this chapter, deamidation *via* direct hydrolysis of the two side chains, which occurs at low pH, is very slow[24] and thus insignificant. On the other hand, the degradation of Asn-containing peptides becomes very significant under neutral to basic conditions. Numerous mechanistic studies with model peptides and proteins have demonstrated that the instability is due to the formation of the same succinimide intermediate as that formed from Asp-containing peptides under acidic conditions.[15,25] Under neutral and basic conditions, the succinimide intermediate is quite unstable and susceptible to hydrolysis. Since early studies with model peptides showed that racemization also occurred in the resulting Asp and iso-Asp residues due to deamidation, it was hypothesized that the racemization proceeds through enolization of the succinimide (Scheme 7.2).[24] More recent studies, nonetheless, suggest that the racemization may mainly occur at the tetrahedron intermediate stage,[26] the precursor to the succinimide, rather than from the succinimide intermediate. On the other hand, a study by DeHart and Anderson indicates that the succinimide intermediate can be susceptible to intramolecular attack by the preceding *N*-terminal amino group (in this case, the Asn has to be the second amino acid residue in the sequence), and by intermolecular attack by an amino group from a nearby protein or peptide molecule.[27] The intramolecular attack should result in the formation of a diketopiperazine (DKP) ring, while the intermolecular attack should yield dimerization and oligomerization.

With the exponential growth in the development of monoclonal antibodies (mAbs) as therapeutic agents, knowledge regarding their degradation behavior has also been rapidly accumulated. It has been demonstrated that mAbs

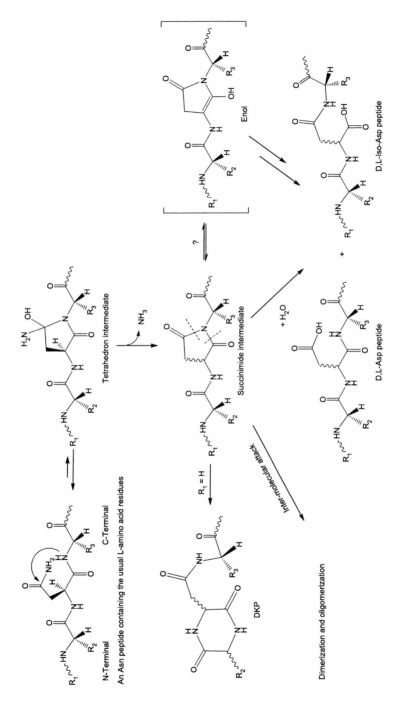

Scheme 7.2

undergo Asn deamidation and Asp isomerization as illustrated in the previous two sections, resulting in the formation of iso-Asp; this degradation pathway is the main reason leading to charge heterogeneity in mAbs.[28–34]

As discussed in Section 7.2.1, Asp-peptides or proteins are susceptible to hydrolysis under acidic pH. Under neutral to basic conditions, Asp-peptides or proteins can form the analogous tetrahedron intermediate and the same succinimide intermediate. Hence, they can also undergo racemization and isomerization. For example, racemization of specific Asp residues was observed in α-a-crystallin from an aged human lens.[35,36] In a study by Shahrokh *et al.*, a succinimide intermediate and the corresponding iso-Asp residue, resulting from the highly flexible sequence of Asp^{15}–Gly^{16}, were detected in the degradation products of basic fibroblast growth factor (bFGF).[37]

7.2.3 Hinge Region Hydrolysis in Antibodies

An antibody consists of two identical heavy (~ 50 kDa each) and two identical light (~ 25 kDa each) chains, respectively, with a total molecular weight of approximately 150 kDa. The two heavy chains are connected by a disulfide bridge and each heavy chain is linked to a light chain by another disulfide bridge. The light and heavy chains can be divided into variable and constant regions, which constitute the antigen binding fragments, Fab. The heavy chains also contain crystallizable fragments, Fc. The whole antibody structure is a Y-shaped tetramer with the Fab fragments as the two open arms of the Y and the Fc fragment as the base. The Fab and Fc fragments of the heavy chains are connected by a flexible hinge region at the converging point of the Y-shaped antibody molecule. The hinge region is responsible for much of the dynamic behavior of immunoglobulins and it has long been known that this highly flexible region is quite susceptible to attack by proteases. In the course of extensive studies of the hydrolytic stability of mAbs in recent years, it has been noticed that non-enzymatic hydrolytic degradation also occurs frequently in the hinge region of these proteins, in particular in the family of immunoglobulin G antibodies (IgGs).[38–41] This hydrolytic degradation has emerged as one of the major degradation pathways of mAbs. The hydrolysis does not appear to be sequence specific and its exact mechanism is not very clear. Some studies suggest that conformation flexibility in the Fab region contributes to increased hydrolysis in the hinge region.[42]

7.2.4 Oxidation of Side Chains of Cys, Met, His, Trp, and Tyr

From the perspective of protein drug stability and degradation, the most relevant oxidation events in proteins are those occurring under transition metal-catalyzed autooxidation and photosensitized oxidation. In both events, molecular oxygen is activated to various reactive oxygen species (ROS), such as H_2O_2, $HO^\bullet$, and singlet oxygen, which then cause oxidation of the proteins. This oxidation usually takes place at or near the site of the ROS generation, where intrinsically oxidizable amino acid residues like Cys, Met, His, Trp, and Tyr, are located. This is because the binding of metal ions or photosensitizers is usually preferred at certain protein sites rather than others; in other words, oxidative degradation of protein drugs takes place typically in a site-specific way.[43] As discussed in extensive detail in Chapter 3, Oxidative Degradation, under transition metal-catalyzed autooxidation conditions, the ROS generated and most responsible for oxidative degradation of drugs are H_2O_2 and $HO^{\bullet}$. As also pointed out by the current author in Chapter 3, the underlying chemistry of ROS generation in the transition metal-catalyzed autooxidation is an Udenfriend reaction rather than a Fenton reaction. The Udenfriend reaction contains three key components, that is, a transition metal ion (typically Fe or Cu in their reduced forms), a chelating agent (which is especially needed in Fe-catalyzed events), and a reducing agent (which can be an excipient or the drug substance itself). The activation of O_2 starts when the reduced form of a transition metal ion transfers one electron to O_2, resulting in the formation of a superoxide anion, $O_2^{-\bullet}$. This is transformed to H_2O_2, which is then converted to $HO^{\bullet}$ by the Fenton reaction. Hence, the Fenton reaction is a component of the Udenfriend reaction in metal-catalyzed autooxidation. Under photo-sensitized oxidation conditions, the ROS generated also include singlet oxygen, $^1O_2$, and carbon-based free radicals, in addition to H_2O_2 and $HO^{\bullet}$, which was discussed in Chapter 6, Photochemical Degradation.

Once various ROS are generated, oxidation tends to occur on the side chains of Cys, Met, His, Trp, and Tyr that are at or near the sites of ROS generation, owing to their susceptibility to oxidation. The oxidative degradation of Cys residue can be divided into two types. First, two thiol groups can be oxidized to form a disulfide bridge, if both are present within a certain distance. This is one of the major pathways that is largely responsible for protein aggregation involving covalent misfolding, dimerization, and oligomerization.[44] Second, the thiol group can be oxidized sequentially to sulfenic, sulfinic, and sulfonic acid.[45,46] At the sulfenic acid stage, it can cyclize with the amino group of the succeeding amino acid residue to form sulfonamide, which can be further oxidized to sulfinamide. The sulfinamide (as well as sulfenamide) is actually a mixed cyclic imide and hence susceptible to hydrolysis, producing cysteine sulfinic acid. Further oxidation of cysteine sulfinic acid gives cysteine sulfonic acid.[45] There is another relevant degradation reaction involving the thiol group. The thiol group can attack a disulfide bridge resulting in the formation of a new disulfide bridge and thiol group. This is an exchange reaction, with the original thiol group being oxidized and part of the original disulfide bridge being reduced. This degradation is another major pathway that is responsible for protein aggregation involving covalent misfolding, dimerization, and olige-merization.[47,48] The oxidative degradation pathways of Cys are summarized in Scheme 7.3.

Met is another sulfur-containing amino acid and its oxidative degradation products are methionine sulfoxide[49] and sulfone as shown in Scheme 7.4. Although the formation of the sulfone does not necessarily have to go through the sulfoxide stage, in particular during photosensitized oxidation as discussed

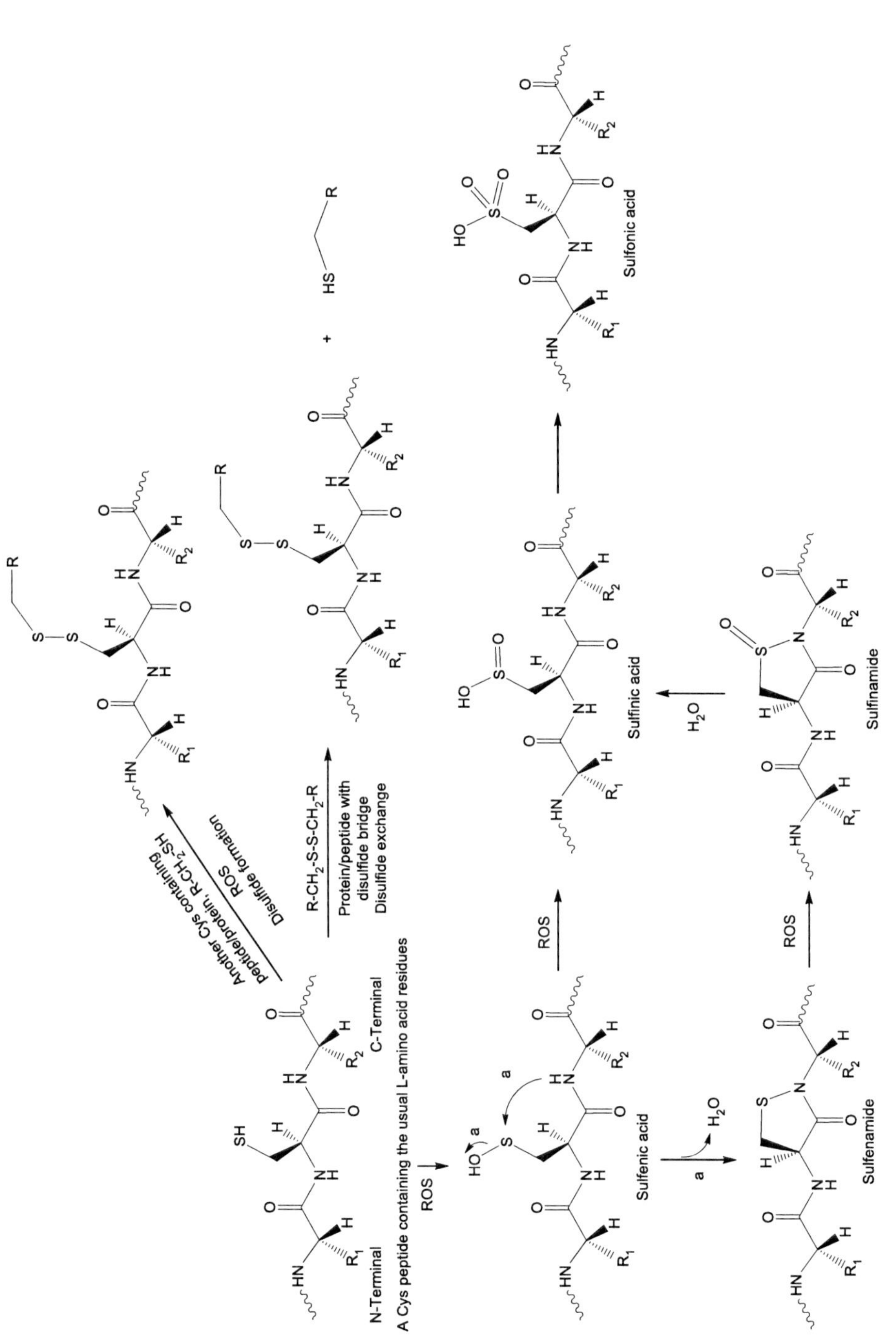
N-Terminal
C-Terminal
A Cys peptide containing the usual L-amino acid residues
ROS
Another Cys containing peptide/protein, R-CH$_2$-SH
ROS
Disulfide formation
R-CH$_2$-S-S-CH$_2$-R
Protein/peptide with disulfide bridge
Disulfide exchange
Sulfenic acid
Sulfinic acid
Sulfonic acid
Sulfenamide
Sulfinamide
H$_2$O
a

Scheme 7.3

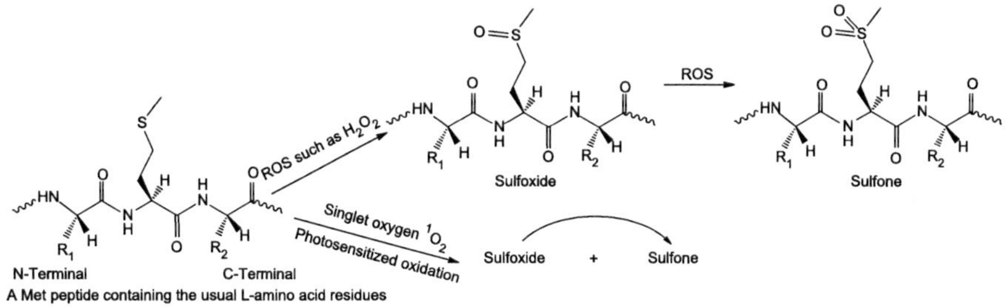

Scheme 7.4

in Chapter 6, the oxidation of Met usually stops at the sulfoxide stage. In an oxidative stress study of recombinant human vascular endothelial growth factor (rhVEGF) under both chemical and photochemical conditions, all six Met residues were found to be oxidized to the corresponding sulfoxides in varying degrees (8–40%); no formation of sulfones was observed.[50] In studies of human growth hormone (hGH) and human insulin-like growth factor I (hIGF-I), the oxidation of Met residues to form Met sulfoxides was also found to be a major degradation pathway for both proteins in the solid state.[51,52] In the latter case, the rates of Met[59] oxidation were measured in both the solution and solid states and were found to be essentially the same under both conditions. When hIGF-I was irradiated under a daylight lamp, the rate of the oxidation increased by 30-fold.[52]

The evidence presented in both studies is consistent with transition metal-catalyzed autooxidation being responsible for the protein degradation. For example, even when the oxygen level was reduced to $\sim 0.05\%$ in the headspace of the product vial, oxidation of Met[14] in hGH remained a significant degradation pathway, which was at least comparable to the degradation caused by deamination.[51] This is typical for a metal-catalyzed, free radical-mediated autooxidation in which the reaction between the free radical of the oxidation substrate and molecular oxygen is not a rate-determining step. As discussed in Chapter 3, this step is extremely fast, close to a diffusion-controlled process.

His, Trp, and Tyr are three amino acid residues that contain oxidizable aromatic side chains. His has an imidazole side chain and the mechanism of its oxidation has been studied with His derivatives, peptides and proteins. A number of oxidative stress studies of His-containing peptides and proteins were conducted using Cu^{2+}/ascorbic acid as a mean of generating hydroxyl radical.[53–56] The Cu^{2+}/ascorbic acid combination can be considered to be a Udenfriend reagent. Under this stress condition, His residues at certain sites were selectively oxidized to 2-oxo-His (2-imidazolone) as illustrated in Scheme 7.5. The His residues that were oxidized appeared to be part of a metal-binding complex involving the Udenfriend reagent. For this reason, the oxidation can be inhibited by metal chelators such as ethylenediamine tetraacetic acid (EDTA),[53] which should disrupt this complex by pulling away the metal ion. In some cases, an intact three dimensional protein structure is necessary to form the metal-binding complex. For example, while His[18] and His[21] residues were

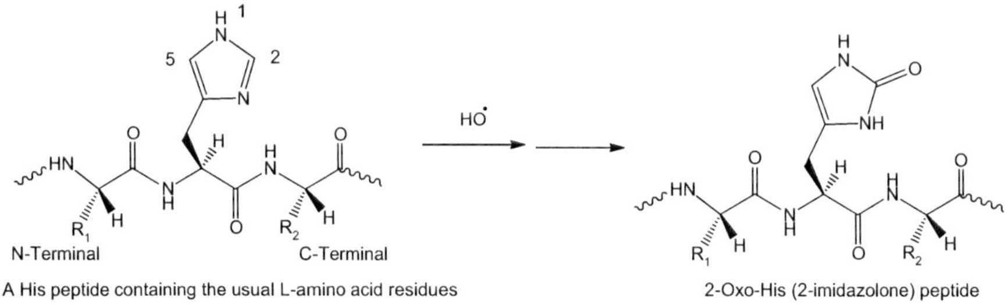

A His peptide containing the usual L-amino acid residues   2-Oxo-His (2-imidazolone) peptide

Scheme 7.5

both oxidized in intact hGH under the stress of the Cu^{2+}/ascorbic acid system, the same two His residues in isolated peptide fragments did not undergo oxidation under identical stress conditions.[56]

Trp contains an indole side chain and the fused pyrrole portion of the indole ring is much more susceptible to oxidative degradation than the phenyl one. Over the past several decades, a great number of studies have been performed in order to determine the oxidative degradation products of Trp residue in proteins and peptides under various stress conditions. Several oxidative degradants have been identified: *N*-formylkynurenine (NFK),[57-61] oxindolylalanine (OIA),[57,58,61] kynurenine (KYN),[58-61] and degradants resulting from further decomposition of these.[57] Nonetheless, it appears that no clear formation mechanisms for these degradants have been provided. Interpretation of results in the early days was usually hampered by further degradation products caused by strong acid or base hydrolysis or Edmon degradation chemistry after the oxidation of Trp occurred. Characterization of Trp degradation products was made easier, after mass spectrometry was used in the studies. Nowadays, the above-mentioned three degradants are typically detected or identified by their mass difference from Trp: NFK has a mass increase of 32, OIA 16, and KYN 4. Based on the fact that the pyrrole moiety is quite susceptible to oxidation by H_2O_2[62,63] and $^1O_2$ (see also Chapters 3 and 6), the following mechanisms are provided in Scheme 7.6 by the current author.

There have been some reports indicating that oxidation of Trp can also occur during sample handling and preparation, in particular during SDS-PAGE and in-gel proteolysis.[64,65] Therefore, caution must be exercised to minimize the artifactual oxidation caused by sample handling and preparation. Since NFK should be mainly a degradant of photosensitized oxidation, protection from laboratory lighting may help reduce Trp oxidation during these procedures.

Tyr has a phenol side chain. As such, its intrinsic degradation pathways should resemble those of phenol-containing compounds, which include oxidation of the phenol moiety to quinone and dimerization resulting from phenolic free radicals (see Chapter 3). Indeed, under both UV irradiation at 280 nm and metal-catalyzed oxidation, Tyr residues were found to crosslink, giving rise to dityrosine (Di-Tyr), although under the latter condition, formation of Di-Tyr is not a major degradation pathway.[66] In another experiment in which

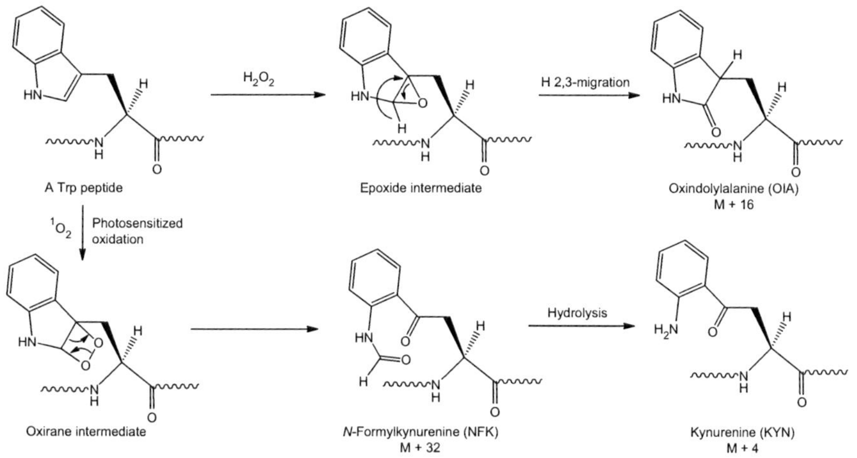

H$_2$O$_2$

H 2,3-migration

A Trp peptide

Epoxide intermediate

Oxindolylalanine (OIA)
M + 16

^{1}O$_2$ | Photosensitized oxidation

Oxirane intermediate

N-Formylkynurenine (NFK)
M + 32

Hydrolysis

Kynurenine (KYN)
M + 4

Scheme 7.6

hemoglobin was stressed with H$_2$O$_2$, Di-Tyr and a number of Tyr oxidative degradants were observed.[67] These key degradation pathways of Tyr residue are summarized in Scheme 7.7.

In the formation mechanism of Di-Tyr proposed by Giulivi *et al.*,[66] a combination of two *ortho*-Tyr radicals was hypothesized (pathway b, Scheme 7.7), leading to Di-Tyr after isomerization. Although such a pathway would be theoretically possible, the probability that they would react with each other is very low for the two short-lived Tyr radicals, which are typically in low quantities especially in metal-catalyzed oxidation. On the other hand, it would be much more likely for the initially formed Tyr radical to react with a nearby Tyr residue to form a Di-Tyr radical (pathway c). This radical can either give up a hydrogen radical (pathway c1) under UV irradiation or first be oxidized by the metal ion in its oxidized form (M^{n+}) and then deprotonated (pathway c2). In the latter scenario, the reduced metal ion M$^{(n-1)+}$ would be put back to work with H$_2$O$_2$, generating more HO$^{\bullet}$ radicals. The HO$^{\bullet}$ radicals generated can also attack the side chain of a Tyr residue directly, resulting in the formation of a dihydroxyl phenyl radical (pathway d). The latter species then converts to Tyr catechol (DOPA) through deprotonation, in the same fashion as the formation of Di-Tyr *via* pathway c. Further oxidation of Tyr catechol should produce Tyr quinone.[66]

7.2.5 Oxidation of Side Chains of Arg, Pro, and Lys

The side chains of Arg, Pro, and Lys are not intrinsically susceptible to oxidative degradation. However, if these residues are located in proximity to or are part of metal-binding sites within a defined three dimensional structure of a protein, their side chains may be oxidized by HO$^{\bullet}$,[68] a very powerful oxidant, generated *in situ* under transition metal-catalyzed Fenton or Udenfriend

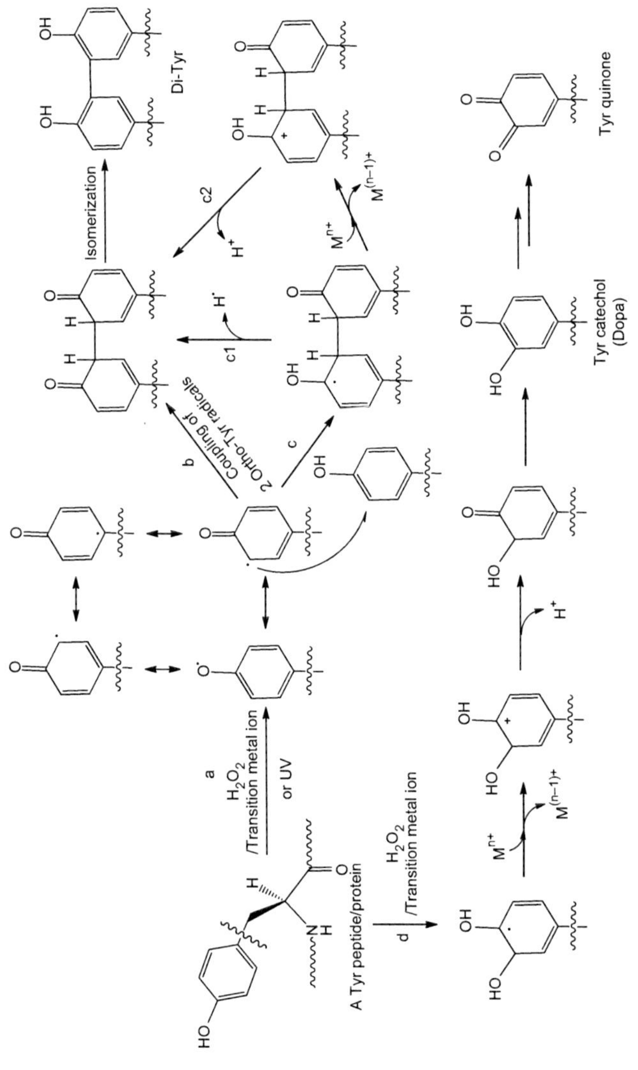
A Tyr peptide/protein
a
H2O2
/Transition metal ion
or UV
b
Coupling of
2 ortho-Tyr radicals
c
c1
c2
Isomerization
Di-Tyr
d
H2O2
/Transition metal ion
Tyr catechol
(Dopa)
Tyr quinone

Scheme 7.7

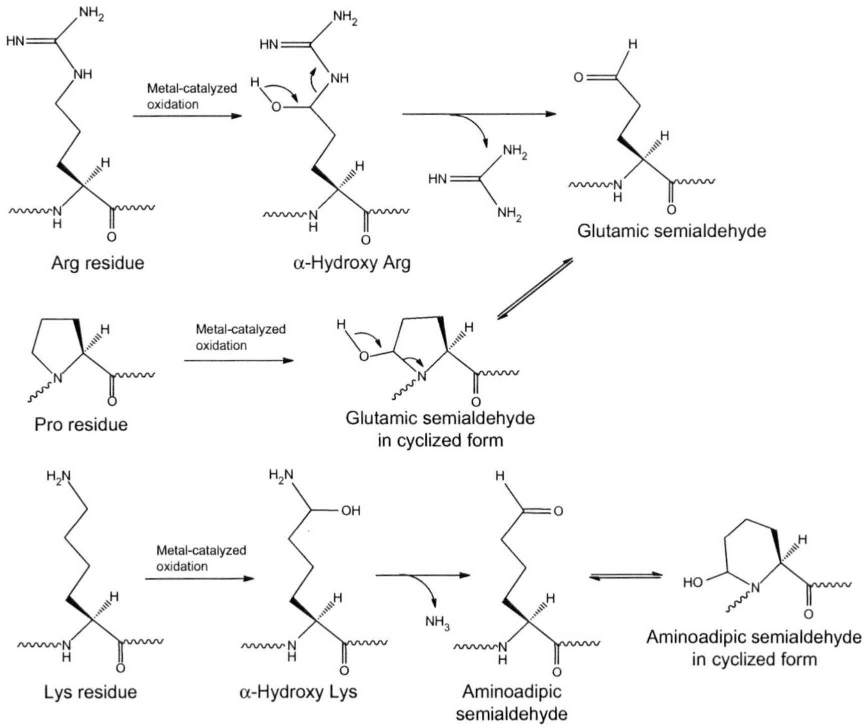

Scheme 7.8

conditions. While the Fenton reagent, such as Fe^{2+}/H_2O_2 or Cu^{2+}/H_2O_2, represents stress or forced degradation conditions, the Udenfriend reagent, for example, Fe^{2+}/chelator/reducing agent or Cu^{2+}/reducing agent, more resembles protein oxidation under a real life scenario. Oxidation by the $HO^{\bullet}$ generated *in situ* occurs on the carbon α to the amino (as in the cases of Pro and Lys) or guanadino (as in the case of Arg) functionality. The resulting α-hydroxy amino acid residues can readily decompose to produce glutamic and aminoadipic semialdehydes,[68,69] respectively, as shown in Scheme 7.8.

In a metal-catalyzed oxidation study of proteins including glutamine synthetase, bovine serum albumin, RNase, and lysozyme, significant amounts of glutamic and aminoadipic semialdehydes were detected in all the proteins studied under metal-catalyzed stress conditions.[69] The aldehydes were also detected in these proteins prior to oxidative stress.

7.2.6 β-Elimination

It has been observed that recombinant human IgG1 monoclonal antibodies are susceptible to non-enzymatic cleavage in the hinge region when placed in long-term storage in solution. The cleavage leads to the formation of Fab and Fc-Fab fragments. The main cleavage appears to occur at the Ser_{219}-Cys_{220} bond

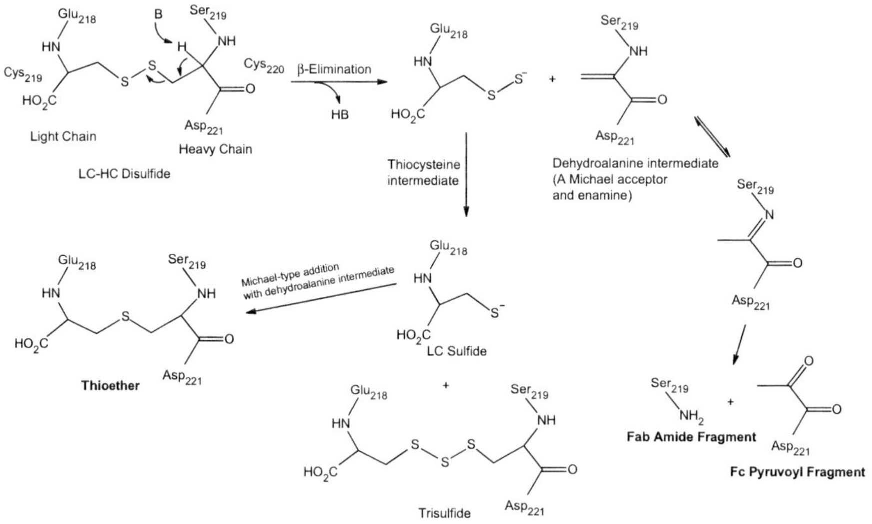

Scheme 7.9

in the heavy chain.[70] In an accelerated stability study of IgG1 formulated in phosphate buffered saline at 45 °C, Cohen *et al.* confirmed the cleavage site and also presented evidence indicating the cleavage is caused by a β-elimination mechanism.[71] According to this mechanism (Scheme 7.9), the cleavage starts with the removal of the α-proton of the heavy chain Cys_{220}, which results in the elimination of the whole disulfide bridge, that is, the one between the light chain Cys_{219} and heavy chain Cys_{220} (LC Cys_{219}–HC Cys_{220}). This mechanism implies that the β-elimination should increase in a higher pH range as the removal of the α-proton should be favored under such conditions. Indeed, this cleavage was significantly higher in an elevated pH range.[71] The resulting dehydroalanine intermediate is both a Michael acceptor and a substituted enamine. As the former, it can react with a free LC Cys to form a thioether linkage, which was also observed by both Cohen *et al.* and a previous study.[72] As the latter, it can isomerize to the imine which would be readily hydrolyzed to produce Fab amide and Fc-Fab pyruvoyl fragments. The β-elimination induced degradation of IgG1 is summarized in Scheme 7.9.

The above degradation pathways are largely based on a study by Cohen *et al.*,[71] but with additional detail. Most notably, in the originally proposed mechanism, the origin of the LC sulfide was not elaborated. Based on the chemistry of sulfur, it is not inconceivable that the thiocysteine intermediate can attack the LC-HC disulfide to release the LC sulfide, with the concomitant formation of the trisulfide. The LC sulfide can then react with the dehydro-alanine intermediate to yield the thioether; the amount of thioether-linked IgG1 was found to increase significantly at higher pH.[71]

Other proteins also undergo β-elimination induced degradation. For example, in a stability study of lyophilized insulin, a significant portion of

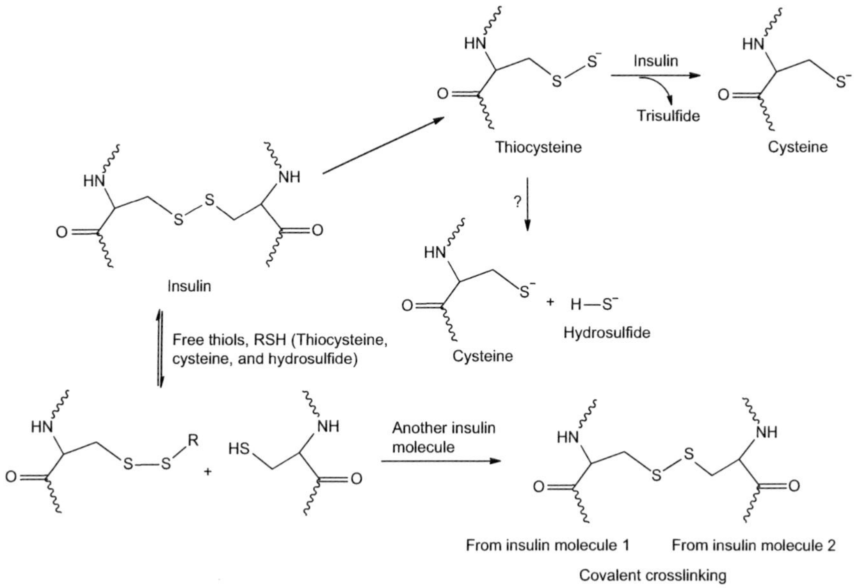

Scheme 7.10

the covalent aggregation of the protein was found to be due to intermolecular thiol-catalyzed disulfide exchange.[73] Insulin contains three pairs of disulfide bridges but no free cysteines. Hence, the free thiols are very likely to be generated during and/or in the aftermath of the β-elimination event. The authors suspected the formation of cysteine and hydrosulfide from thiocysteine formed during the β-elimination. Nevertheless, thiocysteine itself should also be capable of catalyzing disulfide exchange, leading to the covalent aggregation of insulin as shown in Scheme 7.10.

7.2.7 Crosslinking, Dimerization, and Oligomerization

Reactive or unstable intermediates are produced in quite a few of the protein degradation pathways that have been discussed so far in this chapter, for example, succinimide in deamidation, phenolic radical in Tyr oxidation, and thiocysteine and dehydroalanine in β-elimination. These intermediates are capable of reacting or crosslinking with functional groups from the same protein molecule (intramolecule) or from another one (intermolecule). Intermolecular crosslinking leads to the formation of covalent protein dimers and oligomers, which typically causes protein aggregation and precipitation. A few specific examples of protein crosslinking have been given.[27,47,48,66,71–73]

As discussed in Section 7.2.4, oxidation of the His side chain leads to the predominant formation of 2-oxo-imidazole ring. In a study with model compounds, Liu *et al.* demonstrated that the 2-oxo-imidazole moiety can be further oxidized to form an dehydroimidazole intermediate, an electrophilic species, which can be intercepted by the amino group of Lys residues

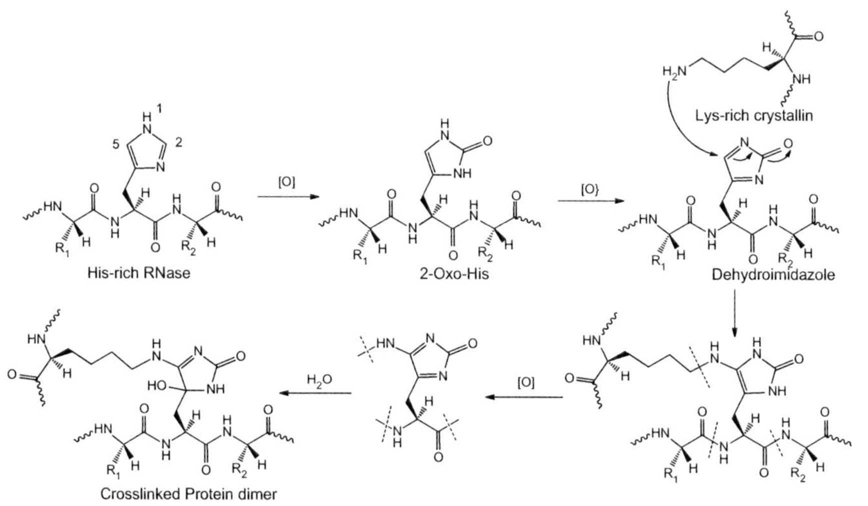

Scheme 7.11

(Scheme 7.11).[74] The results of this study were used to explain the cross-linking observed between a His-rich RNase and Lys-rich crystallin under oxidative conditions.[75]

Among all the crosslinking types, the one linkage caused by disulfide exchange (or disulfide scrambling) can be cleaved through reduction by a few reducing agents such as dithiothreitol (DTT) and tris(2-carboxyethyl)phosphine (TCEP), while others usually cannot be readily disconnected.

7.2.8 The Maillard Reaction

In Chapter 5 Drug–Excipient Interactions and Adduct Formation, we discussed the degradation pathways *via* the Maillard reaction in detail. Since proteins and peptides contain an *N*-terminal amino group and a side chain amino group of Lys residues, both groups can undergo degradation *via* the Maillard reaction in the presence of reducing sugars. This degradation can occur *in vitro*[76] as well as *in vivo* and the latter process has been generally attributed to degenerative disorders and aging.[77] Hence, formulation of protein drugs with reducing sugars is usually avoided.[15] Nonetheless, non-reducing sugars, such as sucrose, can contain reducing sugars as impurities or degradants, while some non-reducing sugars, for example, trehalose, are more resistant to hydrolytic degradation, in which reducing sugars may be generated, than others.[78] Some examples of protein and peptide degradation caused by the Maillard reaction, which is usually referred to as glycation, include recombinant human Mab,[79] recombinant human DNase,[80] human relaxin,[81] and lysine vasopressin.[82] In a stability study of lysine vasopressin in the presence of reducing sugars, the nonapeptide was found to readily undergo the Maillard reaction and the *N*-terminal amino group was much more reactive than the ε-amino group of Lys in a pH range of 3.0–8.5.

7.2.9 Degradation *via* Truncation of a *N*-Terminal Dipeptide Sequence through DKP Formation

Incubation of clinical grade human serum albumin (HAS) solutions at 37 °C caused significant cleavage of the *N*-terminal Asp-Ala dipeptide from the protein. Mass spectrometric analysis, *N*-terminal sequencing, and nuclear magnetic resonance (NMR) measurement suggested that the cleavage was caused by the attack of the *N*-terminal amino group toward the alanyl carbonyl group, resulting in breakage of the peptide bond and concomitant release of a DKP fragment (Scheme 7.12).[83] This attack is likely to be facilitated by the neighboring His$_3$ residue, the side chain of which could donate a proton to the alanyl carbonyl, as suggested by the authors, rendering the alanyl carbonyl more susceptible to nucleophilic attack by the *N*-terminal amino group. Additionally, the side chain of the terminal Asp could act as a general acid promoting the leaving of the His amino group. Comparison of HSA with serum albumins from other species indicates that this cleavage is specific to HSA, as others do not contain the same *N*-terminal sequence.

N-Terminal sequences that contain NH$_2$-X-Pro-, especially NH$_2$-Gly-Pro-,[84] are also prone to the DKP degradation pathway. Recombinant human growth hormone, which has the *N*-terminal sequence of NH$_2$-Phe-Pro-, also undergoes this degradation pathway.[85] In such cases, the susceptibility to DKP formation may be due to the particular conformation, caused by the Pro residue, which makes the cyclization easier to proceed.

7.2.10 Miscellaneous Degradation Pathways

If the *N*-terminal residue of a protein is a Glu, the protein becomes very susceptible to cyclization in which the *N*-terminal amino group forms a five-membered lactam with its side chain (Scheme 7.13). This degradation is a dehydration process as one molecule of water is eliminated and the resulting lactam is a pyroglutamic acid residue, or pyroGlu. Most recombinant monoclonal antibodies contain *N*-terminal Glu in their light chains and thus are susceptible to this degradation pathway.[86–88]

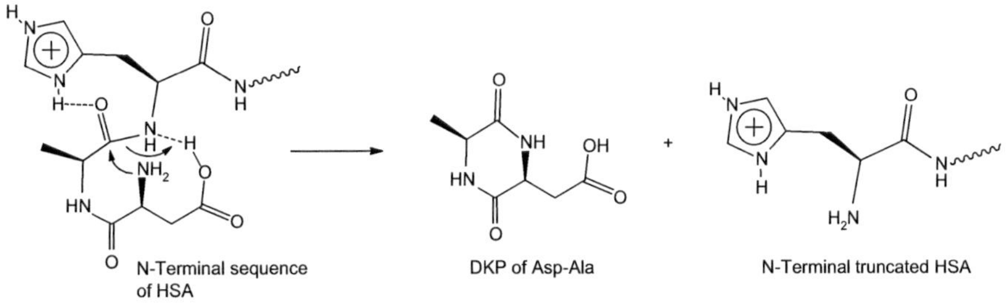

N-Terminal sequence   DKP of Asp-Ala   N-Terminal truncated HSA
of HSA

Scheme 7.12

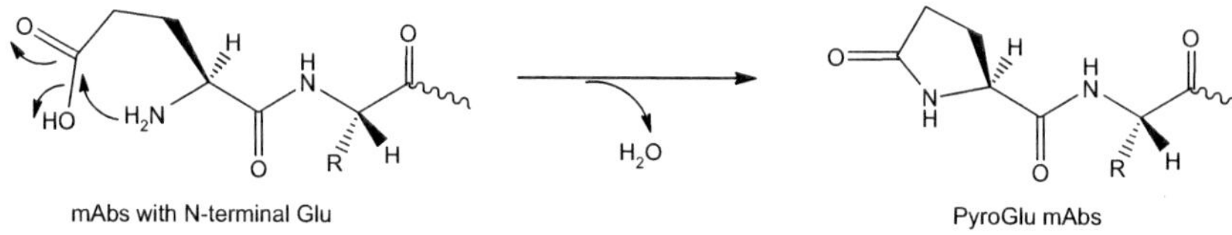

mAbs with N-terminal Glu   PyroGlu mAbs

Scheme 7.13

7.3 Degradation of Carbohydrate-based Biological Drugs

As mentioned in the beginning of this chapter, heparin and hyaluronan may be among the very few carbohydrate-based drugs that fall into the scope of our discussion. Both heparin and hyaluronan are polymeric anionic glycosaminoglycans, with the former containing both sulfate and carboxylate groups and the latter only carboxylate groups. Heparin is a natural product, which is usually extracted from mucosal tissues of porcine intestine. Pharmaceutical grade heparin is essentially free of peptide content and, chemically, it is a highly sulfated glycosaminoglycan prepared in its sodium salt format with an average molecular weight of $\sim 12\,000$. More recently, low molecular weight heparin (LMWH), obtained through both enzymatic and controlled chemical cleavage, has become available, which demonstrates improved bioavailability and therapeutic profile.[89]

Heparin is said to have the highest negative charge density among all the known biomolecules,[90] thanks to the densely populated sulfate and carboxylate groups throughout its linear chains. Heparin is heterogeneous in nature and it contains a variety of repeating sulfated disaccharide units, the most common of which consists of residues of α-L-idopyranosyluronic acid 2-sulfate and 2-deoxy-2-sulfamino-α-D-glucopyranose 6-sulfate. This repeating disaccharide unit is linked through position 4 and accounts for up to 85% of heparin from beef lung and about 75% of from porcine intestinal mucosa.[91,92] Despite its long clinical use history since 1937, few studies are available regarding its chemical stability. In a stress study of heparin in acidic, basic, and neutral solutions, Jandik *et al.* found that heparin, although remarkably stable overall, can undergo hydrolysis at the glycosidic linkages (in 0.1 N HCl), β-elimination and desulfation (in 0.1 N NaOH), and an increase in reducing capacity (in pH 7 phosphate buffer), especially under elevated temperatures (60 °C and 100 °C) over extended periods (*e.g.* > 500 hours).[93] The main degradation pathways of heparin are summarized in Scheme 7.14.

As shown above, the major degradation pathways under acidic and neutral conditions are cleavage at the glycosidic linkages and desulfation at both the sulfonyl and sulfamyl bonds. All of these degradation pathways are hydrolytic in nature. The activation energy for acidic hydrolysis at pH 2 of a glycosidic linkage in heparin should be at least $\sim 25\,\text{kcal}\,\text{mol}^{-1}$

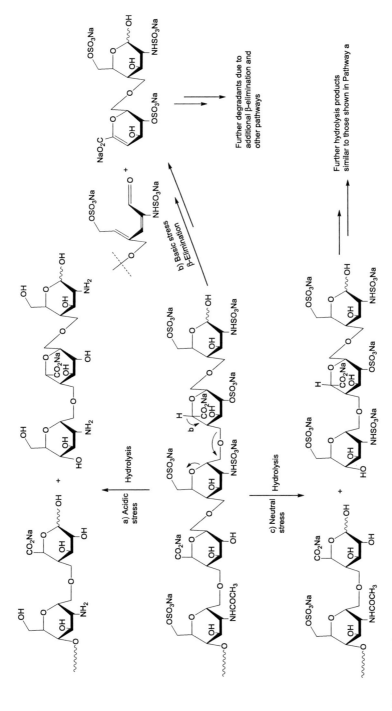

Scheme 7.14

$(103\,\text{kJ}\,\text{mol}^{-1})$.[94] On the other hand, the activation energies for hydrolysis of the sulfonyl and sulfamyl bonds should be in a similar range to the glycosidic linkages, based on the discussion in Chapter 2, Hydrolytic Degradation. These reasonably high activation energies explain the remarkable stability of heparin in solutions at temperatures of $60\,^{\circ}\text{C}$ and below. Nonetheless, appreciable degradation can occur at very high temperatures. In a recent study of formulated heparin sodium solutions under autoclave conditions of $121\,^{\circ}\text{C}$, more than one-third of its activity was lost after 30 minutes autoclaving.[95] Chondroitin sulfate, also a sulfated glycosaminoglycan with structural features closely resembling those of heparin, displays very similar stability and degradation pathways as heparin.[96] It has now been widely used as a dietary supplement for alleviating symptoms of osteoarthritis.

Hyaluronan is a ubiquitous biopolymer, which chemically is an anionic, non-sulfated glycosaminoglycan with a molecular weight in the range of millions of Daltons. This high molecular weight material has long been utilized in ophthalmic surgery as a lubricant; as such, it was approved by the FDA as a surgical device in 1980. While high molecular weight hyaluronan has anti-inflammatory, anti-angiogenic, and immunosuppressive properties, intermediate molecular weight hyaluronan fragments have the opposite effects. Depolymerization of hyaluronan is mainly achieved by chemical means, such as treatment at high temperature, acid hydrolysis, microwave-assisted digestion, and oxidative degradation.[97–100] A few studies examining the effectiveness and structural impact of these chemical methods on the resulting fragments have been reported.[98,100–102] Under non-oxidative degradation conditions, it appears that depolymerization is caused mainly by cleavage at the glycosidic linkages. Under free radical-mediated oxidative degradation conditions, depolymerization is likely to be initiated by the formation of a carbon-centered radical on the C1 of the glucuronate unit as shown in Scheme 7.15.[103]

7.4 Degradation of DNA and RNA Drugs

DNA and RNA molecules comprise repeating units of nucleotides and each of these in turn consists of a nucleobase, pentose, and phosphate group. In DNA, the nucleobases are adenine (A), thymine (T), guanine (G), and cytosine (C) and the pentose is deoxyribose. RNA molecules have the same structural features as DNA, except that RNA uses uracil (U) in place of thymine (T) and ribose in place of deoxyribose, respectively. There are two major degradation pathways of DNA and RNA (Figure 7.1): hydrolysis of the phosphodiester bonds that connect the repeating nucleotide units and oxidation of nucleobases.[104] To a less extent, the pentose unit can also be degraded *via* metal-catalyzed oxidation.[105]

7.4.1 Hydrolytic Degradation of Phosphodiester Bonds

In general, a phosphodiester bond is quite resistant to hydrolysis, as evidenced by its reasonably high activation energy of hydrolysis (see Chapter 2,

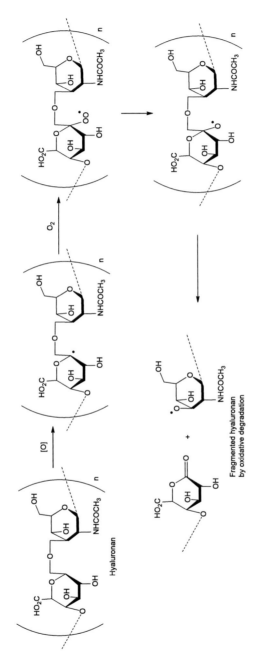

Scheme 7.15

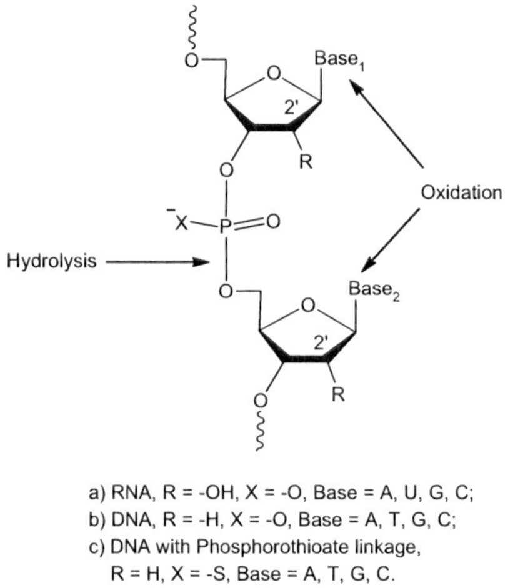

a) RNA, R = -OH, X = -O, Base = A, U, G, C;
b) DNA, R = -H, X = -O, Base = A, T, G, C;
c) DNA with Phosphorothioate linkage,
  R = H, X = -S, Base = A, T, G, C.

Figure 7.1 Major degradation pathways of DNA and RNA.

Hydrolytic Degradation). This explains why DNA is usually very stable in aqueous solution. For example, no measurable hydrolysis of a DNA molecule was observed in 1 M NaOH at 50 °C.[106] On the other hand, despite the fact that non-enzymatic hydrolysis of RNA is still very sluggish under physiological conditions, the rate of its hydrolysis is several orders of magnitude (10^5–10^6 fold) higher than DNA, owing to the presence of a hydroxyl group at the 2'-position of the ribose unit.[106] Under either acid- or base-catalyzed hydrolysis, it has been generally accepted that 2'-hydroxyl or its deprotonated form attacks the phosphodiester functional group, resulting in the formation of a pentavalent phosphorus transition state or intermediate. Decomposition of the latter leads to cleavage of the phosphodiester linkage (Scheme 7.16).[107]

Fomivirsen is the first and remains the only approved DNA-based drug,[2] which was designed based on the concept of "antisense".[3] It is a synthetic, modified 21-member oligonucleotide with the following sequence: 5'-GCG TTT GCT CTT CTT CTT GCG-3'. In this modified oligonucleotide, one of the two non-bridging oxygen atoms in all the phosphodiester bonds is replaced by a sulfur atom (Figure 7.1). The resulting oligonucleotide now contains all phosphorothioate linkages and displays enhanced stability toward hydrolysis by nucleases.[3]

7.4.2 Oxidative Degradation of Nucleic Acid Bases

The nucleobases are vulnerable spots susceptible to oxidative degradation owing to the electron richness of the purine and pyrimidine rings.[108] There are two commonly observed types of oxidative degradation: the first is mediated by transition metal ion-catalyzed processes, in which hydroxyl free

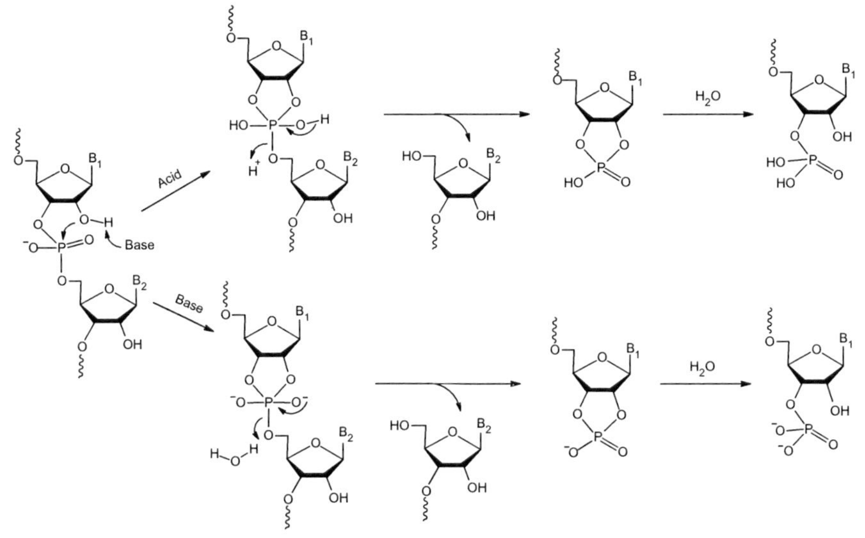
Acid
Base
Base
H_2O
H_2O

Scheme 7.16

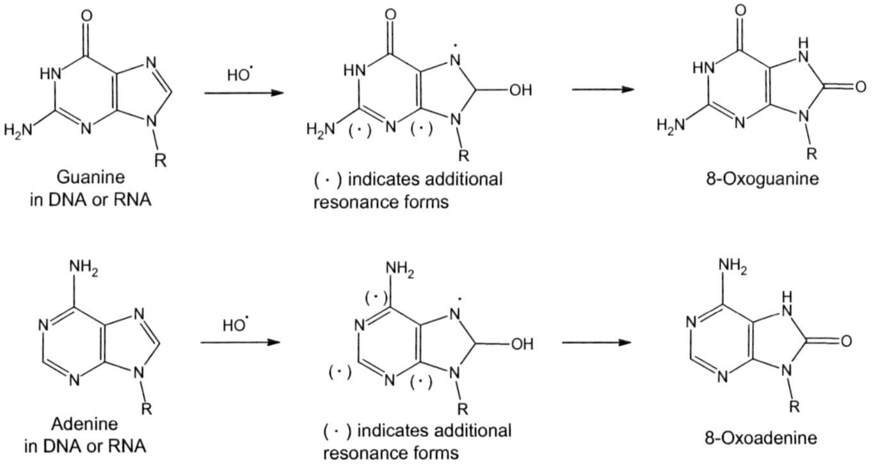
HO•
Guanine
in DNA or RNA
(•) indicates additional
resonance forms
8-Oxoguanine
HO•
Adenine
in DNA or RNA
(•) indicates additional
resonance forms
8-Oxoadenine

Scheme 7.17

radical, HO•, usually plays a pivotal role in the oxidation. The second type is caused by photosensitized oxidation in which singlet oxygen, $^1O_2$, is the key oxidant.

In transition metal ion-catalyzed oxidation, guanine and adenine are oxidized in a very similar way: HO• mainly attacks the 8-positions giving rise to 8-oxoguanine and 8-oxoadenine, respectively (Scheme 7.17).[108] This oxidation also occurs *in vivo* and 8-oxoguanosine has been used as a biomarker for cellular oxidative damage.[109]

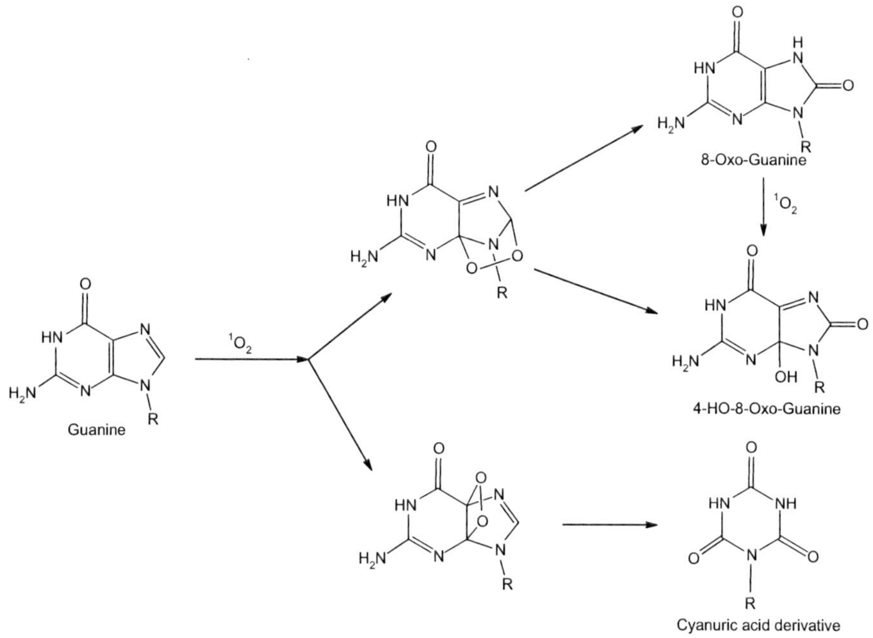

Scheme 7.18

Another major cause of the oxidative degradation of nucleobases is due to photosensitized oxidation (see Chapter 6, Photochemical Degradation). During type I photosensitized oxidation, HO$^•$ and other radicals such as peroxide radicals are generated. Owing to its high reactivity, oxidation by HO$^•$ would be expected to play a major role in type I category reactions and its main degradation mechanism should be the same as illustrated in Scheme 7.17. During type II photosensitized oxidation, the main oxidant is singlet oxygen, $^1O_2$, and among the nucleosides, it has been demonstrated that $^1O_2$ preferentially reacts with guanine,[110–112] probably through dioxygen cyclized intermediates.[113,114] Three immediate degradants resulting from this oxidation are 8-oxoGuanine, 4-HO-8-oxoGuanine, and a cyanuric acid derivative (Scheme 7.18).

References

1. http://pharma.about.com/od/BigPharma/tp/Big-Ten-Branded-Drug-Blockbusters-Of-2010.htm, last accessed 18 Feb, 2012.
2. S. D. Patil, D. G. Rhodes and D. J. Burgess, *The AAPS Journal*, 2005, **7**(1) Article 9 (http://www.aapsj.org).
3. E. Uhlmann and A. Peyman, *Chem. Rev.*, 1990, **90**, 543.

4. L. Poliseno, A. Mercatanti, L. Citti and G. Rainaldi, *Curr. Pharm. Biotechnol.*, 2004, **5**, 361.
5. N. R Wall and Y. Shi, *Lancet*, 2003, **362**, 1401.
6. L. Bonetta, *Cell*, 2009, **136**, 581.
7. E. Y. Chi, S. Krishnan, T. W. Randolph and J. F. Carpenter, *Pharm. Res.*, 2003, **20**, 1325.
8. W. Wang, *Int. J. Pharm.*, 2005, **289**, 1.
9. S. Frokjaer and D.E. Otzen, *Nat. Rev. Drug Discovery*, 2005, **4**, 298.
10. S. J. Shire, Z. Shahrokh and J. Liu, *J. Pharm. Sci.*, 2004, **93**, 1390.
11. H.-C. Mahler, W. Friess, U. Grauschopf and S. Kiese, *J. Pharm. Sci.*, 2009, **98**, 2909.
12. J. Schulz, *Methods Enzymol.*, 1967, **11**, 255.
13. A. S. Inglis, *Methods Enzymol.*, 1983, **91**, 324.
14. C. Oliyai and R. T. Borchardt, *Pharm. Res.*, 1993, **10**, 95.
15. M. C. Manning, D. K. Chou, B. M. Murphy, R. W. Payne and D. S. Katayama, *Pharm. Res.*, 2010, **27**, 544.
16. S. Clarke, *Int. J. Pept. Protein Res.*, 1987, **30**, 808.
17. F. Marcus, *Int. J. Pept. Protein Res.*, 1985, **25**, 542.
18. D. Piszkiewicz, M. Landon and E. L. Smith., *Biochem. Biophys. Res. Commun.*, 1970, **40**, 1173.
19. G. Chen, B. M. Warrack, A. K. Goodenough, H. Wei, D. B. Wang-Iverson and A. A. Tymiak, *Drug Discovery Today*, 2011, **16**, 58.
20. Y. Ge, B. G. Lawhorn, M. ElNaggar, E. Strauss, J.-H. Park, T. P. Begley and F. W. McLafferty, *J. Am. Chem. Soc.*, 2002, **124**, 672.
21. R. A. Kenley and N. W. Warne, *Pharm. Res.*, 1994, **11**, 72.
22. J. A. Schrier, R. A. Kenley, R. Williams, R. J. Corcoran, Y. Kim, R. P. Northey, D. D'Augusta and M. Huberty, *Pharm. Res.*, 1993, **10**, 933.
23. M. Sandor, A. Riechel, I. Kaplan and E. Mathiowitz, *Biochim. Biophys. Acta*, 2002, **1570**, 63.
24. M. C. Manning, K. Patel and R. T. Borchardt, *Pharm. Res.*, 1989, **6**, 903.
25. T. Geiger and S. Clark, *J. Biol. Chem.*, 1987, **262**, 785.
26. B. Li, R. T. Borchardt, E. M. Topp, D. Vander Velde and R. L. Schowen, *J. Am. Chem. Soc.*, 2003, **125**, 11486.
27. M. P. DeHart and B. D. Anderson, *J. Pharm. Sci.*, 2007, **96**, 2667.
28. D. J. Kroon, A. Baladin-Ferro and P. Lalan, *Pharm. Res.*, 1992, **9**, 1386.
29. P. K. Tsai, M.W. Bruner, J. I. Irwin, C. C. Ip, C. N. Oliver, R. W. Nelson, D. B. Volkin and C. R. Middaugh, *Pharm. Res.*, 1993, **10**, 1580.
30. J. Cacia, R. Keck, R. Presta and L. G. J. Frenz, *Biochemistry*, 1996, **35**, 1897.
31. Y. Mimura, K. Nakamura, T. Tanaka and M. Fujimoto, *Electrophoresis*, 1998, **19**, 767.
32. M. Perkins, R. Theiler, S. Lunte and M. Jeschke, *Pharm. Res.*, 2000, **17**, 1110.
33. R. J. Harris, B. Kabakoff, F. D. Macchi, F. J. Shen, M. Kwong, J. D. Andya, S. J. Shire, N. Bjork, K. Totpal and A. B. Chen, *J. Chromatogr., B: Biomed. Sci. Appl.*, 2001, **752**, 233.
34. W. Zhang and M. J. Czupryn, *J. Pharm. Biomed. Anal.*, 2003, **30**, 1479.

35. N. Fujii, S. Muraoka, K. Satoh, H. Hori and K. Harada, *Biomed. Res. (Tokyo)*, 1991, **12**, 315.
36. N. Fujii, Y. Momose, N. Ishii, M. Takita, M. Akaboshi and M. Kodama, *Mech. Ageing Dev.*, 1997, **107**, 347.
37. Z. Shahrokh, G. Eberlein, D. Buckley, M. V. Paranandi, D. W. Aswad, P. Stratton, R. Mischak and Y. J. Wang, *Pharm. Res.*, 1994, **11**, 936.
38. W. Jiskoot, E. C. Beuvery, A. A. de Koning, J. N. Herron and D. J. Crommelin, *Pharm. Res.*, 1990, **7**, 1234.
39. A. J. Alexander and D. E. Hughes, *Anal. Chem.*, 1995, **67**, 3626.
40. M. Paborji, N. L. Pochopin, W. P. Coppola and J. B. Bogardus, *Pharm. Res.*, 1994, **11**, 764.
41. A. J. Cordoba, B.-J. Shyong, D. Breen and R. J. Harris, *J. Chromatogr. B: Anal. Technol. Biomed. Life Sci.*, 2005, **818**, 115.
42. T. Xiang, E. Lundell, Z. Sun and H. Liu, *J. Chromatogr. B: Anal. Technol. Biomed. Life Sci.*, 2007, **858**, 254.
43. E. R. Stadtman, *Free Radicals Biol. Med.*, 1990, **9**, 315.
44. B. L. Chen, T. Arakawa, C. F. Morris, W. C. Kenney, C. M. Wells and C. G. Pitt, *Pharm. Res.*, 1994, **11**, 1581.
45. V. Shetty and T. A. Neubert, *J. Am. Soc. Mass. Spectrom.*, 2009, **20**, 1540.
46. V. Shetty, D. S. Spellman and T. A. Neubert, *J. Am. Soc. Mass Spectrom.*, 2007, **18**, 1544.
47. W. R. Liu, R. Langer and A. M. Klibanov, *Biotechnol. Bioeng.*, 1991, **37**, 177.
48. G. M. Jordan, S. Yoshioka and T. Terao, *J. Pharm. Pharmacol.*, 1994, **46**, 182.
49. J. L. Jensen, C. Kolvenbach, S. Roy and C. Schöneich, *Pharm. Res.*, 2000, **17**, 190.
50. E. T. Duenas, R. Keck, A. DeVos, A. J. S. Jones and J. L. Cleland, *Pharm. Res.*, 2001, **18**, 1455.
51. G. W. Becker, P. M. Tackitt, W. W. Bromer, D. S. Lefeber and R. M. Riggin, *Biotechnol. Appl. Biochem.*, 1988, **10**, 326.
52. J. Fransson, E. Florin-Robertsson, K. Axelsson and C. Nyhlen, *Pharm. Res.*, 1996, **13**, 1252.
53. K. Uchida and S. Kawakishi, *Arch. Biochem. Biophys.*, 1990, **283**, 20.
54. K. Uchida and S. Kawakishi, *J. Biol. Chem.*, 1994, **269**, 2405.
55. F. Zhao, E. Ghezzo-Schoneich, G. I. Aced, J. Hong, T. Milby and C. Schöneich, *J. Biol. Chem.*, 1997, **272**, 9019.
56. C. Schöneich, *J. Pharm. Biomed. Anal.*, 2000, **21**, 1093.
57. L. A. Holt, B. Milligan, D. E. Rivett and F. H. C. Stewart, *Biochim. Biophys. Acta*, 1977, **499**, 131.
58. K. Itakura, K. Uchida and S. Kawakishi, *Chem. Res. Toxicol.*, 1994, **7**, 185.
59. J. D. Kanner and O. Fennema, *J. Agric. Food Chem.*, 1987, **35**, 71.
60. M. K. Krogull and O. Fennema, *J. Agric. Food Chem.*, 1987, **35**, 66.
61. J. A. Ji, B. Zhang, W. Cheng and Y. J. Wang, *J. Pharm. Sci.*, 2009, **98**, 4485.
62. X. Zhang and C. S. Foote, *J. Am. Chem. Soc.*, 1993, **115**, 8867.

63. M. Li, B. Conrad, R. G. Maus, S. M. Pitzenberger, R. Subramanian, X. Fang, J. A. Kinzera and Holly J. Perpall, *Tetrahedron Lett.*, 2005, **46**, 3533.
64. J. M. Froelich and G. E. Reid, *Proteomics*, 2008, **8**, 1334.
65. I. Perdivara, L. J. Deterding, M. Przybylski and K. B. Tomer, *J. Am. Soc. Mass Spectrom.*, 2010, **21**, 1114.
66. C. Giulivi, N. J. Traaseth and K. J. A. Davies, *Amino Acids*, 2003, **25**, 227.
67. C. Giulivi and K. J. A. Davies, *J. Biol. Chem.*, 2001, **276**, 24129.
68. E. R. Stadtman, *Free Radicals Biol. Med.*, 1990, **9**, 315.
69. J. R. Requena, C.-C. Chao, R. L. Levine and E. R. Stadtman, *Proc. Natl. Acad. Sci.*, 2001, **98**, 69.
70. A. J. Cordoba, B.-J. Shyong, D. Breen and R. J. Harris, *J. Chromatogr., B: Anal. Technol. Biomed. Life Sci.*, 2005, **818**, 115.
71. S. L. Cohen, C. Price and J. Vlasak, *J. Am. Chem. Soc.*, 2007, **129**, 6976.
72. G. I. Tous, Z. Wei, J. Feng, S. Bilbulian, S. Bowen, J. Smith, R. Strouse, P. McGeehan, J. Casas-Finet and M. A. Schenerman, *Anal. Chem.*, 2005, **77**, 2675.
73. H. R. Costantino, R. Langer and A. M. Klibanov, *Pharm. Res.*, 1994, **11**, 21.
74. Y. Liu, G. Sun, A. David and L. M. Sayre, *Chem. Res. Toxicol.*, 2004, **17**, 110.
75. P. Guptasarma, D. Balasubramanian, S. Matsugo and I. Saito, *Biochemistry*, 1992, **31**, 4296.
76. M. C. Lai and E. M. Topp, *J. Pharm. Sci.*, 1999, **88**, 489.
77. A. C. Cerami, H. Vlassara and M. Brownlee, *Sci. Am.*, 1987, **256**, 90.
78. J. O'Brien, *J. Food Sci.*, 1996, **61**, 679.
79. B. Zhang, Y. Yang, I. Yuk, R. Pai, P. McKay, C. Eigenbrot, M. Dennis, V. Katta and K. C. Francissen, *Anal. Chem.*, 2008, **80**, 2379.
80. C. P. Quan, S. Wu, N. Dasovich, C. Hsu, T. Patapoff and E. Canova-Davis, *Anal. Chem.*, 1999, **71**, 4445.
81. S. Li, T. W. Patapoff, D. Overcashier, C. Hsu, T.-H. Nguyen and R. T. Borchardt, *J. Pharm. Sci.*, 1996, **85**, 873.
82. E. Tarelli, P. H. Corran, B. R. Bingham, H. Mollison and R. Wait, *J. Pharm. Biomed. Anal.*, 1994, **12**, 1355.
83. B. Chan and N. Dodsworth, J. Woodrow, A. Tuchker and R. Harris, *Eur. J. Biochem.*, 1995, **227**, 524.
84. C. Goolcharran, M. Khossravi and R. T. Borchardt, in *Pharmaceutical Formulation Development of Peptides and Proteins*, ed. S. Frokjaer and L. Hovgaard, CRC Press, New York, 2000, pp. 70–88.
85. J. E. Battersby, W. S. Hancock, E. Canovadavis, J. Oeswein and B. O'Connor, *Int. J. Pept. Protein Res.*, 1994, **44**, 215.
86. L. Yu, R. L. Remmele Jr. and B. He, *Rapid Commun. Mass Spectrom.*, 2006, **20**, 3674.
87. H. Liu, G. Gaza-Bulesco and J. Sun, *J. Chromatogr. B: Anal. Technol. Biomed. Life Sci.*, 2006, **837**, 35.

88. D. Chelius, K. Jing, A. Lueras, D. S. Rehder, T.M. Dillion, A. Vizel, R. S. Rajan, T. Li, M. J. Treuheit and P. V. Bondarenko, *Anal. Chem.*, 2006, **78**, 2370.

89. R. J. Linhardt, D. Loganathan, A. Al-Hakim, H. M. Wang, J. M. Walenza, D. Hoppensteadt and J. Fareed, *J. Med. Chem.*, 1990, **33**, 1639.

90. M. Cox and D. Nelson, Lehninger Principles of Biochemistry, W. H. Freeman & Co., New York, 2005, p. 1100.

91. A. S. Perlin, B. Casu, G. R. Sanderson and L. F. Johnson, *Can. J. Chem.*, 1970, **48**, 2260.

92. A. S. Perlin, M. Mazuek, L. B. Jaques and L. W. Kavanagh, *Carbohydr. Res.*, 1968, **7**, 369.

93. K. A. Jandik, D. Kruep, M. Cartier and R. J. Linhard, *J. Pharm. Sci.*, 1996, **85**, 45.

94. A. Karlsson and S. K. Singh, *Carbohydr. Polym.*, 1999, **38**, 7.

95. J. M. Beaudet, A. Weyers, K. Solakyildirim, B. Yang, M. Takieddin, S. Mousa, F. Zhang and R. J. Linhardt, *J. Pharm. Sci.*, 2011, **100**, 3396.

96. N. Volpi, A. Mucci and L. Schenetti, *Carbohydr. Res.*, 1999, **315**, 345.

97. H. Bothner, T. Waaler and O. Wik, *Int. J. Biol. Macromol.*, 1988, **10**, 287.

98. Y. Tokita and A. Okamoto, *Polymer Degrad. Stab.*, 1995, **48**, 269.

99. S. A. Galema, *Chem. Soc. Rev.*, 1997, **26**, 233.

100. L. Soltes, G. Kogan, M. Stankovska, R. Mendichi, J. Rychly, J. Schiller and P. Gemeiner, *Biomacromolecules*, 2007, **8**, 2697.

101. K. Kubo, T. Nakamura, K. Takagaki, Y. Yoshida and M. Endo, *Glycoconjugate J.*, 1993, **10**, 435.

102. E. Drimalova, V. Velebny, V. Sasinkova, Z. Hromadkova and A. Ebringerova, *Carbohydr. Polym.*, 2005, **61**, 420.

103. J. Rychly, L. Soltes, M. Stankovskab, I. Janigova, K. Csomorova, V. Sasinkova, G. Kogan and P. Gemeiner, *Polym. Degrad. Stab.*, 2006, **91**, 3174.

104. D. Pogocki and C. Schöneich, *J. Pharm. Sci.*, 2000, **89**, 443.

105. W. K. Pogozelski and T. D. Tullius, *Chem. Rev.*, 1998, **98**, 1089.

106. M. Komiyama, N. Takeda and H. Shigekawa, *Chem. Commun. (Cambridge, U.K.)*, 1999, 1443.

107. R. Bredow and M. Labelle, *J. Am. Chem. Soc.*, 1986, **108**, 2655.

108. C. J. Burrows and J. G. Muller, *Chem. Rev.*, 1998, **98**, 1109.

109. K. B. Beckman and B. N. Ames, *J. Biol. Chem.*, 1997, **272**, 19633.

110. H. I. Simon and H. Van Vunakis, *J. Mol. Biol.*, 1962, **4**, 488.

111. A. Kornhauser, N. I. Krinsky, P.-K. Huang and D. C. Clagett, *Photochem. Photobiol.*, 1973, **18**, 63.

112. J. Cadet and R. Teoule, *Photochem. Photobiol.*, 1978, **28**, 661.

113. J.-L. Ravanatt and J. Cadet, *Chem. Res. Toxicol.*, 1996, **8**, 379.

114. J. Cadet, M. Berger, C. Decarroz, J. R. Wagner, J. E. Van Lier, Y. M. Ginot and P. Vigny, *Biochimie*, 1986, **68**, 813.

CHAPTER 8

Strategies for Elucidation of Degradant Structures and Degradation Pathways

8.1 Overview

The ultimate goal of understanding the chemistry of drug degradation is to prevent, minimize, or control drug degradation starting from selection of a drug candidate, proceeding through various stages of process and pharmaceutical development, to commercial scale manufacturing and stability testing. In order to comprehend the degradation chemistry of a drug, it is necessary first to elucidate the structures of its degradants. The predominant analytical methodology utilized in the pharmaceutical industry to ensure adequate potency of a drug and to monitor the impurities and degradants generated during its manufacturing process and throughout its shelf-life is high performance liquid chromatography (HPLC). The structures and root cause (*i.e.* formation mechanism) of unknown degradants need to be identified once the degradants exceed certain specified levels, as required by regulatory guidelines such as the International Conference on Harmonisation (ICH) guidelines.[1,2] Rapid identification of these degradants observed at various stages of stability studies including those under accelerated conditions, is essential to a clear understanding of the quality attributes of a new drug candidate. For commercialized drug products, rapid identification of degradants formed during long term stability storage conditions, registered with various regulatory/health authorities, is also critical for a timely, adequate toxicological evaluation and decisive market action, if warranted, based on the structures and levels of the degradants determined.

Liquid chromatography-mass spectrometry (LC-MS^n, n is typically 1 to 4) is most frequently utilized as the first analytical tool for identification of degradants as well as other impurities at trace levels. First of all, this is due to the natural

RSC Drug Discovery Series No. 29
Organic Chemistry of Drug Degradation
By Min Li
© Min Li 2012
Published by the Royal Society of Chemistry, www.rsc.org

synergy between HPLC and MS, in particular in cases where the HPLC method is MS compatible, that is, the mobile phase of the method does not contain any non-volatile components. In these cases, a mass spectrometer can be simply regarded as an additional detector (MS detector), connected after a conventional or a photodiode array UV-Vis (PDA) detector, for the HPLC method. Second, the sensitivity of a MS detector is usually at least as sensitive as a typical UV detector. In many cases, an MS detector can be much more sensitive than a typical UV detector. Third, with the capability of multi-stage tandem mass spectrometry (MS^n, $n \geq 2$), far more structural information can be obtained from the fragmentation pathways of the parent ion of an unknown degradant. Lastly, with high resolution MS detection technologies, such as time-of-flight, Obitrap, and Fourier transform ion cyclotron resonance (FT-ICR), the molecular formulae of an unknown degradant and its fragments can be readily attainable,[3,4] greatly facilitating the structure elucidation of the unknown degradant.

Despite the strong capability of LC-MS^n, especially LC-high resolution MS^n technology, complementary analytical techniques such as nuclear magnetic resonance (NMR) spectroscopy are required from time to time for unambiguous structure identification of unknown impurities and degradants, especially in cases where the degradation chemistry of the drug is not well understood. The exercise of structure elucidation of an unknown degradant, typically present between 0.1–0.5% levels in a drug substance or product, can be very challenging for a few reasons. The most encountered challenge is usually the availability of a purified sample of the unknown degradant: a limited amount will usually be available, in particular for degradants present in a drug product, for isolation and subsequent structure elucidation *via* NMR spectroscopy. The sample quantity required for structure determination by modern NMR methodology can be as low as hundreds of micrograms or even lower. For example, with a 1.3 mm cryogenic NMR probe, heteronuclear ^{1}H–^{13}C 2D NMR experiments may be performed on a sample as low as $\sim 20\,\mu$g for a small molecule drug.[5] Nevertheless, this capability may not be readily accessible to everyone. On the other hand, LC-NMR has been utilized for structure elucidation of pharmaceutical impurities and degradants, which has the advantage of not performing a tedious off-line sample purification.[6–8] However, it is usually very challenging to acquire satisfactory heteronuclear 2D NMR signals on impurity samples at trace levels with the current LC-NMR capability, owing to insufficient sensitivity. Frequently, heteronuclear ^{1}H–^{13}C 2D NMR measurements such as heteronuclear single-quantum correlation spectroscopy (HSQC) and heteronuclear multiple-bond correlation spectroscopy (HMBC) are needed for high confidence structure determination, in addition to 1D (^{1}H and ^{13}C) and 2D homonuclear NMR measurements (correlation spectroscopy; COSY, total correlation spectroscopy; TOCSY, and nuclear Overhauser effect spectroscopy; NOESY/rotating frame nuclear Overhauser effect spectroscopy; ROESY). As Sharman and Jones concluded, based on the results of their study, "despite the undoubted advantages of LC–NMR in many situations, for the identification of very low-level impurities it may not always be the most efficient overall approach when all factors are considered".[9]

In this chapter, we will demonstrate a strategy that combines LC-MS^n with a mechanism-based approach for design of effective stress studies (forced degradation), which should greatly facilitate the identification of drug degradants. Through this approach, a sufficient amount of an unknown degradant can usually be made available within a short period for further structure analysis by 1D and 2D NMR spectroscopy. Besides, this approach is also very helpful in understanding the degradation pathways of drug substances and their formulated products. If the chemistry (*e.g.* a particular type of oxidation) used in the designed stress study is able to regenerate the degradant of interest, it suggests that this chemistry may resemble the underlining degradation chemistry of the drug occurring in a real life scenario, especially in cases where a similar degradation profile can also be generated. In contrast, if the designed stress study is unable to regenerate the degradant of interest, it implies that the underlining degradation chemistry may be different from the one employed in the stress study. This general strategy, as applied in several case studies presented below, can be readily utilized as a general strategy for structure elucidation of unknown degradants with a high confidence level and high probability of success.

We will also demonstrate the concept and utility of LC-MS^n molecular fingerprinting for structure elucidation of unknown degradants. The use of LC-MS^n molecular fingerprinting can afford very rapid structure elucidation of unknown process impurities and degradants with a high confidence level (without resorting to NMR) in cases where comparison of MS^n fingerprints with those of authentic samples can be made directly or indirectly (with indirect comparison done *via* either gas phase chemistry or wet chemistry). Even in cases where indirect comparison is not feasible, structure predication of unknown impurities or degradants may be possible based on a close but not identical match of MS^n fingerprints, in conjunction with additional knowledge such as perceived degradation mechanisms and known process chemistry. Another advantage of using LC-MS^n molecular fingerprinting is that once the LC-MS^n fingerprint, in particular the MS^n fingerprint, of an impurity or degradant is established in a database, it is not be necessary to maintain a physical collection of this impurity or degradant as a reference material for future positive verification of its identity. Although the case studies presented in this chapter are all from steroidal drugs, the concept and strategy of LC-MS^n molecular fingerprinting is generally applicable to drug molecules of other structures.

8.2 Practical Considerations of Employing LC-MS^n for Structural Elucidation of Degradants at Trace Levels

There is no intention here to give a comprehensive overview of LC-MS technology. Readers who are interested in this topic can refer to the third edition of *Mass Spectrometry, Principles and Applications*.[10] Instead, a few practical issues that have been frequently encountered during typical applications of LC-MS to structure elucidation of pharmaceutical impurities at trace levels are discussed here.

8.2.1   Conversion of MS-unfriendly HPLC Methods to LC-MS Methods

A great number of HPLC methods employ non-volatile buffers such as phosphate buffers to control the pH of their mobile phases. Control of the pH is not only necessary for ionizable drug molecules, for example, those containing acid and/or base functionalities, but may also be necessary for non-ionizable drug molecules if they have ionizable impurities and/or degradants. Hence, in these cases, the non-volatile buffers need to be replaced by volatile buffers with similar pH buffering ranges. In pH range ~ 5.5 to 8, the most commonly used volatile buffer appears to be ammonium acetate. Typically, a concentration of no more than 10 mM of ammonium acetate should be used and, at this concentration, the pH of the aqueous solution of ammonium acetate is approximately 6.8. The pH of the ammonium acetate solution can be adjusted down with acetic acid or up with ammonia solution.

For HPLC methods with a mobile phase pH between ~ 3.5 and 5, formic acid and acetic acid can substitute for the corresponding buffers at the low and high pH ends, respectively. Many HPLC methods utilize trifluoroacetic acid (TFA), typically at a $\sim 0.1\%$ concentration in their mobile phases. Although TFA can suppress ionization in the gas phase, which results in reduced mass spectrometric response, it can still be used in many cases with satisfactory results. For HPLC methods using non-volatile strong acids, such as methanesulfonic acid, use of TFA to replace these strong acids may be mandatory, as other volatile organic acids are not strong enough to yield a comparable pH at a range below 3. If suppression of ionization becomes an issue, one can try to reduce the concentration of TFA to 0.05% or lower. In general, if the original pH can be largely maintained in the corresponding LC-MS methods, these can usually generate very similar elution profiles to those by the original methods. In addition, the UV-Vis profiles of those analytes that have ionizable chromophores should remain the same or be very similar.

In other cases, ion-pairing agents are utilized in reserved phase chromatography to impart peak retention to those analytes that are otherwise not properly retained. For example, sodium hexanesulfonate is commonly used to retain positively charged analytes under acidic conditions. The ion-pairing agents used are usually non-volatile compounds and hence need to be replaced by volatile ion-pairing agents; a commonly used volatile ion-pairing agent in LC-MS is heptafluorobutyric acid.[11] Commonly used volatile buffers, acids and ion-pairing agents in LC-MS are summarized in Table 8.1.

8.2.2   Nomenclature, Ionization Modes and Determination of Parent Ions

Historically, the first ionization mode in MS is electron ionization which used to be called electron impact (ionization). This mode of ionization is made possible by the impact of an electron beam, typically accelerated at 70 eV, toward the analyte. Under this highly energetic ionization, an electron is taken

Table 8.1 Commonly Used Volatile Buffers, Acids and Ion-pairing Agents in LC-MS.

HPLC mobile phase pH/ion pairing agent	Volatile buffer/Ion pairing agent replacement	Note
5.5–8	NH$_4$OAc	10 mM solution has pH ~6.8, which can be adjusted down with CH$_3$CO$_2$H or up with NH$_3$.
4–5	CH$_3$CO$_2$H	0.1%
~3.5	HCO$_2$H	0.1%
<3	TFA	0.05–0.1%
Sodium hexanesulfonate	Heptafluorobutyric acid	0.1%

away from the neutral analyte (M) by the incoming electron beam, forming a radical cation (M$^{+\bullet}$), in addition to the formation of fragment ions. This type of radical cations have been referred to as "molecular ions" since the early days of MS. These radical ions have odd numbers of electrons, which are obviously quite different from the ions generated by the two most utilized modes of ionization in LC-MS, that is, electron spray ionization (ESI) and atmosphere pressure chemical ionization (APCI). In these two cases, non-radical, proto-nated ions ([M + H$^+$]) with even numbers of electrons are usually produced under positive ionization (from both ESI and APCI), although occasionally sodiated, potassiated, or ammoniated ions ([M + Na$^+$], [M + K$^+$], or [M + NH$_4^+$]) can be produced as well. Under negative ionization from both ESI and APCI, deprotonated ions can be formed for analytes that can lose a proton. Under negative ESI, acidic compounds are readily ionized to produce deprotonated ions.

Under negative APCI, compounds that are much less "acidic", including aldehydes,[12] hydrazones,[13] and certain hydroxyl-containing compounds,[14] may also deprotonate. Although the protonated, sodiated, and deprotonated ions generated from ESI and APCI are sometimes also called molecular ions in analytical and MS journals,[15–17] the terminology is considered technically incorrect according to some experts in the field.[4] According to them the term "molecular ions" should always be reserved for the radical cation (M$^{+\bullet}$) as discussed above. Other terms such as "parent ions" and "precursor ions" have been used for the ions generated directly from the analyte molecule *via* pro-tonation, deprotonation, or other processes such as association of the analyte molecule with sodium, potassium, and ammonium ions in the gas phase. In our discussion, we choose to use the term "parent ions".

In general, positive ionization is used in a majority cases in both ESI and APCI MS. However, negative ionization is much preferred for acidic analytes. Additionally, negative ionization usually gives less background interference. Once an LC-MS measurement of a pharmaceutical sample containing unknown degradants is complete, the first step in analysis is typically trying to determine the *m/z* values for the parent ions of the degradants so that their formulae can be deduced. At the typical degradant levels, the mass signals for these degradants may not be obvious in the total ion chromatogram. If this situation occurs, one needs to uncover the m/z values by searching the regions in the total ion chromatogram where they are expected to elute based on the corresponding UV chromatograms. Since the mass detector is almost always connected after the UV-Vis detector in LC-MS analysis of pharmaceutical impurities, there is some delay in determining the mass signals with respect to the corresponding UV-Vis signals. This delay can be used as an opportunity to eliminate interfering mass signals eluting closely with the peaks of interest. Once a possible m/z signal has been found, the single ion chromatogram based on that m/z value must be found; the peak in this chromatogram should have a similar profile to that of the corresponding UV-Vis chromatogram, if there is no interference. In addition, the single ion peak should have the expected retention time delay.

Other ways of locating or confirming the m/z value of a parent ion include the use of so called "mass defect" or "fractional mass" filtering, when high resolution MS is employed. It is based on the fact that the "fractional mass" (or "mass defect"), that is, the mass value after the integer part of a monoisotopic molecular weight, should be consistent between the degradants and the active pharmaceutical ingredient (API) for each established or expected type of chemical transformation. For example, since the theoretical accurate m/z value of the protonated anti-AIDS drug molecule, indinavir, is 614.3701, the accurate m/z value of its monohydroxylated metabolite would be $614.3701 + 15.9949 = 630.3650$. In this case, the two theoretical fractional masses, 0.3701 and 0.3650, are very close to each other owing to the fact that the accurate mass of oxygen is very close to the integer value of 16. In other words, if one sets the mass defect filtering parameter to 0.37 ± 0.01, only monooxidized metabolites will show up in the spectra, while interfering species with a nominal M + 16 mass value will be filtered out. This methodology has been widely used in metabolite identification in commercial software packages.[18–20] To identify pharmaceutical impurities and degradants, this principle can also be applied to assist the correct assignment of a parent ion.

The characteristic isotope abundances of drug molecules that contain elements with relatively high abundant isotopes such as sulfur and chlorine (*e.g.* ^{34}S and ^{37}Cl) can also be used to help identify the parent ions. Although ESI and, to a somewhat lesser degree, APCI, are considered "soft" ionization techniques, the parent ion of an analyte may not always be obvious or even observed under these ionization conditions, particularly for analytes that tend to form electronically stabilized species in the gas phase. In some cases, parent ions can be observed in one type of ionization but not in the other; hence, analyses under both positive and negative ionization may be needed to verify the assignment of parent ions.

Sometimes, an analyte solution can be directly injected into the mass detector, that is, by "direct infusion" mode. This way of analysis bypasses the capability of the LC part of LC-MS. Based on the experience of the current author, this practice should be strongly discouraged in LC-MS analysis of pharmaceutical impurities. This is because (1) direct infusion usually requires much more sample volume than a typical LC-MS analysis and (2) direct infusion may bring about the problem of mass signal suppression for the analyte of interest by minor, but strongly ionizable, impurities present in the sample.

8.2.3 Fragmentation and LC-MS^n Molecular Fingerprinting

Once an LC-MS method becomes available, the first step in the analysis is usually to collect the relevant LC-PDA-MS^n properties (UV profile, molecular weight/molecular formula, and multiple fragmentation patterns) of the unknown degradant. Collectively these properties can be defined as the LC-MS^n molecular fingerprint of a compound, while the multiple stage MS^n fragmentation patterns can be defined as its MS^n molecular fingerprint. This is due to the observation that an LC-MS^n fingerprint (and in most cases an MS^n fingerprint alone) is unique for each chemical entity, including its diastereo-meric isomers. Hence, a quick comparison of the LC-MS^n fingerprint (or the MS^n fingerprint alone) of an unknown impurity to the fingerprints of known compounds should reveal the structure of this unknown impurity immediately, if a match is found. This is analogous to the use of gas chromatography-mass spectrometry (GC-MS) in elucidation of the structure of unknown volatile impurities. However, the biggest limitation on the use of LC-MS^n or MS^n fingerprint comparisons is that the MS^n fingerprints are dependent upon the configuration or make of a particular mass spectrometer. For example, MS^2 fingerprints obtained on an ion-trap MS instrument are not directly compar-able to those obtained on a triple quadrupole instrument.[21] On the other hand, the fragmentation patterns generated by GC-MS under electron ionization (typically with a 70 eV electron beam) are universal across all GC-MS instru-ments from different manufacturers.

Nevertheless, the wide availability of LC-MS equipped with triple quadru-pole or ion-trap mass spectrometers has rendered the acquisition of LC-MS^n fingerprints for compounds of interest a relatively easy undertaking. As such, comparison of the LC-MS^n fingerprint of an unknown impurity with those of relevant, known compounds obtained on the same type of mass spectrometers can be readily performed. Hence, use of LC-MS^n molecular fingerprints remains a very effective and convenient tool in the structure elucidation of unknown chemical entities. This approach to identifying and/or verifying the structure of an unknown compound based on its LC-MS^n molecular fingerprint is called LC-MS^n molecular fingerprinting.

The most reliable element of an LC-MS^n molecular fingerprint is its MS^n fingerprint. This is because structurally similar diastereomers have identical molecular weight/formulae that cannot be differentiated by high resolution MS. Furthermore, they tend to have very similar chromatographic retention times and thus identification based on their retention properties may not be straightforward. Nevertheless, these structurally similar diastereomers can display subtle but distinguishable differences in their MS^n fingerprints.[21-24]

One good example is the case of dexamethasone and betamethasone. The two drug substances are epimeric diastereomers of each other and their only structural difference lies in the orientation of the 16-methyl group: dexamethasone has an α-orientation while betamethasone has a β-orientation. Yet, the two drug molecules and their corresponding esters display reproducible and distinguishable differences in their MS^n fingerprints as illustrated in Figure 8.1.[22,23]

Another example is the two dehydration degradants of betamethasone: Z- and E-betamethasone enol aldehydes. Although the majority of the MS^2 fragments of the two regioisomers are the same, the overall profiles of the fragments (that is MS^2 fingerprints) are different (Figure 8.2).[25]

In some cases, the difference between MS^2 fingerprints of different diastereomers may not be obvious owing to the lack of sufficient numbers of fragments at the MS^2 level. In such a case, molecular fingerprinting at higher MS^n levels may be performed. This scenario was encountered in the case of four diastereomers of betamethasone 17-deoxy-20-hydroxy-21-oic acid. While the MS^2 fingerprints of the m/z 393 parent ions of the four species display no obvious differences, MS^3 fingerprints *via* a significant product ion at m/z 355 (m/z 393 → m/z 353 →) display distinguishable differences between the four diastereomers (Figure 8.3).[26]

These examples clearly demonstrate that structurally very similar species may be individually identified by their own unique LC-MS^n molecular fingerprints, in particular their MS^n fingerprints. In other words, LC-MS^n molecular fingerprinting may be used for rapidly identifying and/or verifying the structures of unknown compounds and is particularly useful in distinguishing structurally similar diastereomers. Although the examples shown here are steroidal molecules, the concept of MS^n fingerprinting is readily applicable to molecules of other structural varieties. For example, gentiobiose and cellobiose are isomeric disaccharides both with two glucose units. Their difference lies in the glycosidic linkages: the former has a β(1 → 6) linkage, while the latter a β(1 → 4) linkage (Figure 8.4). Yet, the MS^2 of the lithium cationized ions of the two isomers displays distinctively different fragmentation patterns (or fingerprints).[27]

Quite often, the LC-MS^n or MS^n fingerprints of a reference compound may not be available for exact fingerprint match during degradant identification. In such a case, structure prediction may be possible by leveraging the available MS^n data according to the degree of similarity between the unknown degradant and the reference compound(s). Alternately, the unknown compound may be converted to a known chemical entity that has an established LC-MS^n fingerprint. The conversion may be achieved *via* gas phase chemistry (CID) or wet chemistry. An example will be given in Section 8.5.7 below to illustrate how the structure prediction is performed based on a similar match of MS^n fingerprints.

A typical practice in structure elucidation of unknown pharmaceutical impurities by LC-MS is to obtain the MS^n fragmentation pathways of the unknown species and then compare them with those of the API. During this comparison, emphasis is placed on (1) the differences between the

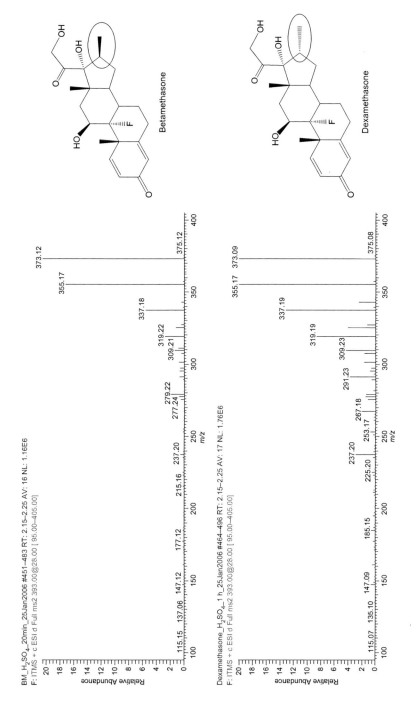

Figure 8.1 MS^2 molecular fingerprint of betamethasone (top) *versus* that of dexamethasone (bottom).[23]

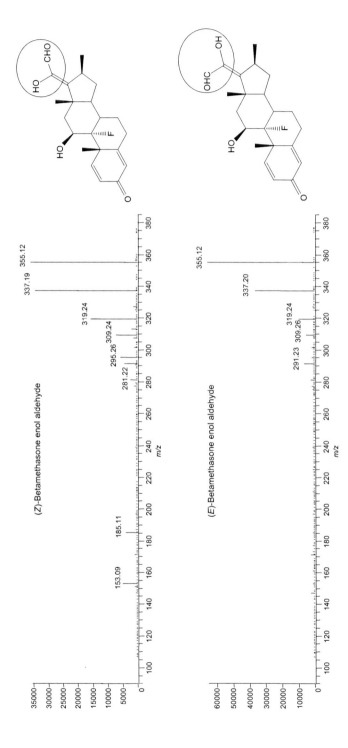

Figure 8.2 MS^2 molecular fingerprint of (*Z*)-betmamethasone enol aldehyde (top) *versus* that of (*E*)-betmamethasone enol aldehyde (bottom).[25]

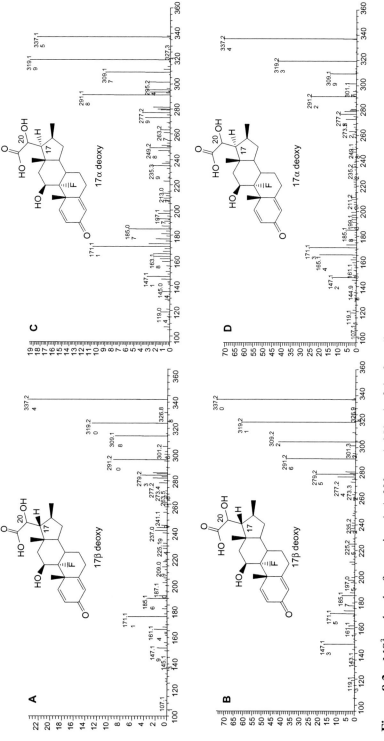

Figure 8.3 MS³ molecular fingerprints (m/z 393 → m/z 353 →) of the four diastereomers of betamethasone 17-deoxy-20-hydroxy-21-oic acid. The exact stereochemistry at C-20 could not be determined for each specific diastereomer.

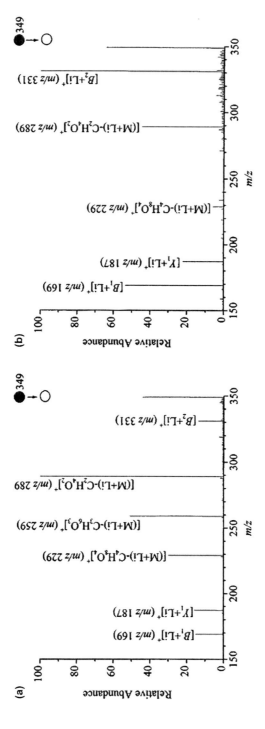

Figure 8.4   MS² fingerprints of lithium cationized parent ions of gentiobiose (a) and cellobiose (b). Reproduction from Reference 27 with permission.[27]

corresponding fragments of the impurities and API, respectively, and (2) the neutral losses during the key fragmentation steps. The latter steps are the ones involving moieties that are different from those of the API owing to side reactions in the process (for process impurities) or degradation (for degradants) of the API. During this exercise, the capability of MS^n molecular fingerprinting as demonstrated here is not utilized. An additional advantage of using MS^n molecular fingerprinting is that it expands the capability of a unit resolution mass spectrometer with an MS^n function. Since MS^n molecular fingerprinting does not require high resolution mass measurement, a unit resolution mass spectrometer is also capable of rapidly identifying or verifying the structure of an unknown/suspected impurity with a high confidence level, based on straightforward comparison of the MS^n fingerprints. Lastly, once the MS^n fingerprints of known impurities are recorded in a database, no physical collection of the reference materials is necessary in future structure verification.

8.3 Brief Discussion of the Use of Multi-dimensional NMR in Structure Elucidation of Trace Level Impurities

As mentioned previously in this chapter, the most encountered challenge in structure elucidation of pharmaceutical impurities at trace levels is usually the availability of a purified sample in "a sufficient quantity". The definition of "a sufficient quantity" depends on the type of NMR probes used. In general, sub 3-mm microprobes are much preferred and ideally in a cryogenic format. In cases where only conventional 5-mm probe NMR technology is available, it would still be advantageous to perform NMR measurements in a 3-mm tube in the 5-mm probe.[28] With cryogenic microprobes, samples of $\sim 100\,\mu g$ (~ 0.25 μmol for a molecular weight of 400) of small molecule impurities can usually afford satisfactory heteronuclear 2D ^1H–^{13}C NMR results.[28,29] For process impurities, isolation of such an amount typically would not be a problem, as ample supplies of drug substances are usually available. Nevertheless, in particular for degradants of potent drugs (which are usually in low doses), an amount of $\sim 100\,\mu g$ can often be a substantial challenge to obtain. For potent drug products on commercial stability programs, there may not be enough sample to isolate such a quantity in many cases.

Once an adequate amount of a purified impurity/degradant sample is available, various 2D NMR experiments are performed. Typically homonuclear experiments are performed first, before the much more time-consuming heteronuclear ones, although a reverse preference is also adopted. In Table 8.2 are summarized the most frequently used 2D NMR experiments along with the structural information revealed by these experiments. Those who would like to learn in more detail about the capability of modern NMR spectroscopy can refer to a book chapter by G.E. Martin, in which a systematic approach is outlined for elucidation of pharmaceutical impurities using modern NMR spectroscopy.[28]

Table 8.2   Common 2D NMR Experiments.

2D NMR Experiment	Structure information revealed
COSY (correlation spectroscopy)	^1H–^1H through-bond coupling; cross peaks correspond to coupling observed in 1D ^1H NMR.
NOESY (nuclear Overhauser effect spectroscopy)	^1H–^1H through-space coupling; cross peaks correspond to NOE effects observed in multiple 1D NOE experiments.
ROESY (rotating frame nuclear Overhauser effect spectroscopy)	ROESY is a variation of NOESY, which is used in place of NOESY for molecules whose NOE effect is too weak to detect.
HSQC (heteronuclear single-quantum correlation spectroscopy); HMQC (heteronuclear multiple-quantum correlation spectroscopy)	Cross peaks indicate directly coupled (one-bond) ^1H–^{13}C correlations. HMQC provides similar results for molecules of low and intermediate molecular weights, while HSQC provides better results for high molecular weight molecules.
HMBC (heteronuclear multiple-bond correlation spectroscopy)	Cross peaks indicate ^1H–^{13}C coupling through multiple (2–4) bonds.

8.4   Performing Meaningful Stress Studies

Stress study is also referred to as stress testing, forced degradation, or simply stress. Even in International Conference on Harmonisation (ICH) guidelines, different nomenclature is used: in Q1A(R2), "Stability testing of new drug substances and products", the term "Stress testing" is used,[30] while in Q1B, "Stability testing: Photostability testing of new drug substances and products", the term "forced degradation testing studies" is used.[31] Why are stress studies necessary or what is the purpose of a stress study and its scope? A simple answer can be found in ICH guideline Q1A(2R), which stipulates: "Stress testing of the drug substance can help identify the likely degradation products, which can in turn help establish the degradation pathways and the intrinsic stability of the molecule and validate the stability indicating power of the analytical procedures used."

Understanding the "real life" degradation chemistry of a new chemical entity, which is typically represented by the degradation chemistry occurring under ICH long term stability storage conditions, may be facilitated by forced degradation/stress studies under various degrading conditions and by accelerated stability studies (*e.g.* 40 °C/75% RH, or relative humidity). Since accelerated stability studies usually take a few months to complete, short term (from a few hours to a few days or weeks) stress studies are much preferred as a quick tool to "predict" the "real life" degradation chemistry of the new chemical entity. Nevertheless, design and interpretation of the stress studies is critical to capture the degradation pathways that are most likely to occur under long term storage conditions, while removing or minimizing "artificial" degradants from consideration. ICH guidelines certainly recognize the limitation of stress studies with regard to artificial degradants, that is, the ones that are not likely to form under long term stability storage conditions. Guideline Q1A stipulates that although the information from stress testing is "useful in establishing degradation pathways and developing and validating suitable analytical procedures", "it may not be necessary to examine specifically for certain degradation products if it has been demonstrated that they are not formed under accelerated or long term storage conditions."[30]

The nature of stress studies will depend on individual drug substances and the type of drug products involved. ICH guidelines provide no specific procedures on how to perform stress studies or details of stress testing conditions in general, except for defining the standard experiment conditions including exposure limit (minimum 1.2 million lux hours) in the "confirmatory testing" of photostability testing.[31] Nonetheless, ICH guidelines for drug substances indicate that stress testing should include the effect of temperatures (in 10 °C increments above that for accelerated testing), humidity (*e.g.* 75% RH or greater), and where appropriate, oxidation and photolysis on the drug substance.[30] The guideline also requires the evaluation of drug substance susceptibility to hydrolysis across a wide range of pH values. ICH guidelines do not have specific languages for stress testing drug products. This is understandable because knowledge of drug degradation pathways in various formulations is usually gained during the compatibility study in pharmaceutical development of the drug candidate.

In this section, we will discuss how to perform meaningful stress studies with regard to generating relevant degradation profiles and their use in elucidation of degradant structures and establishment of degradation pathways.

8.4.1 Generating Relevant Degradation Profiles

The topic of stress studies or forced degradation has been discussed and reviewed in the literature with regard to the following objectives: (1) to help elucidate drug degradation pathways,[32,33] (2) to predict drug stability (shelf-life),[34] and (3) to provide specificity mixtures for developing stability-indicating analytical methods.[35,36] Nevertheless, the related topic of how to design and in particular how to evaluate the results of stress testing does not appear to have been appropriately addressed, except in few cases.[32,35] Quite often, emphasis was placed on generating degradants but did not question if these degradants are "relevant or real", that is, if they can occur under ICH accelerated and long term storage conditions. The most common industry practice is to use 0.1 N HCl, 0.1 N NaOH, and 3% H_2O_2 for the acidic, basic, and oxidative stresses, respectively, as a first line testing protocol for non-photochemical stress studies. These stress studies may be carried out with different reagent concentrations and at ambient or elevated temperatures in order to achieve certain levels in the decomposition of the drug molecule. Commonly used stress/forced degradation conditions employed by practitioners in the pharmaceutical industry are summarized in Table 8.3.

Table 8.3 Common Industry Practice in Performing Stress Studies.

Stress condition	Common industry practice	Expected main degradation	Comments
Heat	50 °C and up to 150 °C under ambient moisture.	Various degradations including dehydration, isomerization, oxidation by molecular oxygen.	Usually carried out with solid drug substance samples (below melting point) in open air.
Heat + Moisture	50 °C and up to 150 °C under controlled moisture (usually with minimum 75% RH).	Various degradations including hydrolysis, isomerization, oxidation by molecular oxygen.	Usually carried out with solid drug substance samples in stability chamber or enclosure with controlled moisture.
Acid	0.1 N HCl, ambient temperature up to reflux	Hydrolysis	More dilute (0.01 N) or concentrated (1 N) solutions are also used. Either in suspension or solution (with organic co-solvent such as acetonitrile or methanol)
Base	0.1 N NaOH, ambient temperature up to reflux	Hydrolysis	More dilute (0.01 N) or concentrated (1 N) solutions are also used. Either in suspension or solution (with organic co-solvent such as acetonitrile or methanol)
Oxidative	3% H_2O_2, ambient temperature up to 80 °C	Oxidation	Higher concentrations up to 30% are also used. Higher concentrations and higher temperature tend to give more artificial degradants.
Photo	(1) Solution, Irradiation in UV with mercury lamp (with 254 nm and 365 nm irradiation) and visible light under laboratory fluorescent lighting	Photochemical degradation	During the initial "stress testing" stage, photolysis is usually carried out in solution and the amount of exposure can be more flexible. In many cases, the exposure does not have to reach Q1B exposure limit.
	(2) Solid API sample, light exposure defined by ICH Q1B		Quantitative light exposure defined in Q1B is required during the "confirmatory testing" stage. Photolysis is usually carried out in solid state (for API and solid dosage forms) with and without product packaging materials (primary and secondary).

In order to probe the real-life degradation chemistry of a drug molecule within a relatively short period, the key points are rational design of stress studies and assessment of the results under various relevant degrading conditions. A mechanism-based approach is needed, that is, to choose a stress condition (or conditions) under which the observed degradation pathways would more resemble the degradation behavior of the drug molecule in its native state (usually solid) as well as in formulated forms, based on the initial assessment of the structure of the molecule and its formulations.

Chemical compounds with functional groups and structure moieties have established degradation pathways. For example, tertiary amines and thioethers tend to degrade into *N*-oxides and sulfoxides, which are usually formed *via* a nucleophilic mechanism. Hence, it would be adequate to use hydrogen peroxide as an oxidant in stress testing of compounds containing tertiary amines and thioethers. On the other hand, autooxidative degradation through free radical mediation is a common drug degradation pathway for drugs containing allylic and benzylic type moieties. A number of stressing systems/conditions mimicking the free radical-mediated autooxidation processes, for example, azobisisobutyronitrile (AIBN), have been reported in the literature,[32,37,38] and a few examples have been discussed in Chapter 3 of this book (Sections 3.5.2 and 3.5.10). Therefore, for drugs containing allylic and benzylic type moieties, the latter stress systems tend to give a degradant profile that would more resemble the one observed under the long term stability conditions caused by free radical-mediated oxidation.

For drug molecules containing more than one oxidizable moiety, oxidative stress testing may have to be carried out under multiple oxidative stress conditions in order to generate all the significant oxidative degradants. Interestingly, although free radical initiators such as AIBN have been frequently utilized to assess the oxidizibility of a drug candidate during the early phase of pharmaceutical development, it has rarely been used to generate relevant oxidative degradants in specificity mixtures during the development of stability-indicating methods. On the other hand, hydrogen peroxide is ubiquitously employed for this purpose without a question of whether the generated degradants are relevant or not. In order to increase the likelihood of generating a degradation profile that is more similar to the real one (and hence to help better probe the underlying degradation pathways), the following additional points also need to be considered.

(1) If the main purpose of a stress study is to help elucidate a drug degradation pathway or provide a meaningful specificity mixture for developing a stability-indicating analytical method (most likely an HPLC method), the extent of degradation of the drug molecule needs to be controlled at an appropriate level, for example, no more than $\sim 10\%$. One of the main problems encountered in stress testing is excessive degradation, which dramatically increases the chance of producing artificial degradants owing to secondary degradation and beyond. On the other hand, if the main purpose of a stress study is to generate a particular unknown degradant in large enough quantities for analyses by 1D and 2D NMR, which requires more sample than LC-MS, it is obvious that the yield of the desired degradant needs to be maximized.

(2) In a similar vein, if the purpose of a stress study is to help elucidate a drug degradation pathway or provide a meaningful specificity mixture for developing a stability-indicating analytical method, too high temperatures in particular for stress studies in solution should be avoided. In general, solution stress at elevated temperatures increases the chance of generating artificial or irrelevant degradants resulting from side reactions, secondary degradation or beyond. An industrial discussion group, Impurity Profiling Group (IPG), even recommends not using elevated temperatures as a means of accelerating degradation in solution, except in special cases such as evaluating the stability of a solution during autoclaving.[35]

In oxidative stress using hydrogen peroxide, high temperatures can cause the degradation mechanism to change. For example, if the oxidative stress using hydrogen peroxide is carried out at a temperature greater than 50 °C, hydrogen peroxide tends to form free radicals *via* homolytic decomposition, which would be likely to promote the formation of free radical-mediated artificial degradants of drug molecules whose main degradation pathways would be mediated *via* nucleophilic oxidation. Likewise, the concentration of hydrogen peroxide used in oxidative stress should probably not exceed 3% concentration. Higher concentration is more likely to induce artificial degradants. A recent example using 30% hydrogen peroxide in the oxidative stress of pentoxifylline, a ketone-containing drug, *gem*-dihydroperoxide was found to be one of the major degradants.[39] This degradant was formed by the initial attack on the keto group by hydrogen peroxide, followed by condensation with a second molecule of hydrogen peroxide (Scheme 8.1).

Such a degradant is highly unlikely to occur under accelerated or normal stability storage conditions. Another example is the formation of dealkylated degradants from tertiary amines during oxidative stress using hydrogen peroxide, particularly if over-stressing occurs. Dealkylation, especially demethylation and de-ethylation, is a common degradation pathway in drug metabolism which is catalyzed by oxidative enzymes. Nevertheless, such a degradation pathway does not appear to be significant in the chemical degradation of drugs under accelerated and long term stability storage conditions.

(3) There is no specific requirement for stress testing a drug product in the ICH guidelines other than "confirmatory testing" in photostability testing. From a practical point of view, however, it would make sense to perform stress testing of the drug product, if deemed necessary, under heat and/or heat with

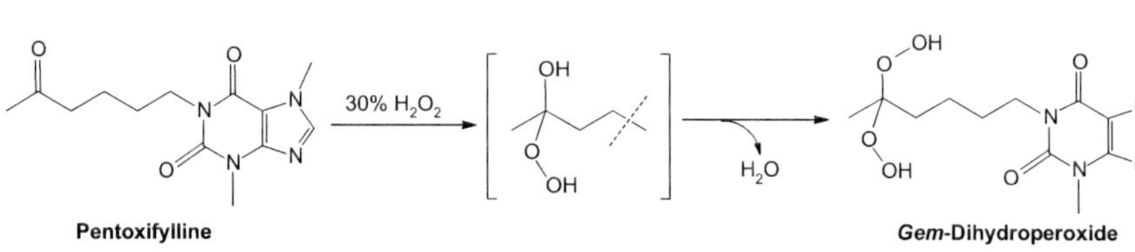

Pentoxifylline   *Gem*-Dihydroperoxide

Scheme 8.1

moisture in non-photostability testing, because these are the two most likely scenarios for non-photochemical decomposition to occur (for a finished drug product).

(4) One should be aware that a particular type of stress may induce another type of degradation. For example, hydrolytic stress under basic conditions may cause certain compounds to undergo base-catalyzed autooxidation.[40,41] In solution stress studies, organic co-solvents such as methanol and acetonitrile are frequently used. Both the solvents may interfere with the degradation profiles as methanol can form esters and ethers with carboxyl- and arylhalide-containing drug molecules, respectively, while acetonitrile can promote oxidative stress by hydrogen peroxide *via* the intermediacy of peroxycarboximidic acid (see Chapter 3, Section 3.3.1).[42–44]

(5) To determine if a stress-generated degradant is real or not, one needs to examine the degradation profiles, obtained from the compatibility studies during the pharmaceutical development stage and ultimately during the accelerated and long term stability studies, for the presence of that particular degradant. Unfortunately, a large number of "stability-indicating" methods were developed solely based on stress-generated degradants, many of which may be artificial ones. For example, several "stability-indicating" methods for finasteride drug substance and formulated products were reported in the literature.[45–48] None of the studies spent any effort in trying to understand the nature of the degradation, nor did they try to evaluate if stress-generated degradants were actually forming in the drug substance and products that were subject to various stability studies.

8.5 Effective Use of Mechanism-based Stress Studies in Conjunction with LC-MS^n Molecular Fingerprinting in Elucidation of Degradant Structures and Degradation Pathways: Case Studies

8.5.1 Outline of General Strategy

Prior to discussing the case studies, the general strategy for the effective use of mechanism-based stress studies in elucidation of degradant structures and degradation pathways is outlined in Scheme 8.2. The critical steps are briefly discussed below.

8.5.2 Proposing Type of Degradation Based on LC-MS^n Analysis

In the vast majority cases, a UV detector is used in tandem with a MS detector. For structural elucidation, it is much preferred to use a photodiode array UV/Vis detector, as it provides results complementary to those generated by the MS detector. Typically, the first goal of structure elucidation is to

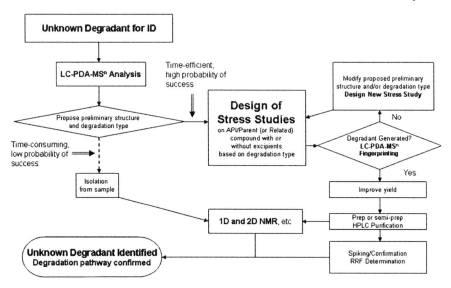

Scheme 8.2

determine the molecular weight and preferably the molecular formula of the unknown degradant. With high resolution MS, determination of the molecular formula can be readily achieved, except in cases where parent ions are not observed. With unit resolution mass spectrometry, the molecular formula may be postulated according to the molecular weight difference between the unknown degradant and the API, in conjunction with the presumed degradation chemistry of the drug based on its structural features.

Secondly, the UV-Vis profile and MS^n fragmentation behavior of the unknown degradant are obtained and then compared with those of the API. Based on their differential behavior under the LC-PDA-MS^n conditions, for example, changes in chromophores, molecular weights, and MS^n fingerprints of degradants compared with those of the API, initial knowledge regarding the type of the degradation may be inferred. For example, a molecular weight increase of 16 (determined by unit resolution MS) usually indicates oxidative degradation by incorporation of an oxygen, although the oxidation could be due to one of the following transformations: *N*-oxidation, hydroxylation of certain allylic carbons or α-keto methylene groups, epoxidation of a double bond, or oxidation of a sulfur. On the other hand, a loss of 18 usually suggests degradation by dehydration of the API, although defluorination can also result in a nominal mass loss of 18. In cases where the molecular weight of a degradant remains the same, degradation *via* isomerization has occurred. A brief summary of commonly observed molecular weight changes compared with potential degradation types is illustrated in Table 8.4. Obviously, with high resolution MS, the exact molecular formula change can be readily determined.

Table 8.4   Commonly observed molecular weight (MW) changes with regard to types of degradation.[a]

MW Difference	Type of degradation	Note[b]
0	Isomerization	
+14	Oxidation of -CH$_2$-protons to form ketone.	Usually degradant formed *via* peroxide intermediate.[c] The latter also leads to the formation of a hydroxyl group on the methylene. An increase of 14 may also suggest incorporation of a CH$_2$ moiety, implicating degradation through interaction with formaldehyde.[d]
+16	Various oxidations including: *N*-oxidation, hydroxylation, epoxidation, Baeyer–Villiger oxidation, and oxidation of sulfur.	Oxidation at multiple sites may be possible. In such cases, MW differs in the multiples of 16. See also below for LC-MS behavior of activated aldehyde groups.
+32	Peroxidation (if not oxidation on two sites).	Peroxides formed are usually not stable. However, they may be observed by LC-MS.
–2	Oxidation of alcohol to form aldehyde.	Certain aldehyde groups may be activated by neighboring electron-withdrawing groups, *e.g.* an α-keto group. In such cases, the hydrated aldehyde form (hemi-acetal) may be observed in LC-MS and MW difference would be $(-2 + 18) = 16$, corresponding to net incorporation of an oxygen.
–18	Dehydration	In rare cases of fluorine-containing compounds, loss of 18 may also represent defluorination (–F + H).
–44	Decarboxylation	

[a]A combination of these types of degradation may occur within a single molecular entity.
[b]High resolution MS can readily differentiate the different scenarios discussed in this column, brought upon by nominal mass difference.
[c]Refer to Chapter 3, Section 3.2.
[d]Refer to Chapter 4, Section 4.7, Dimerization.

8.5.3   Design of Stress Studies According to Presumed Degradation Type

Based on the presumed degradation type according to the LC-PDA-MS[n] analysis outlined above, proper stress studies can be designed accordingly, that is, mechanism-based stress studies can be designed. For example, for a degradation that is perceived to be a dehydration, forced degradation reactions can usually be performed under acidic conditions. This is based on the knowledge that dehydration of a hydroxyl compound typically proceeds well under acidic conditions. Nevertheless, selection of a particular acid as a catalyst to induce dehydration may be different, depending upon the nature of the API involved.

Likewise, for different types of oxidative degradation, selection of different oxidation reagents needs to be evaluated and/or tried. For example, for drug molecules containing tertiary amine moieties, hydrogen peroxide is usually capable of generating degradation profiles, owing to N-oxidation, that are likely to mimic real degradation. Nevertheless, 3-chloroperoxybenzoic acid (mCPBA) would probably be a better reagent for regenerating an epoxide degradant or degradants mediated *via* an epoxide intermediate, in particular when such degradants are needed in large quantities for NMR characterization.[49]

Whenever possible, one should take advantage of reactions or mechanisms known in the literature, which are specific for certain types of molecular transformation. One good example is autooxidation of the 3-hydroxyacetone side chain of certain corticosteroids which can be catalyzed either by certain metal ions or by bases. When catalyzed by cupric ion, the 21-hydroxyl group of the corticosteroids is specifically oxidized into the corresponding 21-aldehyde. This reaction has been used to aid the rapid identification of several 21-aldehyde degradants in betamethasone[50] and related compounds.[23] On the other hand, base-catalyzed autooxidation of the corticosteroids gives rise to a few other oxidative degradants as well.[51] These degradants have been observed in commercial products.[52]

8.5.4 Tracking and Verification of Unknown Degradants Generated in Stress Studies Using LC-MS^n Molecular Fingerprinting

In the general strategy presented here, LC-MS^n molecular fingerprinting is used as a key analytical tool to track and verify if the unknown degradants are formed in the designed stress studies and throughout the isolation process of these degradants. As we have demonstrated above in Section 8.2.3, MS^n fragmentation behavior revealed by MS^n spectra, which are reproducible under identical MS^n conditions, provides reliable molecular fingerprinting and can be used to distinguish between molecules with similar structural features.

8.5.5 Case Study 1: Elucidation of a Novel Degradation Pathway for Drug Products Containing Betamethasone Dipropionate and Similar Corticosteroidal 17,21-Diesters

A drug product containing betamethasone dipropionate as the active ingredient was found to contain a previously unknown peak at a relative retention time (RRT) of 0.67 at a level above the identification threshold.[23] This drug has two known degradants, betamethasone 17-propionate and betamethasone 21-propionate, both of which result from the hydrolysis of the API. Since the hydrolysis does not have an impact on the chromophore of the drug molecule, the two degradants display the same UV profile as the API. This unknown species co-eluted with the later part of the betamethasone 21-propionate peak. A PDA/UV scan of the region showed two distinctively different UV

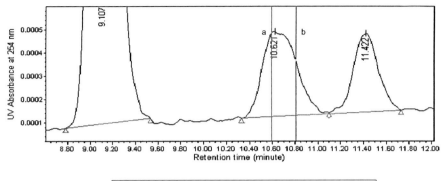

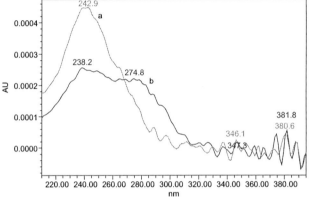

Figure 8.5 Top: HPLC-PDA/UV analysis of the drug product sample containing an RRT 0.67 unknown peak that co-eluted with a known degradant of the product, betamethasone 21-propionate (10.62 min). Bottom: UV spectra of betamethasone 21-propionate (a) *versus* the unknown at RRT 0.67 (b).[23]

chromophores, with the first one belonging to betamethasone 21-propionate and second one to the unknown species (Figure 8.5).

LC-MS analysis of the drug sample revealed that the unknown species showed two characteristic mass signals: the one at *m/z* 375 was determined to be its protonated ion, while the one at *m/z* 355 appeared very likely to be the fragment ion of the unknown species caused by in-source fragmentation. The loss of HF (−20) during in-source fragmentation is a common phenomenon in this particular type of mass spectrometer for betamethasone and related steroids containing the 9α-hydroxy-11β-fluoro moiety. These results suggested that (1) this unknown is structurally related to the API as evidenced by the characteristic loss of HF, and (2) the difference of 18 between the molecular weight of the unknown (374) and that of betamethasone (392) implies that this unknown might be a direct dehydration degradant of betamethasone (but not the API, betamethasone dipropionate). Betamethasone itself is a minor degradant of the product as it is a secondary degradant formed *via* the intermediacy of either 17-propionate or 21-propionate of betamethasone.

A search of the literature indicated that dehydration of betamethasone or related compounds can be effected under acidic stress conditions.[53–58] Hence, stress of betamethasone was performed in a mixture of acetonitrile and water containing ~10% sulfuric acid. Under these conditions, not only was the unknown species generated, as demonstrated by LC-MS, but also, in a somewhat less amount, an isomer of the unknown species was formed which has a very similar UV profile but a different MS^2 fingerprint.[25] At a temperature of 60 °C, the total yield of the two isomeric degradants reached about 40% in 20 hours. The two isomers were isolated by HPLC and then analyzed by high resolution MS and 2D NMR spectroscopy. Based on the high resolution MS and various 2D NMR experiments (NOESY, HMQC, HMQC–TOCSY and HMBC), the RRT 0.67 unknown species and its isomer were identified as the *E*- and *Z*-isomers of betamethasone enol aldehyde (structures shown in Scheme 8.3), respectively. Reanalysis of the product sample using a modified method revealed that the product also contains *Z*-betamethasone enol aldehyde, albeit at a lower level. In the product method, the *Z*-isomer completely co-eluted with the major degradant, betamethasone 21-propionate. At this point, with the use of LC-MS in conjunction with the acid-catalyzed dehydration of betamethasone, which is known as the Mattox rearrangement,[53–55] the structure of the RRT 0.67 unknown degradant was resolved. Additionally, a hidden degradant (the *Z*-isomer) was also uncovered. Thus, it appeared that this investigation could be closed.

Nevertheless, based on the available degradation mechanism (the Mattox rearrangement) and pathways, the formation of the two enol aldehyde degradants would be *via* a linear pathway that is mediated consecutively through two reasonably stable intermediates, that is, either of the monopropionates of betamethasone and betamethasone. Acidic stress of the API, betamethasone dipropionate, indeed showed the sequential formation of the two monopropionates and betamethasone prior to the formation of the two enol aldehydes as the stress continued (Scheme 8.3).

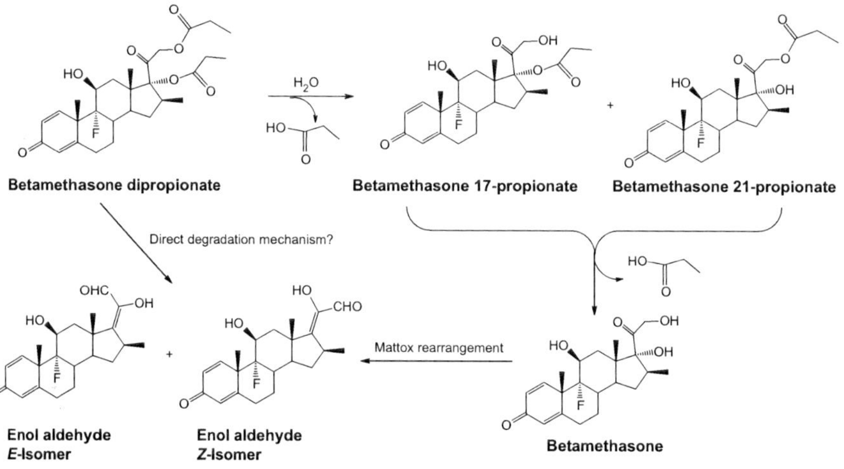

Scheme 8.3

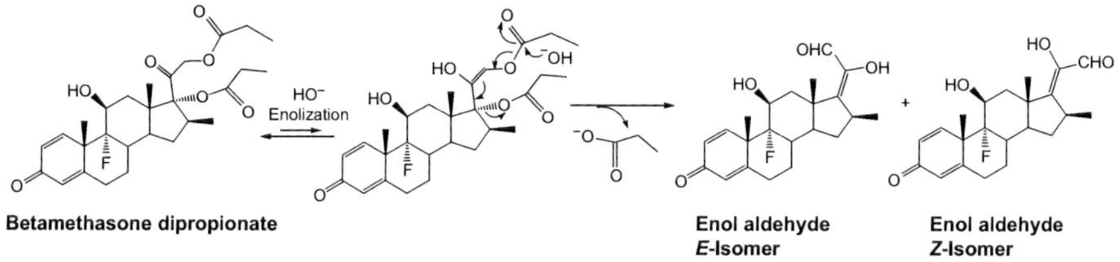

Betamethasone dipropionate   Enol aldehyde   Enol aldehyde
                                                                                        E-Isomer   *Z*-Isomer

Scheme 8.4

Thus, the two enol aldehydes would be tertiary degradants according to this linear degradation mechanism. In other words, the formation of these two degradants would be negligible in this product under the usual stability conditions. Since they are present in the product at levels greater than the identification threshold, the formation of these two degradants must proceed through a direct, yet unknown mechanism. With this rationale in mind, stress of the API, betamethasone dipropionate, under other conditions was performed. When a solution of the API in acetonitrile was stressed with a small amount of sodium hydroxide solution, the *E*- and *Z*-isomers of betamethasone enol aldehyde were quickly formed in approximately 10% and 30% yields within 20 minutes at room temperature. This experiment demonstrated that a direct degradation pathway from betamethasone dipropionate to the two enol aldehyde degradants does exist. A mechanism for this direct pathway is illustrated in Scheme 8.4.[23]

According to this mechanism, the 21-propionate activates the drug molecule (so that a nucleophilic attack can occur on the 21-ester functionality), while the 17-propionate provides a good leaving group for the masked 17-hydroxyl group. In other words, this mechanism, which can be considered a variation of the original Mattox rearrangement, requires both ester groups in order to proceed. Indeed, stress of either betamethasone 17-propionate or 21-propionate under identical basic stress conditions did not generate the enol aldehydes. Although this direct degradation mechanism was demonstrated under alkaline conditions, which is not a typical environment surrounding the API in this drug product, it indicates that the 17, 21-diester of betamethasone and related drug molecules should be susceptible to attack by a general base or nucleophile, which would result in the formation of the corresponding enol aldehydes. It has since been found that this new, direct degradation mechanism is generally applicable to formulated products containing betamethasone dipropionate and similar diesters.[59]

8.5.6   Case Study 2: Rapid Identification of Three Betamethasone Sodium Phosphate Isomeric Degradants – Use of Enzymatic Transformation When a Direct MS" Fingerprint Match is not Available

Betamethasone sodium phosphate (BSP) is a water-soluble pro-drug of betamethasone, typically used in injectable formulations. Despite its long history of clinical usage, little was known about its degradation chemistry until recently.[24,60,61] The case study discussed below is among the work by the current author's research group that has led to a clear understanding of the degradation chemistry of this drug molecule.

In an amorphous sample of BSP, three unknown degradant peaks were observed at RRT 0.55, 0.71, and 0.81, respectively. Analysis by LC-MS indicated that they are isomers of BSP as shown in Figure 8.6.[61]

During several previous studies, we elucidated the structures of a large number of isomeric degradants of betamethasone and hence their LC-MS^n fingerprints were available. Since BSP is the 21-phosphate ester of betamethasone, it was logical to hypothesize that the three isomeric degradants of BSP could be the corresponding phosphate derivatives of some of the known betamethasone isomeric degradants. If this hypothesis is correct for at least one of the three degradants, the MS^n molecular fingerprint of its dephosphorylated form should be identical to that of a known betamethasone isomeric degradant. In other words, this unknown BSP isomeric degradant may be identified based on matching the MS^n fingerprint of its dephosphorylated form to that of the known betamethasone isomeric degradant. Hence, the key to making the MS^n molecular fingerprinting strategy work in this case is to obtain the MS^n fingerprints of the dephosphorylated form of the BSP degradant. In a previous study, we had been able to confirm the structure of betamethasone 9,11-epoxide by comparing its MS^2 fingerprint (m/z 373 →) to the MS^3 fingerprint of betamethasone (m/z 393 → m/z 373 →).[23] In that case, betamethasone undergoes dehydrofluorination (–HF) in the gas phase and the resulting fragment ion, m/z 373, is structurally identical to the parent ion (m/z 373) of betamethasone 9,11-epoxide (Scheme 8.5).

By following the successful example of betamethasone 9,11-epoxide illustrated above, we initially tried to cleave the phosphate moiety of the three BSP degradants from the steroid core structures in the gas phase by collision induced fragmentation (CID). Unfortunately, this approach did not work in this case because no meaningful amount of the expected m/z 393 ion was generated in the gas phase. The results suggested that the phosphate bond in these steroid molecules is more stable than some others in the gas phase. Therefore, the only alternative would be to use solution chemistry to remove the phosphate group. Since simple acid- or base-catalyzed forced degradation of these BSP degradants may not be selective, it occurred to us that the task could be achieved specifically *via* enzymatic hydrolysis with a phosphatase. Hence, an inexpensive acidic phosphatase from wheat germ was tried in aqueous solutions of the three isolated BSP degradants and in all the three cases the phosphate was cleaved from the BSP isomeric degradants. The resulting betamethasone isomers thus formed were analyzed by LC-MS^n and their MS^3 molecular fingerprints (m/z 393 → m/z 355 →) were compared to those of the known betamethasone degradants. An exact match was found (Figure 8.7) between the RRT 0.81 degradant and a betamethasone isomeric degradant resulting from D-homoannular ring expansion,[62] indicating that RRT 0.81 degradant has an identical steroid core structure to the latter betamethasone

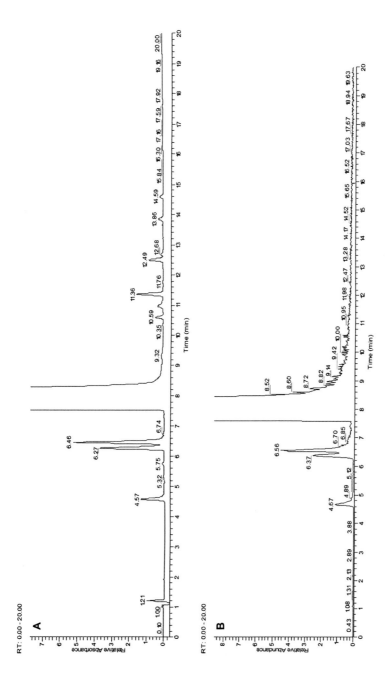

Figure 8.6   Top: UV chromatogram at 240 nm of an amorphous betamethasone sodium phosphate (BSP) sample containing three unknown degradant peaks at RRTs 0.55, 0.71 and 0.81 (in the retention region of 4 to 7 minutes). Bottom: the corresponding single ion chromatogram at *m/z* 473.[60]

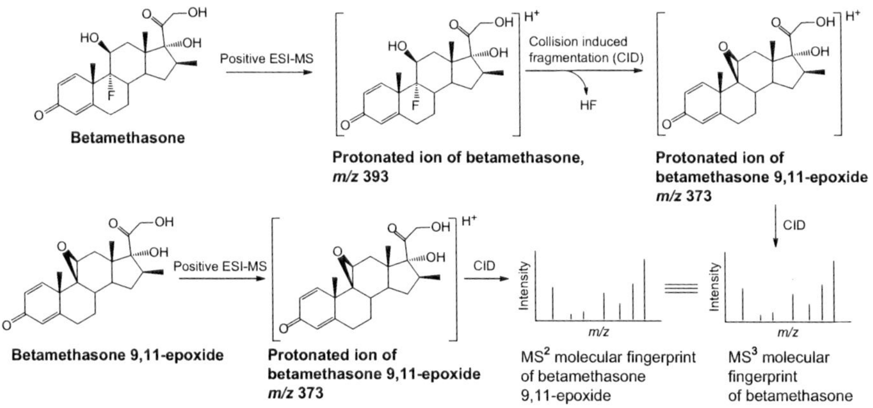

Scheme 8.5

degradant. Therefore, based on MS^3 fingerprinting, the RRT 0.81 BSP degradant was identified as 9-fluoro-11β,17aβ-dihydroxy-17a-[(phosphonooxy)-methyl]-D-homoandrosta-1,4-diene-3,17a-dione.

Likewise, the MS^3 fingerprints of the dephosphorylated species, generated enzymatically from RRT 0.55 and RRT 0.71 BSP degradants respectively, were obtained and then compared with the MS^3 fingerprints of the known betamethasone degradants. Unfortunately, no exact match was found unlike the previous case. Nevertheless, upon further examination of Figure 8.7, the MS^3 fingerprints of the m/z 393 species converted from the RRT 0.55 and RRT 0.71 degradants appeared similar to that of the m/z 393 species converted from the RRT 0.81 degradant in varying degrees. This observation suggested that both RRT 0.55 and RRT 0.71 species would also be likely to be degradants caused by D-homoannular ring expansion, as a maximum of four degradants could be formed during this isomerization step (see also Chapter 4, Section 4.5.6): two sets of positional isomers, each having two epimers (Scheme 8.6).

Based on the observation that the MS^3 fingerprint resulting from RRT 0.55 degradant has a higher degree of similarity to that of RRT 0.81, the RRT 0.55 degradant was assigned as the epimer of RRT 0.81 degradant. According to the same rationale, the RRT 0.71 degradant was assigned as one of two epimers among the other set of positional isomers, assuming α-orientation for the 17-(phosphonooxy)methyl group.

In the current case, the structures of three unknown degradants of BSP were elucidated with reasonably high confidence within a matter of only a few days, based on the effective use of LC-MS^n molecular fingerprinting and knowledge of the perceived degradation mechanism. The MS^n fingerprints were acquired on a unit resolution ion trap mass spectrometer. In order to validate the strategy of using LC-MS^n molecular fingerprinting, the structures of RRT 0.71 and 0.81 degradants were later confirmed unequivocally by NMR analysis (1D and 2D) of the samples isolated *via* semi-preparative HPLC.[62]

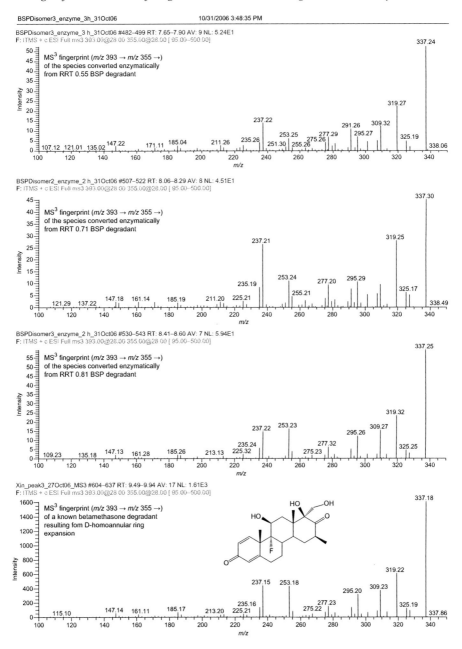

Figure 8.7 Comparison of the MS^3 (m/z 393 → m/z 355 →) fingerprints of the m/z 393 species converted enzymatically from BSP degradants at RRT 0.55, 0.71, and 0.81 with that of a known betamethasone degradant resulting from D-homoannular ring expansion, 9-fluoro-11β,17β-dihydroxy-17-hydroxy-methyl-D-homoandrosta-1,4-diene-3,17a-dione.[60]

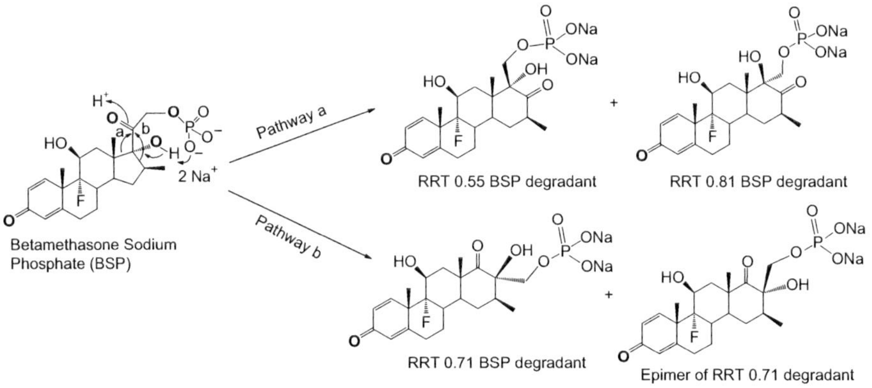

Scheme 8.6

8.5.7 Case Study 3: Identification of an Impurity in Betamethasone 17-Valerate Drug Substance – Structure Prediction When an Exact MS^n Fingerprint Match is not Available

An unknown impurity at RRT 0.90 was observed in betamethasone 17-valerate drug substance at a level above the ICH identification threshold. LC-MS analysis indicated that RRT 0.90 impurity is an isomer of betamethasone 17-valerate as both show the same parent ion at m/z 477.[21] According to the process chemistry of the drug substance, RRT 0.90 impurity appeared to be another valeryl derivative of betamethasone or one of its isomers. Since the structures of a large number of betamethasone isomers had been elucidated and their MS^n molecular fingerprints are available, this is another case where MS^n molecular fingerprinting could be used to resolve the identity of RRT 0.90 impurity quickly. The first step towards that goal would be to cleave the valeryl group from the steroid core of the unknown impurity to allow the MS^n fin-gerprint of the core to be recorded. By comparing the MS^n fingerprint with those of betamethasone and its isomers, the structure of the steroid core could be deduced. Hence, RRT 0.90 impurity was isolated by semi-preparative HPLC and then treated with a small aliquot of 1 N NaOH in acetonitrile in an attempt to hydrolyze the valeryl group under basic catalysis. Upon treatment with NaOH, the majority of RRT 0.90 impurity disappeared instantaneously (Figure 8.8), with the concurrent occurrence of a major product peak eluting at 8.13 minutes and a minor product peak eluting at 3.10 minutes.

   The major product showed a parent ion at m/z 477 and the minor product showed a parent ion at m/z 393. This result suggested that a small portion of RRT 0.90 impurity was hydrolyzed as expected, while the majority underwent isomerization unexpectedly. The MS^2 fingerprint of the minor product at 3.10 minutes was recorded and it was found to match the MS^2 fingerprint of dexamethasone, indicating the steroid core of RRT 0.91 impurity is

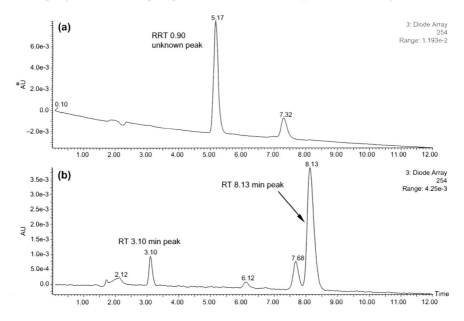

Figure 8.8   Chromatograms of the isolated RRT 0.90 unknown species at 254 nm before treatment with NaOH (a) and immediately after treatment with NaOH (b).

dexamethasone. In other words, RRT 0.91 impurity itself should be a valeryl derivative of dexamethasone.

The next step would be to determine the position of the valeryl ester. Since the 17- and 21-positions are the two most likely positions for the valery ester group, based on the process chemistry of betamethasone and related compounds, RRT 0.91 impurity would most likely be either dexamethasone 17-valerate or dexamethasone 21-valerate. Unfortunately, their MS^n fingerprints were not available for comparison, nor were authentic samples of the two dexamethasone valerates. On the other hand, authentic samples of betamethasone 17-valerate and betamethasone 21-valerate were readily available. Therefore, MS^2 fingerprints of betamethasone 17-valerate and betamethasone 21-valerate were recorded, respectively, and were used as surrogates for the MS^2 fingerprints of the two corresponding dexamethasone valerates. As shown in Figure 8.9, the MS^2 fingerprint of RRT 0.90 impurity resembles that of betamethasone 17-valerate, while the MS^2 fingerprint of the 8.13 minute peak resembles that of betamethasone 21-valerate. As a result, RRT 0.90 impurity was assigned to dexamethasone 17-valerate and the 8.13 minute peak, its isomeric product formed under alkaline conditions, was assigned to dexamethasone 21-valerate. The structure assignments were later confirmed by 1D and 2D NMR analysis of the two isolated compounds.

In this case, MS^n molecular fingerprinting was performed using structurally closely related compounds as surrogates owing to the unavailability of the

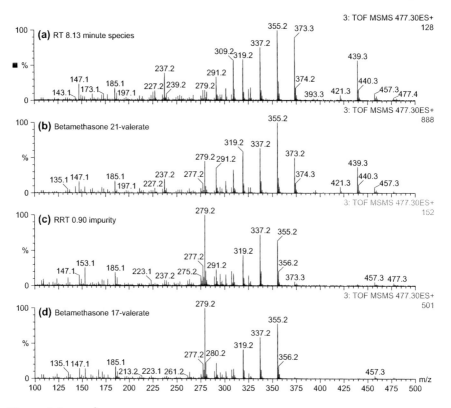

Figure 8.9   MS^2 fingerprint of the ions at m/z 477 from (a) RT 8.13 minute species, the isomer formed *via* NaOH treatment of the isolated RRT 0.90 impurity, (b) betamethasone 21-valerate, (c) RRT 0.90 impurity, and (d) beta-methasone 17-valerate.[21]

desired authentic compounds as well as their MS^n fingerprints. The structure of the unknown impurity was correctly predicted within a very short period based on a similar match of the relevant MS^2 fingerprints.

References

1. International Conference on Harmonisation, *ICH Harmonised Tripartite Guideline: Impurities in New Drug Substances Q3A(R2)*, dated 25 October 2006.
2. International Conference on Harmonisation, *ICH Harmonised Tripartite Guideline: Impurities in New Drug Products, Q3B(R2)*, dated 2 June 2006.
3. D. H. Russell and R. D. Edmondson, *J. Mass Spectrom.*, 1997, **32**, 263.
4. A. G. Marshall and C. L. Hendrickson, *Ann. Rev. Anal. Chem.*, 2008, **1**, 579.
5. Gary E. Martin, personal communication.
6. S. X. Peng, B. Borah, R. L. M. Dobson, Y. D. Liu and S. Pikul, *J. Pharm. Biomed. Anal.*, 1999, **20**, 75.

7. J. C. Lindon, J. K. Nicholson and I. D. Wilson, *J. Chromatogr. B: Biomed. Sci. Appl.*, 2000, **748**, 233.
8. P. Novak, P. Tepes, I. Fistric, I. Bratos and V. Gabelica, *J. Pharm. Biomed. Anal.*, 2006, **40**, 1268.
9. G. J. Sharman and I. C. Jones, *Magn. Reson. Chem.*, 2003, **41**, 448.
10. E. de Hoffmann and V. Stroobant, *Mass Spectrometry, Principles and Applications*, 3rd edn, John Wiley & Sons, Chichester, 2007.
11. Y. Zhu, P. S. H. Wong, M. Cregor, J. F. Gitzen, L. A. Coury and P. T. Kissinger, *Rapid Commun. Mass Spectrom.*, 2000, **14**, 1695.
12. V. Sewram, J. J. Nair, T. W. Nieuwoudt, N. L. Leggott and G. S. Shephard, *J. Chromatogr., A*, 2000, **897**, 365.
13. S. Kölliker, M. Oehme and C. Dye, *Anal. Chem.*, 1998, **70**, 1979.
14. Y. Zhou, S.-X. Huang, L.-M. Li, J. Yang, X. Liu, S.-L. Peng, L.-S. Ding and H.-D. Sun, *J. Mass Spectrom.*, 2008, **43**, 63.
15. B. Schilling, R. H. Row, B. W. Gibson, X. Guo and M. M. Young, *J. Am. Soc. Mass. Spectrom.*, 2003, **14**, 834.
16. A. S. Weiskopf, P. Vouros and D. J. Harvey, *Anal. Chem.*, 1998, **70**, 4441.
17. W. Chai, A. M. Lawson and V. Piskarev, *J. Am. Soc. Mass. Spectrom.*, 2002, **13**, 670.
18. H. Zhang, D. Zhang and K. Ray, *J. Mass Spectrom.*, 2003, **38**, 1110.
19. H. Zhang, M. Zhu, K. L. Ray, L. Ma and D. Zhang, *Rapid Commun. Mass Spectrom.*, 2008, **22**, 2082.
20. K. P. Bateman, J. Castro-Perez, M. Wrona, J. P. Shockcor, K. Yu, R. Oballa and D. A. Nicoll-Griffith, *Rapid Commun. Mass Spectrom.*, 2007, **21**, 1485.
21. M. Li, M. Lin and A. Rustum, *J. Pharm. Biomed. Anal.*, 2008, **48**, 1451.
22. K. E. Arthur, J.-C. Wolff and D. J. Carrier, *Rapid Commun. Mass Spectrom.*, 2004, **18**, 678.
23. M. Li, B. Chen, M. Lin and A. Rustum, *Am. Pharm. Rev.*, 2008, **11**(1), 98.
24. M. Li, X. Wang, B. Chen, T.-M. Chan and A. Rustum, *J. Pharm. Sci.*, 2009, **98**, 894.
25. M. Li, B. Chen, M. Lin, T.-M. Chan, X. Fu and A. Rustum, *Tetrahedron Lett.*, 2007, **48**, 3901.
26. M. Li, X. Wang, B. Chen, T.-M. Chan and A. Rustum, *J. Pharm. Sci.*, 2009, **98**, 894.
27. M. R. Asam and G. L. Glish, *J. Am. Soc. Mass. Spectrom.*, 1997, **8**, 987.
28. G. E. Martin in *Analysis of Drug Impurities*, ed. R. J. Smith and M. L. Webb, Blackwell Publishing, Oxford, 2007, Chapter 5.
29. B. D. Hilton and G. E. Martin, *J. Nat. Prod.*, 2010, **73**, 1465.
30. International Conference on Harmonisation, *ICH Harmonised Tripartite Guideline: Stability Testing of New Drug Substances and Products, Q1A(R2)*, dated 6 February 2003.
31. International Conference on Harmonisation, *ICH Harmonised Tripartite Guideline: Stability Testing: Photostability Testing of New Drug Substances and Products, Q1B*, dated 6 November 1996.

32. G. Boccard, in *Pharmaceutical Stress Testing*, ed. S.W. Baertschi, Taylor & Francis, Boca Raton, 2005, Chapter 7.
33. S. W. Baertschi and K. M. Alsante, in *Pharmaceutical Stress Testing*, ed. S. W. Baertschi, Taylor & Francis, Boca Raton, 2005, Chapter 3.
34. K. C. Waterman and R. C. Adami, *Int. J. Pharm.*, 2005, **293**, 101.
35. S. Klick, P. G. Muijselaar, J. Waterval, T. Eichinger, C. Korn, T. K. Gerding, A. J. Debets, C. Sänger-van de Griend, C. van den Beld, G. W. Somsen and G. J. De Jong, *Pharm. Technol.*, 2005, **2**, 48.
36. D. W. Reynolds, K. L. Facchine, J. F. Mullaney, K. M. Alsante, T. D. Hatajik and M. G. Motto, *Pharm. Technol.*, 2002, **2**, 48.
37. E. D. Nelson, P. A. Harmon, R. C. Szymanik, M. G. Teresk, L. Li, R. A. Sebrug and R. A. Reed, *J. Pharm. Sci.*, 2006, **95**, 1527.
38. P. A. Harmon, K. Kosuda, E. Nelson, M. Mowery and R. A. Reed, *J. Pharm. Sci.*, 2006, **95**, 2014.
39. M. K. Mone and K. B. Chandrasekhar, *J. Pharm. Biomed. Anal.*, 2010, **53**, 335.
40. P. A. Harmon, S. Biffar, S. M. Pitzenberger and R. A. Reed, *Pharm. Res.*, 2005, 1716.
41. M. Spangler and E. Mularz, *Chromatographia*, 2001, **54**, 329.
42. G. B. Payne, P. H. Deming and P. H. Williams, *J. Org. Chem.*, 1961, **26**, 659.
43. Y. Sawaki and Y. Ogata, *Bull. Chem. Soc. Jpn*, 1981, **54**, 793.
44. S. W. Hovorka, M. J. Hageman and C. Schöneich, *Pharm. Res.*, 2002, **19**, 538.
45. A. A. Syed and M. K. Amshumali, *J. Pharm. Biomed. Anal.*, 2001, **25**, 1015.
46. G. Srinivas, K. K. Kumar, Y. R. K. Reddy, K. Mukkanti, G. V. Kanumula1 and P. Madhavan, *J. Chem. Pharm. Res.*, 2011, **3**, 987.
47. A. I. Segall, M. F. Vitale, V. L. Perez, M. L. Palacios and M. T. Pizzorno, *J. Liq. Chromatogr. Relat. Tech.*, 2002, **25**, 3167.
48. M. Xie, *Zhongguo Yiyao Gongye Zazhi*, 2002, **33**, 341.
49. M. Li, B. Conrad, R. G. Maus, S. M. Pitzengburger, R. Subramanian, X. Fang, J. A. Kinzer and H. J. Perpall, *Tetrahedron Lett.*, 2005, **46**, 3533.
50. Q. Fu, M. Shou, D. Chien, R. Markovich and A. M. Rustum, *J. Pharm. Biomed. Anal.*, 2010, **51**, 617.
51. M. Li, B. Chen, S. Monteiro and A. M. Rustum, *Tetrahedron Lett.*, 2009, **50**, 4575.
52. J. Lu, Y. Wei and A. M. Rustum, *J. Chromatogr., A.*, 2010, **1217**, 6932.
53. V. R. Mattox, *J. Am. Chem. Soc.*, 1952, **74**, 4340.
54. H. L. Herzog, M. J. Gentles, H. Marshall and E. B. Hershberg, *J. Am. Chem. Soc.*, 1961, **83**, 4073.
55. M. L. Lewbart and V. R. Mattox, *J. Org. Chem.*, 1964, **29**, 513.
56. T. Hidaka, S. Huruumi, S. Tamaki, M. Shiraishi and H. Minato, *Yakugaku Zasshi*, 1980, **100**, 72.
57. M. L. Lewbart, C. Monder, W. J. Boyko, C. J. Singer and F. Iohan, *J. Org. Chem.*, 1989, **54**, 1332.
58. Z. You, M. A. Khalil, D.-H. Ko and H. J. Lee, *Tetrahedron Lett.*, 1995, **36**, 3303.

59. B. Chen, M. Li, M. Lin, G. Tumambac and A. Rustum, *Steroids*, 2009, **74**, 30.
60. M. Li, X. Wang, B. Chen, M. Lin, A. V. Buevich, T.-M. Chan and A. M. Rustum, *J. Pharm. Sci.*, 2009, **23**, 3533.
61. M. Li, X. Wang, B. Chen, M. Lin, A. V. Buevich, T.-M. Chan and A. M. Rustum, *Rapid Commun. Mass Spectrom.*, 2009, **23**, 3533.
62. L. L. Smith, M. Marx, J. J. Garbarini, T. Foell, V. E. Origoni and J. J. Goodman, *J. Am. Chem. Soc.*, 1960, **82**, 4616.

CHAPTER 9

Control of Drug Degradation

9.1 Overview

In Chapters 2 to 7, we have discussed the organic chemistry of drug degradation *via* various representative drug degradation pathways and mechanisms. In this chapter, we will provide a high level overview of strategies for controlling drug degradation based on the understanding of these degradation pathways and mechanisms. On many occasions during the discussions earlier in this book, we may have mentioned some of these strategies or such strategies may have become obvious as the discussion evolved. Nevertheless, we hope a high level overview of the strategies here will further enhance our understanding and knowledge of the organic chemistry of drug degradation, which should help us to resolve any issues and challenges of drug degradation in the future. The overview is organized through the following discussion topics.

9.2 Degradation Controlling Strategies *Versus* Multiple Degradation Pathways and Mechanisms

It is important to bear in mind that the same degradants can result from multiple degradation pathways and/or mechanisms. For example, the formation of sulfoxide degradants from sulfides can be mediated *via* nucleophilic (non-radical) oxidation,[1,2] radical-initiated oxidation (see Chapter 3, Section 3.5.6),[3] and/or photochemical oxidation through singlet oxygen (see Chapter 6, Section 6.3.3).[4] Hence, use of an antioxidant that is based on the principle of free radical trapping will be unlikely to have a substantial effect in inhibiting the formation of sulfoxides in formulations where nucleophilic oxidation is the major pathway. In a study of control of oxidative degradation of a model protein, parathyroid hormone (PTH), which contains two Met residues, Trp was evaluated as an antioxidant in oxidative stress of PTH with a Fenton reagent and H_2O_2, respectively.[5] Trp was found to greatly reduce oxidation of

RSC Drug Discovery Series No. 29
Organic Chemistry of Drug Degradation
By Min Li
© Min Li 2012
Published by the Royal Society of Chemistry, www.rsc.org

the Met residues by the Fenton reagent. On the other hand, Trp had no effect in protecting Met residues from the stress by H_2O_2. This is consistent with the fact that oxidation of Trp in this case is mediated *via* a free radical mechanism. Therefore, as an antioxidant, Trp is capable of inhibiting the free radical-mediated oxidation of Met by the Fenton reagent but is ineffective in controlling the nucleophilic oxidation of Met by H_2O_2.

9.3 Design and Selection of a Drug Candidate Considering Drug Degradation Pathways and Mechanisms

Whenever possible, during the initial design or selection of a drug candidate in the critical phase after the early development stage, the degradation pathways and mechanisms of the selected candidate should be taken into consideration so that the one selected has a desirable chemical stability profile, while keeping its potency and other desirable properties such as favorable absorption, distribution, metabolism, and excretion (ADME) properties and toxicity profiles. This exercise can be naturally combined with the exercise when the metabolic profile and stability of the drug candidate is considered. As mentioned previously, chemical degradation (*in vitro*) and drug metabolism (*in vivo*) can produce a subset of degradants/metabolites that are common to both the *in vitro* and *in vivo* processes. Special attention should be placed on certain structural moieties that can degrade or metabolize to "structure alerts" – structural moieties that may be potentially genotoxic,[6,7] based on the knowledge reported in the literature and included in certain databases such as the National Institutes of Health (NIH) TOXNET database.[8] One such type of structural moieties are phenolic moieties, in particular the activated ones such as 1,2-, 1,4-di- or multiple hydroxylated phenyl rings. These moieties can degrade or metabolize to redox active as well as electrophilic 1,2- and 1,4-quinoids; a number of drugs have been shown or suspected to display adverse drug effects *via* the intermediacy of such quinoids.[9,10] Regulatory requirements for the evaluation and qualification of potential genotoxic impurities (PGI) have increased dramatically since the start of the millenium.[11]

Once the mechanism for a particular degradation pathway has been elucidated, structural modification of a drug candidate or an existing drug undergoing that pathway may be performed in such a way that the degradation is suppressed or minimized. For example, a family of corticosteroid drugs containing the cross-conjugated 2,5-cyclohexadienone A-ring, which include prednisolone, betamethasone, dexamethasone, and their derivatives (Figure 9.1), undergo a rather facile photoisomerization under both UV-A and UV-B light sources (see Chapter 6, Photochemical Degradation).[12–17]

This photoisomeric degradation takes place at the triplet excited state of the A-ring. When the hydrogen or fluorine at the nearby 9-position of the steroid core is replaced by a heavy atom such as chlorine, the resulting corticosteroids, for example, beclomethasone 17,21-dipropionate, no longer undergo the A-ring

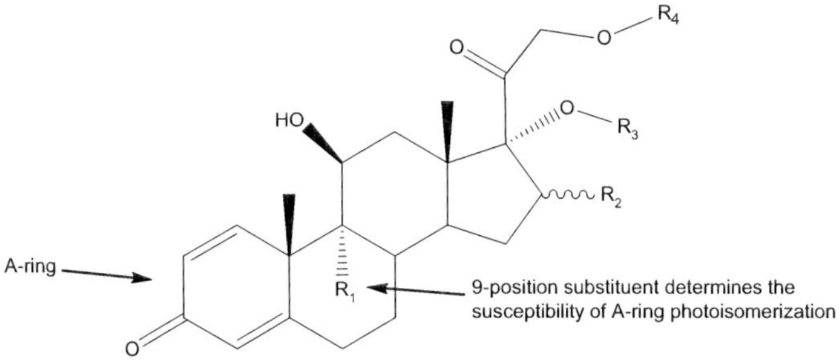

Prednisolone, $R_1 = R_2 = R_3 = R_4 = H$;
Betamethasone, $R_1 = F$, R_2 = methyl at β-position, $R_3 = R_4 = H$;
Betamethasone 17,21-dipropionate, $R_1 = F$, R_2 = methyl at β-position.
$R_3 = R_4$ = propionyl;
Dexamethasone, $R_1 = F$, R_2 = methyl at α-position, $R_3 = R_4 = H$;

Beclomethasone 17,21-dipropionate, $R_1 = Cl$, R_2 = methyl at β-position.
$R_3 = R_4$ = propionyl;

Figure 9.1   Family of corticosteroid drugs containing the cross-conjugated 2,5-cyclohexadienone A-ring.

photoisomerization owing to quenching of the A-ring triplet excited state by the chlorine at the 9-position.[18] On the other hand, corticosteroids containing the 9α-chloro,11β-hydroxy moiety, as in the case of beclomethasone 17,21-dipropionate, tend to undergo elimination to form the 9,11-epoxide degradant in liquid formulations (see Chapter 4, Section 4.1.2). This degradation pathway seems insignificant in corticosteroids containing the 9α-fluoro,11β-hydroxy moiety, as no epoxide degradant was observed in 35 commercial batches of a lotion product containing betamethasone 17,21-dipropionate.[19] The ages of these batches range between four and 32 months, with quite a few batches exceeding the product shelf-life of 18 months. Hence, by replacing 9-F with 9-Cl, one is able to suppress the photodegradation but end up with a structure that more easily undergoes elimination. This case represents a typical dilemma frequently encountered during the process of drug design and development. Ultimately, one should determine which degradant is more critical than others, in terms of its toxicological profile and intended use, so that it needs to be controlled much more tightly.

A typical approach for controlling or preventing degradation in protein- and peptide-based drugs is to change some chemically labile residues by site-directed mutagenesis, when they are present in therapeutically non-critical regions of the proteins and peptide sequences. Insulin is one of the earliest proteins used clinically for the treatment of type I diabetes. It is probably the most studied therapeutic protein in terms of improving its chemical and physical stability by site-directed mutagenesis.[20] For example, Asn residue is known to cause deamidation, which sometimes triggers dimerization in proteins, as we have discussed in Chapter 7. By replacing Asn[B3] with Gln and Asn[A21] with Ala or Gly, the resulting insulin analogs showed between 10 and 30 times improvement in their stability with regard to their respective chemical degradation.[21]

Chemical modification of protein drugs is also performed to improve their stability. The most frequently utilized methodology is to derivatize proteins with polyethylene glycol (PEG), a process known as PEGylation, which has been widely used in a large number of protein drugs on the market.[22] PEGylation can significantly improve *in vivo* stability of biological drugs primarily by slowing down the enzymatic degradation, such as proteolysis, of the drugs, in addition to imparting other favorable properties to the PEGylated drugs.[23–25] Moreover, PEGylation can enhance *in vitro* stability of biological drugs as well.[26,27]

9.4 Implication of the Udenfriend Reaction and Avoidance of a Formulation Design that may Fall into the "Udenfriend Trap"

As we discussed quite extensively in Chapter 3, Oxidative Degradation, the Udenfriend reaction is much more relevant in the autooxidation of drugs than the better known Fenton reaction. The Udenfriend reaction consists of three key elements: a transition redox metal ion (which is usually iron), a chelating agent, and a reducing agent. The combination of the three elements is capable of activating molecular oxygen at its ground state, resulting in the formation of various reactive oxygen species or ROS. Chelating agents and antioxidants are frequently used in the formulation of drug products for the purpose of product preservation and stability. If the formulation of a drug product contains both a chelating agent and a reducing agent (*e.g.* an antioxidant or a preservative with reducing capability), the drug product could potentially be intrinsically vulnerable to autooxidation. This is because a slight increase in a transition redox metal ion in the product, introduced either from the primary packaging, raw materials or during manufacturing, could trigger the Udenfriend process, causing decreased stability of the drug product. Hence, the combination of the three elements of the Udenfriend reaction in a drug formulation can be a "trap", hampering one's effort to develop a robust, quality drug product for the following two reasons. First, during the pharmaceutical development of a drug product, the leachable metal content from the primary packaging, raw materials or manufacturing process may be slightly different from that of commercial scale manufacturing. Hence, the problem may not occur until very late in the development stage or even be carried over to the commercial production stage. The transition metal content needed to trigger the Udenfriend reaction can be as low as hundreds of ppb or even lower. Such low levels are usually not controlled or very difficult to control in primary packaging materials, excipients, and manufacturing equipment. Second, the autooxidative degradation caused by the Udenfriend reaction may only become obvious when the drug product is stored under regular long term stability conditions. Forced or accelerated degradation may not be able to "predict" this degradation pathway, as the drug product may undergo different degradation pathways at the higher temperature and/or moisture of the forced and accelerated degradation conditions.

A formulation study reported by Reed *et al.* is a good example in which the Udenfriend reaction is apparently the root cause of the observed photo instability.[28] This study involved formulating a development stage drug candidate in liquid formulations that contain citrate buffer at pH 6. The drug candidate contains a phenyl ether functionality. Although its full structure was not revealed, it was indicated that the drug molecule does not contain any chromophore that absorbs light at wavelengths greater than 300 nm. Nevertheless, during photostability testing with UV-B and visible light irradiation under the ICH photostability conditions, the formulated drug was found to undergo oxidative degradation. The presence of citrate, iron, and light exposure was necessary for the degradation to proceed. The distribution and type of oxidative degradants strongly suggested that hydroxyl radical was generated in the degradation process. This scenario would be consistent with the fact that Fe(II) can be photochemically generated from the [Fe(III)-citrate] complex,[29] during which process the formation of hydroxyl radical was also reported.[30] Hence, the authors proposed a degradation mechanism in which photochemically generated Fe(II) triggers the sequential formation of superoxide anion, hydrogen peroxide, and hydroxyl radical. The hydroxyl radical would then initiate the oxidation of the drug candidate. The presentation of the following mechanism (Scheme 9.1) is somewhat different from that of the original authors, but is fundamentally consistent with them.

In the mechanism as shown in Scheme 9.1, it is the [Fe(II)-citrate] complex, rather than Fe(II) itself as proposed by the original researchers,[28] that activates molecular oxygen by transferring an electron from the chelated Fe(II) to O_2. At pH 6, Fe(II) should be mostly chelated by citrate in solution. Although not specified by the original researchers, it is evident that this mechanism fits all the key characteristics of the Udenfriend degradation chemistry. The unique point about this example is that the role of the reducing agent, which is played by ascorbic acid in the original Udenfriend reaction, is now played by citrate and photons. In other words, Fe(III), in the form of its citrate complex, is continuously cycled back to the catalytically active Fe(II) through the photo reduction in which the ligand citrate is oxidized. Based on this perspective, this

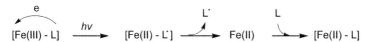

Photo reduction of Fe(III)-Citrate complex to catalytically active Fe(II)-Citrate complex; L = Citrate.

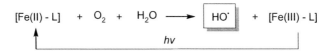

Photo reduction per the step shown above;
Intermediate steps for the formation of $O_2^{-\cdot}$ and H_2O_2 are not shown.

Scheme 9.1

photocatalyzed oxidation can be considered to be the photochemical version of the Udenfriend reaction.

This study also revealed additional challenges that are not uncommon in cases of free radical-mediated oxidative degradation during pharmaceutical development. For example, the oxidation of the drug candidate caused by the photochemical Udenfriend reaction, became noticeable at an iron level as low as 1 ppb. At 50 ppb of iron, more than 20% of the drug was degraded after irradiation by visible light under International Conference on Harmonisation (ICH) photostability conditions. At such low levels, suppression of the photooxidation by lowering the iron content becomes impractical, because of the ubiquitous presence of iron in the drug substance itself, primary packaging materials (glass vials in this case), and excipients.[28] It was also found that iron leached out of the borosilicate glass vials that were used to package the liquid product; the level of iron in the formulation increased from 5 ppb at 3 months to 45 ppb by 23 months. This result indicates that the rate of the oxidation will accelerate over time.

9.5 Control of Oxygen Content in Drug Products

As mentioned previously, the ultimate source of the oxidizing agent is molecular oxygen (O_2) in the vast majority cases of autooxidative degradation. For free radical-mediated autooxidation, a carbon-centered radical usually reacts with molecular oxygen at the diffusion control rate. In such a case, the rate of the autooxidation is usually independent of the concentration of oxygen, as discussed in Chapter 3. Hence, if one tries to reduce drug oxidation by controlling or reducing the oxygen content in the product, the oxygen content needs to be reduced to a really low level, essentially to a level that is close to oxygen-free, in order for this approach to work.

In the study of the photochemical Udenfriend degradation discussed in the previous section, the effect of reducing oxygen content by sparging was evaluated. It turned out that sparging of the product with helium had little effect in inhibiting the photooxidation of the product; the same level (a few percentage points) of oxidative degradation was observed with or without sparging.[28] In this case, the molar ratio of the drug degraded relative to the oxygen remaining in the product solution was only about 1%. In other words, the molarity of the oxygen after sparging was still much higher than the molarity of the portion of the drug substance to be oxidized. Hence, the reaction of the carbon-centered free radical of the drug molecule with oxygen is still not rate determining, which explains the ineffectiveness of sparging. The lesson from this study is that reducing oxygen content tends to be ineffective in suppressing autooxidation, unless one is able to reduce the oxygen in a formulation to a level that is well below the quantity of the drug substance on a molar basis.

On the other hand, purging of drug products in solid dosage forms with nitrogen can be quite effective, if the products are packaged in gas impermeable bottles.[31] This is most likely for the following two reasons. First, the vast majority of oxygen can be readily purged out of the bottles containing the solid products. In contrast, this is usually very difficult to achieve with liquid formulations, as we have seen previously. Second, the gas impermeability ensures an essentially oxygen-free environment inside the bottles as long as the seals of the bottles remain intact.

For solid dosage products packaged in gas permeable bottles, for example, polyethylene (PE) bottles, use of oxygen scavenger can achieve and maintain extremely low levels of oxygen inside the bottles. In a case study by Waterman and Roy,[32] use of an iron-based oxygen scavenger was able to drive down the level of oxygen to 80 ppm within 24 hours. According to the authors, the amount and cost of the scavenger needed to achieve multi-year shelf-life is manageable for the commonly used high density polyethylene (HDPE) bottles.

9.6 Use of Antioxidants and Preservatives

Antioxidants are usually effective in inhibiting free radical-mediated auto-oxidative degradation. Nevertheless, in some cases, use of antioxidants promotes autooxidation. Such seemingly paradoxical scenarios are quite likely to occur in formulations where both an antioxidant and a transition metal ion chelator are present. As we have discussed before, such a combination could easily fall into the Udenfriend trap, that is, combination of a reducing agent, a chelating agent, and transition metal ions, owing to the ubiquitous presence of transition metal ions such as iron. The combination of an antioxidant and chelator may not always cause a problem or the problem may not be obvious initially when the amount of a redox metal ion is still low. Nevertheless, such a combination tends to increase the risk of getting a formulation that is intrinsically vulnerable towards autooxidation under long term stability storage conditions. One also needs to be aware that certain preservatives, such as *m*-cresol and chlorocresol, are also reducing agents. Hence, they may also be capable of reducing Fe(III) in the presence of a chelator, thus triggering the Udenfriend degradation pathway.

9.7 Use of Chelating Agents to Control Transition Metal Ion-mediated Autooxidation

Use of chelating agents is a common way to suppress transition metal-initiated autooxidation in formulation development. The results of this approach have been inconsistent: sometimes it works, sometimes it has no effect, while in other cases, it accelerates the oxidation. Such inconsistent outcomes were perplexing and it appears that few satisfactory explanations have been provided in the literature.[33] Now, with knowledge of Udenfriend degradation chemistry, it should not be difficult in most cases to understand the root causes of the inconsistent outcomes when employing chelating agents in formulation development.

In the case of the antitumor drug 9-hydroxyellipticine, use of ethylenediamine tetraacetic acid (EDTA) dramatically decreased autooxidation of the drug.[34] As discussed in Chapter 3, this may be a Udenfriend type of autooxidation in which the oxidation substrate (the drug) serves as both the chelator

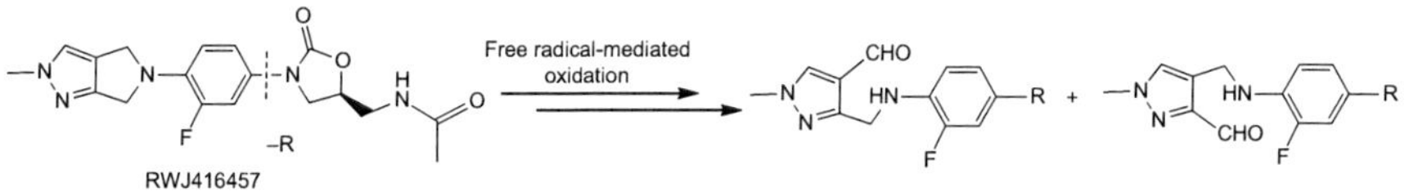

Scheme 9.2

and the reducing agent. Therefore, EDTA should disrupt the complex formed between the drug, transition metal ions, and oxygen, thus inhibiting oxidation. If an oxidation substrate binds to a transition metal ion so tightly that the chelator fails to pull the metal ion away from the substrate, using the chelator would not be able to prevent metal ion-initiated autooxidation.[35] On the other hand, in a preformulation study by Wu *et al.*,[36] use of EDTA was found to promote the oxidation of bortezomib. This suggests that the Udenfriend reaction may be responsible for the autooxidative degradation. In an injectable solution of epinephrine containing metabisulfite, Grubstein and Milano found that addition of EDTA caused the resulting formulation to be photochemically more labile than the original formulation.[37] This appears to be another case of photooxidation caused by the photo Udenfriend reaction.

Dong *et al.* performed degradation studies of an oxazolidinone-based anti-bacterial drug candidate, RWJ416457, and found it underwent mostly oxidative degradation to form two major degradants (Scheme 9.2).[38] Use of a metal chelator alone, citrate or EDTA, stabilized the molecule in pH 7 aqueous solutions. In the presence of antioxidants such as ascorbic acid, BHA, or BHT, however, citrate was found to increase the oxidation of the drug candidate; the pro-oxidation effect of citrate was most evident when ascorbic acid was present. This is clearly another good example of oxidative drug degradation caused by the Udenfriend degradation chemistry.

9.8 Control of Moisture in Solid Dosage Forms

In some limited cases, it appears that certain levels of moisture may be required to maintain the physical or chemical stability of a drug product.[39] Nevertheless, an increase in moisture generally aggravates drug degradation, even in cases where water is not directly involved in the degradation as a reactant. The role of water in the latter cases has been discussed extensively in the literature and, quite often it can be attributed to the plasticizing effect that it has in elevating molecular mobility in the solid state.[39–41] Therefore, control of moisture in solid dosage drug products is important. For moisture-permeable primary packaging materials such as high density polyethylene (HDPE) bottles, control of moisture is generally achieved by using desiccants in the product bottles. The desiccants commonly utilized in the pharmaceutical industry include silica gel, montmorillonite clay, and molecular sieves, which are usually available in prepackaged canisters, cartridges, or sachets.[42] For solid dosage drugs for-mulated in hard shell gelatin capsules, however, desiccants are generally not used because the brittleness of the gelatin starts to increase below a relative

humidity of 30%. Other ways of limiting the impact of moisture on drug stability include use of a moisture-protecting coating[43,44] and melt granulation.[45]
For drugs that need maximum protection from moisture, cold-form aluminum
blisters may be used as the primary packaging material.[46]

9.9   Control of pH

As we have seen previously, pH has a significant impact not only on the
hydrolytic, but also on the oxidative degradation of drugs. The range of
optimal pH for a drug product depends upon the structure of the drug molecule
and the type of the degradation involved. Drugs containing an ester functionality tend to be hydrolyzed more easily at basic pH than acidic pH in
general. For a great number of β-lactam antibiotics possessing both acidic and
basic functional groups, the optimal pH tends to be near the value at which the
zwitterion form of the molecule predominates.[47,48] For drug molecules that
undergo *A*-1 type hydrolysis, strongly acidic pH is generally not favored as it
catalyzes this degradation pathway rather efficiently.[49]

On the other hand, drugs containing amine functional groups, especially
secondary and tertiary amines, are electron-rich nucleophiles in the neutral to
basic pH range. Hence, they are capable of undergoing nucleophilic oxidation
relatively easily. At acidic pH, particularly when it is controlled well below the
pK_a value of its conjugated acid, the nuleophilicity of the amine group is greatly
suppressed owing to protonation of the amine, resulting in a much slower
oxidative degradation.[50] Drugs containing phenol or catechol-type moieties are
strongly electron-rich species at basic pH and thus can undergo various free
radical-mediated oxidative degradation pathways.[34] Therefore, acidic pH is
usually preferred to ensure these drugs are fully protonated, in order to minimize oxidation. Neutral to basic pH can also facilitate the degradation of drugs
that undergo base-catalyzed autooxidation.[51,52]

Although the concept of pH is established in solutions, particularly aqueous
solutions, it has been frequently used in cases of degradation in the solid state,
where it is sometimes referred to as "microenvironment pH".[53] The "pH" in
the solid state can be determined based on the pH of the solution or slurry that
is used to prepare a lyophilized drug product or a regular dosage form. In a
formulation study of a drug containing an ester functionality, Badawy *et al.*
found that simply mixing a solid acid into a formulation for dry granulation is
not as effective as in a wet granulation in terms of preventing undesirable
hydrolytic degradation in the final product.[54] Apparently, controlling the tar-
geted pH (~ 4) uniformly in the microenvironment of the resulting solid dosage,
as rendered by wet granulation, is critical to suppress the hydrolysis.

9.10   Control of Photochemical Degradation Using Pigments, Colorants, and Additives

Control of the degradation of photolabile drug molecules can be achieved using
colorants, pigments, and additives, in addition to photo-protective coatings[55]

and primary packaging.[56] In these three cases, UV and visible irradiation are preferentially absorbed or blocked by the colorants, pigments, and additives. In a formulation study of a tretinoin lotion product, it was found that the photosensitive drug could be stabilized by the yellow colorant, chrysoin, by a factor of 3.5-fold at a concentration of 0.025%.[57] Despite the observation that higher colorant concentrations could further stabilize the drug as much as 8-fold, a concentration of 0.025% was chosen in the final formulation, because this concentration does not color the skin but still provides an acceptable stabilization effect. The stabilization effect is apparently attributable to full UV-Vis spectral overlap between the drug substance and chrysoin.[58]

In a formulation study of molsidomine tablets, aimed at stabilizing the photolabile drug, use of colorants (in the core tablets) and pigments (in both the core tablets and coating) was investigated.[59] At a level of 0.5% of the formulation, yellow iron oxide pigment was able to reduce the degradation from 33% to 5% under the photostability testing conditions; the protection provided by iron oxide was somewhat better than that by titanium oxide pigment and two colorants (azorubine and curcumine). Use of various tablet coatings was also investigated. A hydroxypropylmethylcellulose film coating containing yellow or red iron oxide in combination with titanium oxide was able to completely suppress the photodegradation of molsidomine.

9.11   Variability of Excipient Impurity Profiles

Excipients contain impurities and, quite often, these impurities may not be well controlled by the vendors. As we have mentioned before, excipients such as polyethylene glycol (PEG), polysorbates, and povidone contain varying degrees of peroxides, hydroperoxides including hydrogen peroxide, formaldehyde, and/or formic acid owing to the autooxidation of these excipients.[60–63] Consequently, the impurity profiles of the excipients can vary significantly from vendor to vendor and from lot to lot by the same vendor. In some cases, variability within the same lot of an excipient can occur due to sub-division and subsequent different storage conditions of the lot. Such variability not only poses a challenge during the pharmaceutical development stage but is also a concern in source-of-supply changes that may occur later in the product life cycle. As discussed earlier in this book, especially in Chapters 3 and 5, impurities of the excipients can cause a wide variety of degradation reactions.

9.12   Use of Formulations that Shield APIs from Degradation

In aqueous liquid formulations, one way to control or reduce drug degradation is to segregate the drug substances from water. This can be achieved by using a number of excipients, for example, cyclodextrins,[64] surfactants,[65] and liposomes.[66] The use of these excipients not only imparts favorable properties to the resulting drug products, such as increased solubility, bioavailability and controlled release of the active ingredients, but also improves the drug stability. For example, cyclodextrins, such as 2-hydroxypropyl-β-cyclodextrin and γ-cyclodextrin, can not only increase the solubility of certain hydrophobic drug molecules[67] but can also substantially improve their stability by forming drug–cyclodextrin inclusion complexes. This process is a part of a formulation strategy known as microencapsulation.[68] Reported case studies with cyclodextrin formulations include nitrazepam, mitomycin, and taxol.[69–71]

For acid-sensitive drugs, enteric-coated formulations can be used to prevent drug degradation in the stomach.[72,73] Drug products that contain two chemically incompatible active ingredients can be formulated in bi-layer formulations.[74,75]

9.13 Impact of Manufacturing Process on Drug Degradation

With its formula established, how a drug product is manufactured can have a significant impact on the stability of the drug product. A drug product in tablet formulation, can generally be manufactured by either a wet granulation or direct compression process. In quite a few cases, drug products manufactured *via* wet granulation displayed better stability profiles (*i.e.* slower degradation rates) than those manufactured *via* direct compression.[54,76,77] The stabilization effect can be attributed to a better and more uniform control of the micro-environment pH in the resulting solid dosage forms.

Nevertheless, wet granulation is more likely to induce phase transition,[78] in addition to causing chemical degradation such as hydrolysis during manufacturing processes. In the case of phase transition, if the solid phase impurity formed is less stable than the API, the resulting drug product will show an increased degradation rate, which may have an impact on the product shelf-life. This appears to be the case in the process development of two photolabile drugs, nifedipine and molsidomine, by Aman and Thoma.[79] They found that the products of two wet granulation processes displayed 4% higher degradation than those from an alternative direct compression process. The destabilization was attributed to the formation of an amorphous phase during the wet granulation processes. Therefore the direct compression process would be preferred in this case.

With regard to chemical degradation, wet granulation could cause more process impurities, in particular for moisture-sensitive APIs. However, as we have just discussed, the drug product manufactured *via* wet granulation could have a slower degradation rate than that manufactured *via* the alternative, direct compression process. Hence, if the process impurities are within the control limits, wet granulation would be preferred over the alternative direct compression in such a case, based on consideration of the improved stability profile.

9.14 Selection of Proper Packaging Materials

Primary packaging is the last defense for preventing or minimizing drug degradation after the formulation and process development is complete, as they can block or reduce the three most important elements in drug degradation: oxygen, moisture, and light. Oxygen and light are reagents of drug degradation reactions, while moisture is not only a reagent but also acts as a reaction medium in liquid formulations and a plasticizer facilitating drug degradation in the solid state. Selection of the primary package for a drug product needs to be first evaluated based on the degradation pathways of the drug product. For drugs susceptible to hydrolysis and oxidation, the moisture and oxygen permeability of the package should be an important consideration. For those drugs that are very sensitive to oxygen and moisture, use of impermeable packaging along with oxygen scavengers and desiccants may be necessary. Additionally, these drug products may be packaged under a nitrogen atmosphere further to ensure an adequate product shelf-life.[80]

For photosensitive drugs, the primary packages should be able to protect the products from photodegradation, in particular to ensure that the products pass ICH photostability confirmatory testing.[81]

When selecting a proper primary package, cost is obviously a factor to consider, which needs to be balanced with the intended use and shelf-life of the product.

9.15 Concluding Remarks

Overall, control of drug degradation can be very challenging. Development of a robust, quality drug product relies on a clear understanding of the underlying organic chemistry of the drug degradation. A slight change in the drug substance, excipients, or manufacturing process could trigger an unexpected increase in drug degradation. During pharmaceutical development, one needs to be aware that the role of an excipient may change under certain specific conditions. As we have seen, an antioxidant can actually promote oxidative degradation in cases where the Udenfriend reaction plays a key role in the observed drug degradation.

Another example in this category is given here to illustrate further the complexity of the degradation chemistry that may be encountered during the pharmaceutical development. In a study of the role of mangiferin, a natural product present in mango, in inhibiting ferrous iron-induced lipid peroxidation, Pardo-Andreu *et al.* found that mangiferin promotes the oxidation of 2-deoxyribose by the classic Udenfriend reagent, [Fe(III)-EDTA] plus ascorbate, while it inhibits the oxidation of 2-deoxyribose by a variation of the classic Udenfriend reagent, [Fe(III)-citrate]–ascorbate.[82] In the former case, mangiferin is not able to break up the [Fe(III)-EDTA] complex owing to its weaker affinity for Fe(III) than EDTA. Consequently, it can apparently only act as a reducing agent that facilitates the reduction of [Fe(III)-EDTA] to catalytically reactive [Fe(II)-EDTA]. In the latter case, mangiferin is able to break up the [Fe(III)-citrate] complex owing to its stronger affinity for Fe(III) than citrate, hence inhibiting the oxidation inflicted upon by the [Fe(III)-citrate]–ascorbate system. Such mechanistic complexity would make the drug development process even more challenging.

The ultimate goal of having a clear understanding of the organic chemistry of drug degradation is to put one into a better position to design desirable quality attributes into a drug product by overcoming various challenges that may be encountered during the overall drug development process. In addition, such understanding is also essential to maintaining the stability, efficacy, and safety of the drug product throughout its life cycle. I hope this book has contributed toward this goal.

References

1. G. Modena and P. E. Todesco, *J. Chem. Soc.*, 1964, 4920.
2. J. W. Chu and B. L. Trout, *J. Am. Chem. Soc.*, 2004, **126**, 900.
3. C. Schöneich, A. Aced and K.-D. Asmus, *J. Am. Chem. Soc.*, 1993, **115**, 11376.
4. C. Gu, C. S. Foote and M. L. Kacher, *J. Am. Chem. Soc.*, 1981, **103**, 5949.
5. J. A. Ji, B. Zhang, W. Cheng and Y. J. Wang, *J. Pharm. Sci.*, 2009, **98**, 4485.
6. S. P. Raillard, J. Bercu, S. W. Baertschi and C. M. Riley, *Org. Process Res. Dev.*, 2010, **14**, 1015.
7. S. D. Nelson, *J. Med. Chem.*, 1982, **25**, 753.
8. http://toxnet.nlm.nih.gov. Last accessed 24 April 2012.
9. J. L. Bolton and G. R. J. Thatcher, *Chem. Res. Toxicol.*, 2008, **21**, 93.
10. M. Pirmohamed, A. M. Breckenridge, N. R. Kitteringham and B. K. Park, *Br. Med. J.*, 1998, **316**, 1295.
11. A. Giordani, W. Kobel and H. U. Gally, *E. J. Pharm. Sci.*, 2011, **43**, 1.
12. J. R. Williams, R. H. Moore, R. Li and C. M. Weeks, *J. Org. Chem.*, 1980, **45**, 2324.
13. T. Hidaka, S. Huruumi, S. Tamaki, M. Shiraishi and H. Minato, *Yakugaku Zasshi*, 1980, **100**, 72.
14. O. T. Y. Fahmy, *Generation, isolation, characterization and analysis of some photolytic products of dexamethasone and related steroids*, Doctoral Thesis, University of Mississippi, 1997.
15. M. Lin, M. Li, A. V. Buevich, R. Osterman and A. M. Rustum, *J. Pharm. Biomed. Anal.*, 2009, **50**, 275.
16. G. Miolo, F. Gallocchio, L. Levorato, D. Dalzoppo, G. M. J. Beyersbergen van Henegouwen and S. Caffieri, *J. Photochem. Photobiol., B.*, 2009, **96**, 75.
17. Y. Shirasaki, K. Inada, J. Inoue and M. Nakamura, *Steroids*, 2004, **69**, 23.
18. A. Ricci, E. Fasani, M. Mella and A. Albini, *J. Org. Chem.*, 2001, **66**, 8086.
19. M. Shou, W. A. Galinada, Y.-C. Wei, Q. Tang, R. J. Markovich and A. M. Rustum, *J. Pharm. Biomed. Anal.*, 2009, **50**, 356.
20. J. Brange, *Diabetologia*, 1997, **40**, S48.
21. J. Brange and S. Havelund, *Novel insulin analogues stabilized against chemical modifications. European Patent* 0419504, 1991.
22. J. S. Kang, P. P. DeLuca and K. C. Lee, *Expert Opin. Emerging Drugs*, 2009, **14**, 363.

23. D. H. Na, Y. S. Youn, E. J. Park, J. M. Lee, O. R. Cho, K. R. Lee, S. D. Lee, S. D. Yoo, P. P. Deluca and K. C. Lee, *J. Pharm. Sci.*, 2004, **93**, 256.
24. J. M. Harris and R. B. Chess, *Nat. Rev. Drug Disc.*, 2003, **2**, 214.
25. F. M. Veronese and G. Pasut, *Drug Discovery Today*, 2005, **10**, 1451.
26. N. V. Katre, *Adv. Drug Delivery Rev.*, 1993, **10**, 91.
27. M. A. Croyle, Q.-C. Yu and J. M. Wilson, *Hum. Gene Ther.*, 2000, **11**, 1713.
28. R. A. Reed, P. Harmon, D. Manas, W. Wasylaschuk, C. Galli, R. Biddell, P. A. Bergquist, W. Hunke, A. C. Templeton and D. Ip, *PDA J. Pharm. Sci. Tech.*, 2003, **57**, 351.
29. J. J. Llorens-Molina, *J. Chem. Educ.*, 1988, **65**, 1090.
30. H. B. Abrahamson, A. B. Rezvani and G. Brushmiller, *Inorg. Chim. Acta*, 1994, **226**, 117.
31. S. Maki, S. Ando and C. Nakano, *Oral Preparations Containing Bromhexine or Ambroxol and Stabilization of the Preparations in Oxygen-Free Atmosphere*, Japanese Patent 1,010,1581, 1998.
32. K. C. Waterman and M. C. Roy, *Pharm. Dev. Technol.*, 2002, **7**, 227.
33. S. W. Hovorka and C. Schöneich, *J. Pharm. Sci.*, 2001, **90**, 253.
34. C. Auclair and C. Paoletti, *J. Med. Chem.*, 1981, **24**, 289.
35. S. M. Yatin, S. Varadarajan, C. D. Link and D. A. Butterfield, *Neurobiol. Aging*, 1999, **20**, 325.
36. S. Wu, W. Waugh and V. J. Stella, *J. Pharm. Sci.*, 2000, **89**, 758.
37. B. Grubstein and E. Milano, *Drug Dev. Ind. Pharm.*, 1992, **18**, 1549.
38. J. Dong, S. B. Karki, M. Parikh, J. C. Riggs and L. Huang, *Drug Dev. Ind. Pharm.*, posted online on, January 23, 2012. (doi: 10.3109/03639045.2011.648195)
39. E. Y. Shalaev and G. Zografi, *J. Pharm. Sci.*, 1996, **85**, 1137.
40. C. Ahlneck and G. Zografi, *Int. J. Pharm.*, 1990, **62**, 87.
41. S. R. Byrn, W. Xu and A. W. Newman, *Adv. Drug Delivery Rev.*, 2001, **48**, 115.
42. K. C. Waterman and B. C. MacDonald, *J. Pharm. Sci.*, 2010, **99**, 4437.
43. N. Pearnchob, J. Siepmann and R. Bodmeier, *Drug Dev. Ind. Pharm.*, 2003, **29**, 925.
44. E. M. Rudnic and M. K. Kottke, in *Modern Pharmaceutics*, ed. G. S. Banker and C. T. Rhodes, Marcel Dekker, New York, 3rd edn, 1996, pp. 333–394.
45. J. Kowalski, O. Kalb, Y. M. Joshi and A. T. M. Serajuddin, *Int. J. Pharm.*, 2009, **381**, 56.
46. J. G. Allinson, R. J. Dansereau and A. Sakr, *Int. J. Pharm.*, 2001, **221**, 49.
47. J. P. Hou and J. W. Poole, *J. Pharm. Sci.*, 1969, **58**, 447.
48. R. Chadha, N. Kashid and D. V. S. Jain, *J. Pharm. Pharmacol.*, 2003, **55**, 1495.
49. P. J. Jansen, P. L. Oren, C. A. Kemp, S. R. Maple and S. W. Baertschi, *J. Pharm. Sci.*, 1998, **87**, 81.

50. A. L. Freed, H. E. Strohmeyer, M. Mahjour, V. Sadineni, D. L. Reid and C. A. Kingsmill, *Int. J. Pharm.*, 2008, **357**, 180.
51. J. S. Edmonds, M. Morita, P. Turner, B. W. Skelton and A. H. White, *Steroids*, 2006, **71**, 34.
52. P. A. Harmon, S. Biffar, S. M. Pitzenberger and R. A. Reed, *Pharm. Res.*, 2005, **22**, 1716.
53. A. T. M. Serajuddin, A. B. Thakur, R. N. Ghoshal, M. G. Fakes, S. A. Ranadive, K. R. Morris and S. A. Varia, *J. Pharm. Sci.*, 1999, **88**, 696.
54. S. I. F. Badawy, R. C. Williams and D. L. Gilbert, *J. Pharm. Sci.*, 1999, **88**, 428.
55. S. R. Bechard, O. Quraishi and E. Kwong, *Int. J. Pharm.*, 1992, **87**, 133.
56. K. Thoma and W. Aman, in *Pharmaceutical Photostability and Stabilization Technology*, ed. J. T. Piechocki and K. Thoma, Informa Healthcare USA, New York, 2007.
57. M. Brisaert and J. Plaizier-Vercammen, *Int. J. Pharm.*, 2000, **199**, 49.
58. K. Thoma and R. Klimek, *Int. J. Pharm.*, 1991, **67**, 169.
59. W. Aman and K. Thoma, *J. Pharm. Sci.*, 2004, **93**, 1860.
60. J. W. McGinity, T. R. Patel and A. H. Naqvi, *Drug Dev. Commun.*, 1976, **2**, 505.
61. T. Huang, M. E. Garceau and P. Gao, *J. Pharm. Biomed. Anal.*, 2003, **31**, 1203.
62. W. R. Wasylaschuk, P. A. Harmon, G. Wagner, A. B. Harman, A. C. Templeton, H. Xu and R. A. Reed, *J. Pharm. Sci.*, 2007, **96**, 106.
63. K. C. Waterman, W. B. Arikpo, M. B. Fergione, T. W. Graul, B. A. Johnson, B. C. MacDonald, M. C. Roy and R. J. Timpano, *J. Pharm. Sci.*, 2008, **97**, 1499.
64. T. Loftsson, H. Fridriksdottir and B. J. Olafsdottir, *Acta Pharm. Nord.*, 1991, **3**, 215.
65. V. P. Torchilin, *J. Controlled Release*, 2001, **73**, 137.
66. A. Manosroi, L. Kongkaneramit and J. Manosroi, *Int. J. Pharm.*, 2004, **270**, 279.
67. T. Loftsson, D. Hreinsdottir and M. Masson, *Int. J. Pharm.*, 2005, **302**, 18.
68. *Microencapsulation: Methods and Industrial Applications*, ed. S. Benita, CRC Press, Taylor & Francis Group, Boca Raton, FL, 2nd edn, 2006.
69. S. I. Saleh, A. A. Rahman, A. E. Aboutaleb, Y. Nakai and M. O. Ahmed, *J. Pharm. Belg.*, 1993, **48**, 383.
70. O. Bekers, J. H. Beijnen, M. J. Tank, A. Bult and W. J. Underberg, *J. Pharm. Biomed. Anal.*, 1991, **9**, 1055.
71. H. Montaseri, F. Jamalib, J. A. Rogers, R. G. Micetich and M. Daneshtalab, *Iran. J. Pharm. Sci.*, 2004, **1**, 43.
72. A. Pilbrant and C. Cederberg, *Scand. J. Gastroenterol.*, 1985, **20**(suppl. 108), 113.
73. P. J. Jansen, P. L. Oren, C. A. Kemp, S. R. Maple and S. W. Baertschi, *J. Pharm. Sci.*, 1998, **87**, 81.
74. S. Aryal and N. Skalko-Basnet, *Acta Pharmaceutica.*, 2008, **58**, 299.

75. C. Lacaze, T. Kauss, J.-R. Kiechel, A. Caminiti, F. Fawaz, L. Terrassin, S. Cuart, L. Grislain, V. Navaratnam, B. Ghezzoul, K. Gaudin, N. J White, P. L Olliaro and P. Millet, *Malar. J.*, 2011, **10**, 142.
76. S. Badawy, R. Vickery, K. Shah and M. Hussain, *Pharm. Dev. Technol.*, 2004, **9**, 239.
77. E. A. Zannou, Q. Ji, Y. M. Joshi and A. T. M. Serajuddin, *Int. J. Pharm.*, 2007, **337**, 210.
78. G. G. Z. Zhang, D. Law, E. A. Schmitt and Y. Qiu, *Adv. Drug Delivery Rev.*, 2004, **56**, 371.
79. W. Aman and K. Thoma, *Pharm. Ind.*, 2002, **64**, 1287.
80. R. Mahajan, A. Templeton, A. Harman, R. A. Reed and R. T. Chern, *Pharm. Res.*, 2005, **22**, 128.
81. International Conference on Harmonisation, *ICH Harmonised Tripartite Guideline: Stability Testing: Photostability Testing of New Drug Substances and Products Q1B*, dated 6 November 1996.
82. G. L. Pardo-Andreu, R. Delgado, A. J. Nunez-Selles and A. E. Vercesi, *Pharmacol. Res.*, 2006, **53**, 253.

Subject Index

absorption, distribution, metabolism and excretion (ADME) 263
acetal and hemiacetal groups 40–1
acetaminophen 173
N-acetylcysteine 83
activation energies for hydrolytic degradation (drug molecules) 18–19
active pharmaceutical ingredient (API)
  betamethasone dipropionate 248–51
  clopidogrel bisulfate (Plavix) 64
  decarboxylation 119
  degradation
    counter ions/two APIs 156–7
    excipients 160
    impurities in packaging 161–2
    manufacturing 272
    shielding 271–2
  drug substance 1, 143
  ester and amide linkage 153
  hydralazine HCl 158
  impurities in polymeric excipients 159–60
  lactam degradation 44
  manufacturing process 272
  Maillard reaction 151–2
  meropenem 31, 154
  moisture and manufacturing process 272
  PEG/polysorbate and formaldehyde 143
N,O-acyl migration 132–3
additional reactions of free radicals 56–7

adverse drug reactions (ADRs) 2
"aerial oxidation" term 48
aflatoxin B1 42
albuterol (salbutamol) 94
"allomerization" term 48
allylic/benzylic positions susceptible to hydrogen abstraction by free radicals 62–8
Amadori rearrangement 152
4-amino salicyclic acid 118
5-amino salicyclic acid 119
amide group (hydrolysis) 24–25
amoxicillin 26–7, 140
ampicillin 26–7, 140
amyltryptyline 68
angiotensin-converting-enzyme (ACE) inhibitors
  DKP cyclization 138
  fosinopril sodium 155
  telmisartan 180–2
  thiol (sulfhydro) functionality 83
antifungal drugs 96–8
antioxidants and preservatives (drug products) 268
aromatic rings
  cyclization 180–2
  heterocyclic 96–9
  oxidation 93
  phenols, polyphenols and quinones 92–6
aromatization of 1, 4-dihydropydine drugs 174–6
Arrhenius Equation 9
aryl halides 178–9
2-arylpropionic acid NSAIDs 167–70, 177

aspirin 22, 157, 173
atorvastatin 180, 182, 191–2
"autooxidation" term 48
autooxidative chain reactions and
    kinetic behavior 54–6
Avastin 198
avermectins 66
azithromycin 114
azobisisobutyronitrile (AIBN) 71, 94,
    98, 243
azole antifungal drugs 96, 98

Baeyer–Villager oxidation 81, 87,
    89–91
barzelesin 31
beclomethosone 173, 186
benorylate 173
benoxaprofen 167
benzocaine 22
benzophenone 167
benzylpenicillin 26
betamethasone
    Cannizaro rearrangement 137
    degradation 86, 134, 234
    dehydration 111–112
    esterification 44
    dehydrofluorination 115
    phosphates and
        phosphoramides 32–33
    photoisomerization 172
    pro-drug 138
    rearrangement *via* ring
        expansion 133–5
    retro-aldol reaction 126
    transesterification 44
    see also stress studies: LC-MS^n
        fingerprinting combination: case
        studies
biapenem 30, 141–2
biological drugs (chemical
    degradation)
    carbohydrate-based drugs 216–18
    DNA and RNA Drugs 218–22
    overview 198–9
    protein drugs 199–215
bond dissociation energies (BDE) 87,
    176, 178

book summary 11–14
bortezomib 99–101

calcium channel blockers
    (hypertension) 174
camptothecin 133
Cannizzaro rearrangement 136–7
carbanion/enolate-mediated
    autooxidation 61–2, 83–7
captopril 83
carbamates 30–2
carbapenem series 29, 116, 139–42
carbohydrate-based biological drugs
    (degradation) 216–18
carprofen 167, 170, 177, 185
carzelesin 31–2
cefaclor 28–9
cefepime 28–9
cefotaxime 172
cefpodoxime 128, 131
ceftibuten analogs 131–2
cephalosporins 26–9, 131, 172
chelating agents
    transition metal ion-mediated
        autooxidation 168–9
    Udenfriend reaction 52–3,
        268–9
chloramphenicol 24
chloroacarbazole 177, 180
cholesterol 190–1
chondroitin sulfate 218
chrysoin (yellow colorant) 271
cimetidine 82
ciprofloxacin 182–3
cis-trans isomerization around
    carbon–carbon,
    carbon–heteroatom
    heteroatom–heteroatom double
    bonds 170–2
clinafloxacin 183
clocortolone 115
clopidogrel bisulfate 64–5
collision induced fragmentation
    (CID) 234, 252
control of drug degradation
    antioxidants and preservatives 268
    API shielding 271–2

control of drug degradation
(*continued*)
   chelating agents to control
     transition metal ion-mediated
     autooxidation 268–9
   conclusions 273–4
   design/selection of drug
     candidate 263–5
   excipient impurity profiles 271
   manufacturing process 272
   moisture in solid dosage
     forms 269–70
   overview 262
   oxygen content in drug
     products 267–8
   packaging materials 272–3
   pH 270
   photochemical degradation using
     pigments, colorants and
     additives 270–1
   strategies *versus* multiple
     pathways/mechanisms 262–3
   Udenfriend reaction 265–7, 268
   Udenfriend "trap" 265–7, 268
corticosteroidal drugs
   (degradation) 85–7
crosslinking, dimerization and
   oligomerization (protein
   drugs) 213–14
cyclization
   diketopiperazine 137–8
   other reactions 138–9
   polyaromatic rings 180–2
2,5-cyclodienone rings and
   photoisomerization 172–3
cyclodextrins 271–2
cyclophosphamide 33–4
cyclosporin A 132–3
cytomegalovirus (CMV) 198

D-ring expansion
   (D-homoannulation) in
   corticosteroids 133–5
deamidation and succinimide
   intermediate (protein
   degradation) 202–4

decarboxylation 118–21
degradation reactions
   aldol condensation and retro-aldol
     124–5
   cyclization 137–9
   decarboxylation 118–21
   dimerization/
     oligomerization 139–44
   elimination 110–18
   isomerization and
     rearrangement 127–37
   miscellaneous mechanisms 144–6
   nucleophile/retro-nucleophilic
     conjugate addition 121–4
   retro-aldol 126
dehalogenation of aryl halides 176–8
dehydration elimination 110–14
dehydrohalogenation
   elimination 110, 114–15
denagliptin 138–9
design/selection of drug
   candidate 263–5
desoximetasone 115
dexamethasone 85–6, 89, 111, 115,
   133, 172
diclofenac 43–4, 180
Diels–Alder reaction 144–5, 190
diflusinal 118, 176
1,4-dihydropyridines
   (aromatization) 174–6
Diketopiperazine (DKP)
   biapenem 142
   cyclization 137–8
   deamidation and succinide
     intermediate 202
   dipeptide degradation 114, 215
   β-lactam antibiotics 27–30
dimerization/oligomerization 139–44
direct interaction between drugs and
   excipients (degradation)
   APIs 156–7
   magnesium stearate 154–6
   Maillard reaction 150–3
   ester and amide linkage 153–4
   others 157–8
   transesterification 154

diuretic drugs 34–5
DNA and RNA Drugs
  (chemical degradation)
  hydrolytic degradation of
    phosphodiester bonds 218–20
  oxidative degradation of nuclei
    acid bases 220–2
double bind equivalency (DBE) 110
double-bonds susceptible to addition
  by hydroperoxides 68–71
doxorubicon (adriamycin) 40
drug degradation chemistry
  description 3–4
drug-excipient interaction and adduct
  formation
  degradants of excipients 160–1
  degradation by impurity of
    excipients 158–60
  direct interaction 150–8
  impurities from packaging
    materials 161–2
drugs containing alcohol, aldehyde
  and ketones 87–92
duloxetine 41, 159
dyclonine 124

electron capture degradation
  (ECD) 201
elimination
  dehydration 110–14
  dehydrohalogenation 114–15
  description 110
  Hofmann 110, 116–17
  miscellaneous 117–18
  photochemical 182–4
  protein drugs 211–13
enamines and imines (Schiff
  bases) 79–80
Enbrel 198
epimerization 129
epinephrine (adrenalin) 129, 157
episerone 124
epoxides 41–3, 59–61, 62
ertapenem 140–1
erythromycin A 113
Eschweiler–Clarke reaction 158

ester groups (hydrolysis) 20–3
esterification, transesterification and
  amide linkages 43–4
estramustine 30–1
ethacrynic acid 122, 145
ethers 41–3
etodolac 120
etoposide 129
excipients
  degradants 158–60, 160–1
  formaldehyde 143
  impurity profiles 271
  see also direct interaction between
    drugs and excipients
Eyring equation 6
ezlopitant 66

FDA (Food and Drug
  Administration) in US 1, 218
Fenton, H. J. H. 49
Fenton reaction
  autooxidation 265
  free radicals 49–53, 54
  hydroxyl radical 205, 209–11
  oxidation of aromatic rings 93
  oxidative photochemical
    degradation 187
fluoroquinolone 182–3
flupenthixol 68, 171
fluvoxamine 172
fomivirsen 198, 220
formaldehyde
  degradation (excipient
    impurities) 158–9
  excipients 143, 158–9
  hydrochlorothiazide 142
  irbesartan 159
  packaging materials
    161–2
formic acid (degradation caused by
  excipient impurities) 158–9
fosinopril sodium 155
free-radical mediated autooxidation
  additional reactions 56–7
  autooxidative radical chain
    reaction and kinetics 54–6

free-radical mediated autooxidation
(*continued*)
   Fenton/Udenfriend reactions
     49–53
   homolytic/heterolytic cleavage of
     peroxides 53–4
   radical chain reactions and
     kinetics 54–7

gemcitabine 37–9
ginger (spice) 126
6-gingerol 126
glibenclamide 36

half-lives (drug product shelf-lives)
   7–9
haloperidol 125
heparin 198, 216, 218
heterocyclic aromatic rings 96–9
heterolytic cleavage of peroxides
   formation of epoxides 59–61
   metal ion oxidation 53–4
   oxidation of amines/sulfides
     57–9
high density polyethylene (HDPE)
   bottles 268, 269
hinge region hydrolyis in
   antibodies 204
HMG-CoA reductase inhibitors 67
Hofmann elimination 110, 116–17
homolytic bond dissociation
   energies 57
homolytic cleavage of peroxides
   (thermolysis) 53–4
human growth hormone (hGH) 207
human insulin-like growth factor I
   (hIGF-I) 207
human serum albumin (HAS)
   215
human vascular endothelial growth
   factor (rhVEGF) 207
Humira 198
hyaluran 198, 216, 218
hydralazine HCl 158
hydrochlorothiazide 142
hydrocortisone 86, 126

hydrogen peroxide
   excipient impurities 158–9
   Udenfriend reaction 57–9
   primary/secondary amines 77
   tetrapezam 70
hydrolysis
   reaction 16–18
   rearrangement of peptide
     backbone by Asp residue
     (protein degradation)
     199–202
hydrolytic degradation
   acetal and hemiacetal groups
     40–1
   activation energies 19–20
   carbamates 30–2
   drugs
     amide group 24–5
     ester group 20–3
     lactone group 23–4
     sulfonamides 34–5
   esterification, transesterification
     and amide linkages 43–4
   ethers and ethoxides 41–3
   imides (Schiff bases) and
     deamination 36–9
   imides and sulfonylureas 35–6
   β-lactam antibiotics 26–30
   overview 16–20
   phosphates and
     phosphoramides 32–3
hydrothiazides 34
9-hydroxyellipticine 95, 268
hydroxypropyl methylcellulose
   acetate succinate
   (HPMCAS) 159–60
hydroxypropyl methylcellulose
   phthalate (HPMCP) 159–60

imides 35–6
imines (Schiff bases) 36–9, 79–80, 99,
   101, 158
imipenem 29, 142
impurities
   drugs (importance) 1–2
   excipients (degradation)

hydrogen peroxide, formaldehyde and formic acid 158–9
residual impurities in polymeric excipients 159–60
ICH 3, 1, 227
packaging 161–2
Impurity Profilng Group (IMG) 244
indole rings 69, 98–9
indomethacin 25, 69, 120, 167
indoprofen 167, 188
insulin 264
International Conference on Harmonization (ICH)
impurities/degradants 3, 61, 227
photostability 266–7, 273
stability 2, 3, 74
storage 1, 73–4, 241
stress studies 240–1, 244
intersystem crossing (ISC) 166
irbesartan 159, 182
isomerization and rearrangement
N,O-acyl migration 132–3
cis-trans isomerization 129–31
epimerization 129
intramolecular Cannizzaro rearrangement 136–7
introduction 127
racemization 128–9
ring expansion 133–6
tautomerization 127–8

ketoprofen 167, 189
ketorolac 84, 168
kinetics of chemical reactions 5–7

β-lactam antibiotics
hydrolytic degradation 26–30
nucleophilic attack 139–40
oxime ethers 39
lactone group (hydrolysis) 23
latamoxef (maxalactam) 28
LC-MS^n for structural elucidation of trace levels
MS (unfriendly) HPLC to LC-MS conversion 230

nomenclature, ionization and determination of parent ions 230–3
Lederer–Manasse mechanism 144
lidocaine 25
losartan potassium 139
lovastatin 23, 67
low molecular weight heparin (LMWH) 216
lowest unoccupied molecular orbit (LUMO) 180

magnesium stearate (excipient) 154–6
Maillard, Louis-Camille 150
Maillard reaction 150–3, 214–15
maleic acid 123
mangiferin 273
Mannich reaction 123
manufacturing process and drug degradation 272
Mass Spectrometry, Principles and Applications 229
Mattox process (corticosteroids) 111–13
meclofenamic acid 180
Meisenheimer rearrangement 77
menadione (vitamin K$_3$) 184
meropenem 31, 154
Merrem 30
Michael addition 95, 119, 122–3, 151, 156
metasulfite/bisulfite salts 157, 160–1
miconazole 96, 160
microencapsulation 272
miscellaneous degradation mechanisms
Diels-Alder reaction 144–5
reduction or disproportionation 145–6
moisture
solid dosage forms 269–70
solid state degradation 10–11
molsidomine 271, 272
mometasone furoate 114–15
montelukast 82, 170
morphine 65–6, 96, 139

naproxen 169
National Institutes of Health (NIH)
    TOXNET database 263
nifedipine 272
non-oxidative photochemical
    degradation
    aromatization of 1, 4-
        dihydropyridine drugs 174–6
    cyclization in polyaromatic
        rings 180–2
    dehalogenation of aryl
        halides 176–9
    introduction 166–7
    photochemical elimination 182–4
    photochemistry of ketones: Norris
        type I AND II photo
        reactions 185–7
    photodecarboxylation:
        2-arylpropionic acid 167–8
    photodimerization and
        photopolymerization 184–5
    photoisomerization 170–4
non-radical reactions of peroxides
    (heterolytic cleavage)
        epoxides 59–61
        oxidation 57–9
non-steroidal anti-inflammatory
    (NSAIDs) drugs
    4-aminosalicyclic acid 118
    5-aminosalicyclic acid 119
    benoxaprofen 167
    carprofen 167, 170, 177, 185
    cyclization in polyaromatic
        rings 180
    diclofenac 43–4, 180
    decarboxylation 118–21
    etodolac 120
    indomethacin 25, 69, 120, 167
    indoprofen 167, 188
    ketoprofen 167, 189
    ketorolac 84, 168
    meclofenamic acid 180
    naproxen 169
    photodecarboxylation 121, 167–70
    suprofen 168
    tiaprofenic acid 168, 189

norfloxacin 155, 182–3
nucleobases (DNA and RNA) 218,
    220
nucleophile/retro-nucleophilic
    conjugate addition 121–4

obidoxime 39
ofloxacin 183
olmesartan medomomil 21–2
oxidation pathways of drugs
    allylic/benzylic positions
        susceptible to hydrogen
        abstraction by free radicals 62–8
    aromatic rings: phenols,
        polyphenols and quinones 92–6
    carbanion/enolate mediated
        autooxidation 83–7
    double-bonds susceptible to
        addition by hydroperoxides
        68–71
    drugs containing alcohol, aldehyde
        and ketones 87–92
    enamines/imines (Schiff
        bases) 79–80
    heterocyclic aromatic rings 96–9
    introduction 62
    miscellaneous
        degradations 99–101
    primary/secondary amines 76–9
    tertiary amines 71–6
    thioethers (organic sulfides),
        sulfoxides and thiols 80–3
oxidation of side chains
    Arg, Pro and Lys 209–11
    Cys, Met, His, Trp and Tyr 204–9
oxidative degradation
    carbanion/enolate-mediated
        autooxidation (base-catalyzed
        autooxidation) 61–2
    free-radical mediated
        autooxidation 49–57
    introduction 48–9
    non-radical reactions of
        peroxides 57–61
    oxidation pathways of
        drugs 62–101

oxidative photochemical degradation
introduction 187–8
pathways *via* reaction with singlet
oxygen 190–4
type I photosensitized oxidation:
radical formation/electron
transfer 188–9
type II photosensitized oxidation
by singlet oxygen 189–90
oximes 39, 129–31, 172
oxygen content in drug
products 267–8
oxytocin 153

packaging materials and drug
degradation
impurities 161–2
leachables 1
primary materials 272–3
paliperidone
(9-hydroxyrisperidone) 128
parathyroid hormone (PTH) 262
penicillins 26–8, 139–40
pentoxifylline 2
pH
drug degradation 270
hydrolysis 18
manufacturing process 272
oxidation of thioethers and
sulfoxides 82
solid state 10–11
phenlyephrine 156, 157
phenobarbital 35–6
phenothiazine-derived drugs 72
phenyl rings 93–4
phenylbutazone 86–7
phosphates and phosphoramides
32–3
photo-Fries rearrangement
173–4
photochemical degradation
non-oxidative 166–87
overview 165–6
oxidative 187–94, 222
pigments, colorants and
additives 270–1

photochemical elimination
182–4
photochemistry of ketones: Norris
type I AND II photo
reactions 185–7
photodecarboxylation:
2-arylpropionic acid 167–8
photodimerization and
photopolymerization 184–5
photoisomerization
cis-trans isomerization around
carbon–carbon,
carbon–heteroatom,
heteroatom–heteroatom double
bonds 170–2
drugs containing a
2,5-cyclodienone ring
172–3
photo-Fries rearrangement
173–4
photochemical elimination
182–4
piperazine 24–5
Plavix (clopidogrel bisulfate) 64
polyethylene bottles 268
polyethylene glycol (PEG)
autooxidation 4, 89–90
excipient impurities 271
free radicals 53
formaldehyde 143, 158–9
formic acid 158–9
hydrogen peroxide 158–9
hydroperoxides 80
hydroxyindoles 144
protein drugs 265
polysorbate 4, 90, 143, 158,
162, 271
povidone (polyvinylpyrrolidone) 4,
158, 271
prednisolone 32, 86, 111, 126,
133–5, 172
pregabalin 151
pridinol 112
primary/secondary amines 76–9
procaine 22
prontosil 34

protein drugs (chemical degradation)
crosslinking, dimerization and
oligomerization 213–14
deamidation and succinimide
intermediate 202–4
DNA and RNA Drugs 218–22
β-elimination 211–13, 215
hinge reaction hydrolysis in
antibodies 204
hydrolysis and rearrangement of
peptide backbone by Asp
residue 199–202
introduction 199
Maillard reaction 214
miscellaneous 215–16
N-terminal dipeptide truncation
through DKP formation 215
oxidation of side chains
Arg, Pro and Lys 209–11
Cys, Met, His, Trp and
Tyr 204–9

rabeprazole 145–6
racemization (degradation) 128–9
radical chain reactions and
kinetics 54–7
raloxifene 73
ranitidine 82
reaction orders and half lives (drug
product shelf-lives) 7–9
reactive oxygen species (ROS) 48,
52–3, 188, 204–5, 265
rebeccamycin 143–4
recombinant human interleukin II
(rhIL-II) 202
recombinant human vascular
endothelial growth factor
(rhVEGF) 207
reduction or disproportionation
(degradation) 145–6
Remicade 198
retinoic acid (tretinoin) 171
retro-aldol degradation 126, 134
risperidone 67–8
Rituxin 198
rofecoxib 84–5
roxithromycin 130–1

Schiff bases *see* imines
self-life of drugs 7–9
simvastatin 23, 67
singlet oxygen 176, 180, 189–90, 190–4
solid state degradation
moisture 10–11
pH 10–11
polymorphs 9–10
Solid-State Chemistry of Drugs 4
*Stability of Drugs and Dosage
Forms* 4
stilbene 180
stress studies
artificial degradants 2
degradation profiles 241–5
industry practice 242
introduction 240–1
stress studies/LC-MS^n fingerprinting
combination: case studies
1: betamethasone dipropionate/
corticosteroidal 17, 21
esters 248–51
2: betamethasone sodium
phosphate isomeric degradants
(enzymatic transformation)
251–5
3: betamethasone 17-valerate
(MS^n fingerprint not
available) 256–8
general strategy 245
stress studies according to
presumed degradation
type 247–8
tracking/verification of unknown
degradants 248
type of degradation based on
LC-MS^n analysis 245–7
structures and pathways
(degradation)
fragmentation and LC-MS^n
molecular fingerprinting 233–9
LC-MS^n for structural elucidation
of trace levels 229–39
multi-dimensional NMR in
structure elucidation of trace
level impurities 239–40
overview 227–9

sulfanilimide 34
sulfonamide drugs 34–5
sulfonylureas 35–6
suprofen 168

tamoxifen 180, 182
tautomerization (degradation)
  ceftibuten 131–2
  enamines/imines 79
  imines 37, 79, 127–8,
  isomerization and
    rearrangement 127–8
TEMPO (2,2,6,6-
  tetramethylpiperidin-1-yl)oxyl
  78–9
telmisartan 180–2
tertiary amines
  drugs 75
  free-radical mediated
    autooxidation 75–6
  N-oxides
    decomposition 73–5
    formation 71–3
tetrachlorohydroquinone (TCHQ) 49
tetrazepam 70–1, 80
thiazides 34
thermodynamics of chemical
  reactions 5–7
thioethers (organic sulfides),
  sulfoxides and thiols 80–3
thioxanthine antipsychotic drugs 171
tiagabine 69
tiaprofenic acid 168, 189
tirilazad mesylate 90–1
tobramycin 40, 198
tolperisone 124

triamcinolone 115, 134–5
trimethoprim 34
tylosin 124–5

Udenfriend reaction
  autooxidation of tertiary
    amines 75–6
  chelating agents 52–3, 268–9
  description 205, 265
  free radicals 49–53, 75–6
  formulation and Udenfriend
    "trap" 265–7, 268
  hydrogen peroxide 57–8
  hydroxyl radical 209–211
  oxidation of aromatic rings 93
  oxidative degradation 273
  photochemical degradation 267
  reactive oxygen species 93, 205
Udenfriend, Sydney 50
Udenfriend "trap" 265–7, 268
UV (ultra violet) light
  photodimerization/
    photopolymerization 184–5
  photochemical degradation 165
  photoisomerization 170–2, 172–3,
    174–5

vancomycin 135–6, 198
vitamin D 131
vitamin D_3 154

Weitz-Scheffer reaction 60
World Health Organization
  (WHO) 1

ziprasidone 125–6